LABYRINTH AND PIANO KEY WEIRS II –
PKW 2013

PROCEEDINGS OF THE SECOND INTERNATIONAL WORKSHOP ON LABYRINTH AND PIANO KEY WEIRS 2013, CHATOU, PARIS, FRANCE, 20–22 NOVEMBER 2013

Labyrinth and Piano Key Weirs II – PKW 2013

Editors

Sébastien Erpicum
Laboratory of Engineering Hydraulics – HECE, Université de Liège, Belgium

Frédéric Laugier
Hydro Engineering Center, Electricité de France – EDF-CIH, France

Michael Pfister
Laboratory of Hydraulic Constructions, Ecole Polytechnique Fédérale de Lausanne, Switzerland

Michel Pirotton
Hydraulics in Environmental and Civil Engineering, Université de Liège, Belgium

Guy-Michel Cicero
Hydraulics and Environment National Laboratory – EDF R&D-LHNE, France

Anton J. Schleiss
Laboratory of Hydraulic Constructions, Ecole Polytechnique Fédérale de Lausanne, Switzerland

CRC Press
Taylor & Francis Group
Boca Raton London New York Leiden

CRC Press is an imprint of the
Taylor & Francis Group, an **informa** business
A BALKEMA BOOK

Cover picture:
Malarce dam PKW – France – 2013
Courtesy of Electricite de France (EDF)

CRC Press/Balkema is an imprint of the Taylor & Francis Group, an informa business

Typeset by MPS Limited, Chennai, India
Printed and bound in Great Britain by CPI Group (UK) Ltd, Croydon, CR0 4YY

Published by: CRC Press/Balkema
P.O. Box 447, 2300 AK Leiden, The Netherlands
e-mail: Pub.NL@taylorandfrancis.com
www.crcpress.com – www.taylorandfrancis.com

ISBN: 978-1-138-00085-8 (Hardback)
ISBN: 978-1-135-81290-8 (eBook PDF)

Labyrinth and Piano Key Weirs II – PKW 2013 – Erpicum et al. (eds)
© 2014 Taylor & Francis Group, London, ISBN 978-1-138-00085-8

Table of contents

C. Hydraulic and structural design

D. Planned projects

E. Existing projects

F. Future developments

Labyrinth and Piano Key Weirs II – PKW 2013 – Erpicum et al. (eds)
© 2014 Taylor & Francis Group, London, ISBN 978-1-138-00085-8

Preface

Dam engineering is currently experiencing a strong revival of labyrinth oriented weirs. More and more acute extreme flood issues makes the requirement for performing and reliable spillways becoming higher and higher. Labyrinth weirs, with a repetitive constructional character and an increased specific discharge capacity, are a very good technico-economical compromise. Moreover, as free flow spillways, their operation is quite reliable compared to gated ones. The concept of Piano Key Weir (PKW), with alveoli developed in overhangs from a reduced support area, enables to install non-linear crests at the top of concrete dams. It eliminates thus the main drawback of classical labyrinth weirs and enables their use to rehabilitate numerous existing dams. In addition, innovative fuse plug spillways are often based on labyrinth shaped units

Since the first implementation of PKW by Electricité de France on Goulours dam (France) in 2006, at least eight PKWs have been built in France, Vietnam and Switzerland. Their operation over a few years has already provided the first prototype data. Other projects are under study, construction or planning in France, Vietnam, United Kingdom, Sri Lanka, Australia and South Africa. On another hand, research programs are under progress in France, Belgium, Switzerland, Vietnam, Algeria, United States, India and Iran. The Journals of *Hydraulic Research* (IAHR), *Hydraulic Engineering* (ASCE) and *Irrigation and Drainage Engineering* (ASCE) published in 2011 and 2012 together some 9 papers on PKW.

These projects show that labyrinth and especially piano key weirs are up-to-date true solutions in hydraulic engineering, spreading all over the world. This was the motivation to organize a second workshop dedicated to this specific topic, following the successful event held in Liège (Belgium) in 2011. This second workshop was co-organized by Electricité de France, University of Liege and Ecole Polytechnique Fédérale de Lausanne, which all together have played a significant role in PKW development. The conference took place at Electricité de France Research and Development site of Chatou (Paris) in France, on 20-22 November 2013. The 2 days long workshop was followed by a technical tour in the Ardèche Department (France) where a large PKW was recently built on Malarce dam (EDF). The workshop addressed several themes such as physical and numerical modeling or future developments. Actual design issues were discussed and several planned and existing projects in Europe and abroad were presented.

This book gathers the Proceedings of the second Workshop on Labyrinth and Piano Key Weirs and can be considered as an up-to-date Reference document on the subject.

Sébastien Erpicum, HECE – ULg
Frédéric Laugier, EDF – CIH
Michael Pfister, LCH – EPFL

Labyrinth and Piano Key Weirs II – PKW 2013 – Erpicum et al. (eds)
© 2014 Taylor & Francis Group, London, ISBN 978-1-138-00085-8

Acknowledgements

The organization of a conference is only possible with motivated contributors. We are therefore grateful to practitioners and researchers who allowed taking up the challenge of proposing a workshop on the very specialized domain of Labyrinth and Piano Key Weirs. In order to warrant the high technical and scientific level of the contributions, a serious reviewing process was carried out by the International Scientific Committee, in collaboration with the Committee on Hydraulic Structures of IAHR. All the Reviewers merit our gratitude for the serious of their work and the respect of very short deadlines.

The success of a workshop is also dependent of the support of professional associations and technical and scientific journals that helped to announce the manifestation. Our thanks are devoted to the Société Hydrotechnique de France, the Comité Français des Barrages et Réservoirs, the Swiss Committee on Dams, the International Journal of Hydropower & Dams, the Schweizerischer Wasserwirtschaftsverband, the Committee on Hydraulic Structures of the International Association for Hydro-environment Engineering and Research (IAHR).

The edition of these proceedings has been made possible thanks to the contribution of the workshop sponsors. We greatly acknowledge the help of Artelia, Coyne et Bellier, ISL, Lombardi, Pöyry, Schnabel Engineering and Stucky.

The technical tour in Ardèche Department allowed visiting the Malarce dam owned by EDF, where a PKW spillway was recently built. The tour was guided by competent persons in charge of management and safety of these hydraulic schemes. All our thanks go to EDF-UP Centre representatives for their pleasant availability.

The involvement of three Co-organizing institutions, Electricité de France, University of Liège and Ecole Polytechnique Fédérale de Lausanne was fundamental for the preparation of the workshop and for the edition of the proceedings. Concerning these activities, the organizers want to express their recognition to the publisher for his guidance, to Julie Chieze and Eve Delahaye from EDF-R&D, to Stephane Rouy and Benjamin Sautel from EDF-UP Centre and to Pierre Archambeau from HECE-ULg for their helpful contribution.

The Scientific Secretariat of PKW 2013

Labyrinth and Piano Key Weirs II – PKW 2013 – Erpicum et al. (eds)
© 2014 Taylor & Francis Group, London, ISBN 978-1-138-00085-8

Organization

Members of the International Scientific Committee

Boillat, J.-L.	Ecole Polytechnique Fédérale de Lausanne, Switzerland
Chanson, H.	University of Queensland, Australia
Cicero, G.M.	Electricité de France, France
Crookston, B.	Schnabel Engineering, USA
Delorme, F.	Electricité de France, France
Dewals, B.	Université de Liège, Belgium
Erpicum, S.	Université de Liège, Belgium
Falvey, H.T.	American Society of Civil Engineers, USA
Ho Ta Khanh, M.	VNCOLD & CFRB, France
Laugier, F.	Electricité de France, France
Lempérière, F.	Hydrocoop, France
Lesleighter, E.	Australia
Leite Ribeiro, M.	Stucky, Switzerland
Matos, J.	Technical University of Lisbon, Portugal
Ouamane, A.	University Mohamed Kheider, Biskra, Algeria
Pfister, M.	Ecole Polytechnique Fédérale de Lausanne, Switzerland
Pirotton, M.	Université de Liège, Belgium
Schleiss, A.J.	Ecole Polytechnique Fédérale de Lausanne, Switzerland
Sharma, N.	Indian Institute of Technology, Roorkee, India
Taquet, B.	Electricité de France, France
Tullis, B.	Utah State University, USA
Vermeulen, J.	Electricité de France, France
Vigny, J.-P.	Hydrocoop, France
Violeau, D.	Electricité de France, France

Members of the Organizing Committee

Delahaye, E.	Electricité de France, France
Cicero, G.M.	Electricité de France, France
Laugier, F.	Electricité de France, France
Vermeulen, J.	Electricité de France, France
Erpicum, S.	Université de Liège, Belgium
Pfister, M.	Ecole Polytechnique Fédérale de Lausanne, Switzerland

Labyrinth and Piano Key Weirs II – PKW 2013 – Erpicum et al. (eds)
© 2014 Taylor & Francis Group, London, ISBN 978-1-138-00085-8

Organizing institutions

Labyrinth and Piano Key Weirs 2013 was organized by the following institutions:

Laboratoire National d'Hydraulique et Environnement & Centre d'Ingénierie Hydraulique
Electricité de France

Laboratoire d'Hydraulique des Constructions – HECE
Université de Liège

Laboratoire de Constructions Hydrauliques
Ecole Polytechnique Fédérale de Lausanne

Labyrinth and Piano Key Weirs II – PKW 2013 – Erpicum et al. (eds)
 ISBN 978-1-138-00085-8

Supporting institutions

Labyrinth and Piano Key Weirs 2013 was supported by the following institutions:

Société Hydrotechnique de France

Comité français des barrages et réservoirs

Swiss Committee on Dams

International Journal on Hydropower & Dams

Schweizerischer Wasserwirtschaftsverband

Committee on Hydraulic Structures
International Association for
Hydro-environment
Engineering and Research

Labyrinth and Piano Key Weirs II – PKW 2013 – Erpicum et al. (eds)
© 2014 Taylor & Francis Group, London, ISBN 978-1-138-00085-8

Sponsoring companies

Labyrinth and Piano Key Weirs 2013 was sponsored by the following companies:

ARTELIA — *Artelia*

COYNE ET BELLIER Bureau d'Ingénieurs Conseils; TRACTEBEL Engineering GDF SUEZ — *Coyne et Bellier*
Tractebel Engineering

ISL ingénierie — *ISL Ingénierie*

Lombardi — *Lombardi*

PÖYRY — *Pöyry*

Schnabel ENGINEERING — *Schnabel Engineering*

STUCKY — *Stucky*

Keynote lectures

Labyrinth and Piano Key Weirs II – PKW 2013 – Erpicum et al. (eds)
© 2014 Taylor & Francis Group, London, ISBN 978-1-138-00085-8

The Piano Key Weirs: 15 years of Research & Development – Prospect

M. Ho Ta Khanh
VNCOLD, Vietnam

ABSTRACT: The Piano Key Weir (PKW), invented at the end of the 90's by Hydrocoop, was the subject of fruitful Research & Development during these last 15 years. The keynote summarizes the main results obtained with the physical and numerical models and their applications to existing and planned projects. For the recent projects, the paper indicates the possibilities and advantages of the PKW's utilization and the problem of the choice of the main parameters for the design. The paper presents briefly the past studies concerning some specific issues for a good utilization of the PKW and some proposed alternatives linked to the PKW concept. For planned projects, it is always necessary to carry out in parallel, at the commencement of the design, the hydraulic and structural optimizations, taking into account all the site and construction data. In conclusion, the note suggests some R&D to be carried out in the future to expand the knowledge of PKW in order to improve their utilization and to develop new applications for more safe and economic spillways.

1 INTRODUCTION

1.1 *A brief history of PKW*

The concept of PKW – a variation of traditional labyrinth weirs – was presented almost 15 years ago by Hydrocoop (Blanc & Lempérière, 2001) and Ouamane (Lempérière & Ouamane, 2003). Hydrocoop has promoted and coordinated relevant hydraulic model tests in five countries: France, Algeria, India, China and Vietnam (Hien et al., 2006; Ho Ta Khanh et al., 2011a; Ho Ta Khanh et al., 2012). For this period, the University of Biskra (Algeria) has been deeply associated with Hydrocoop as well (Ouamane, 2006; Ouamane & Lempérière, 2006a & 2006b). All these studies and tests, performed with some PKW of different types and shapes (Fig. 1), have the same practical goals:

- To find a compromise between the hydraulic efficiency, the structural and relevant construction problems and the economic considerations. This point is very important as the best hydraulic solution is not always the optimal solution if one has to take into account the other constraints.
- To elaborate some standard models of PKW to facilitate preliminary designs, cost estimation and comparison with other solutions.
- To adjust the geometrical parameters of a new PKW design by taking into account the results of a "standard model".

Following some years of concept development, interest in PKW spillways augmented internationally in particular in France with Electricité de France (EDF), which has to increase the capacity of some existing spillways (Barcouda et al., 2006), and in Vietnam for some new projects (Hien et al., 2006), the main advantages of the PKW being the possibility to evacuate high discharge with low overflow depth and to be easily placed on the crest of gravity dam in relatively narrow valley.

The design and construction of several PKW by EDF, after the successful example of the Goulours dam (Laugier, 2007), was the origin and the commencement of numerous tests concerning the PKW by the EDF-LNHE (France), HECE-ULg (Belgium) and LCH-EPFL (Switzerland). The interest for PKW increases internationally as reflected by more than 80 publications (some of them are in the Reference), conferences and seminars devoted to this topic during the last 10 years.

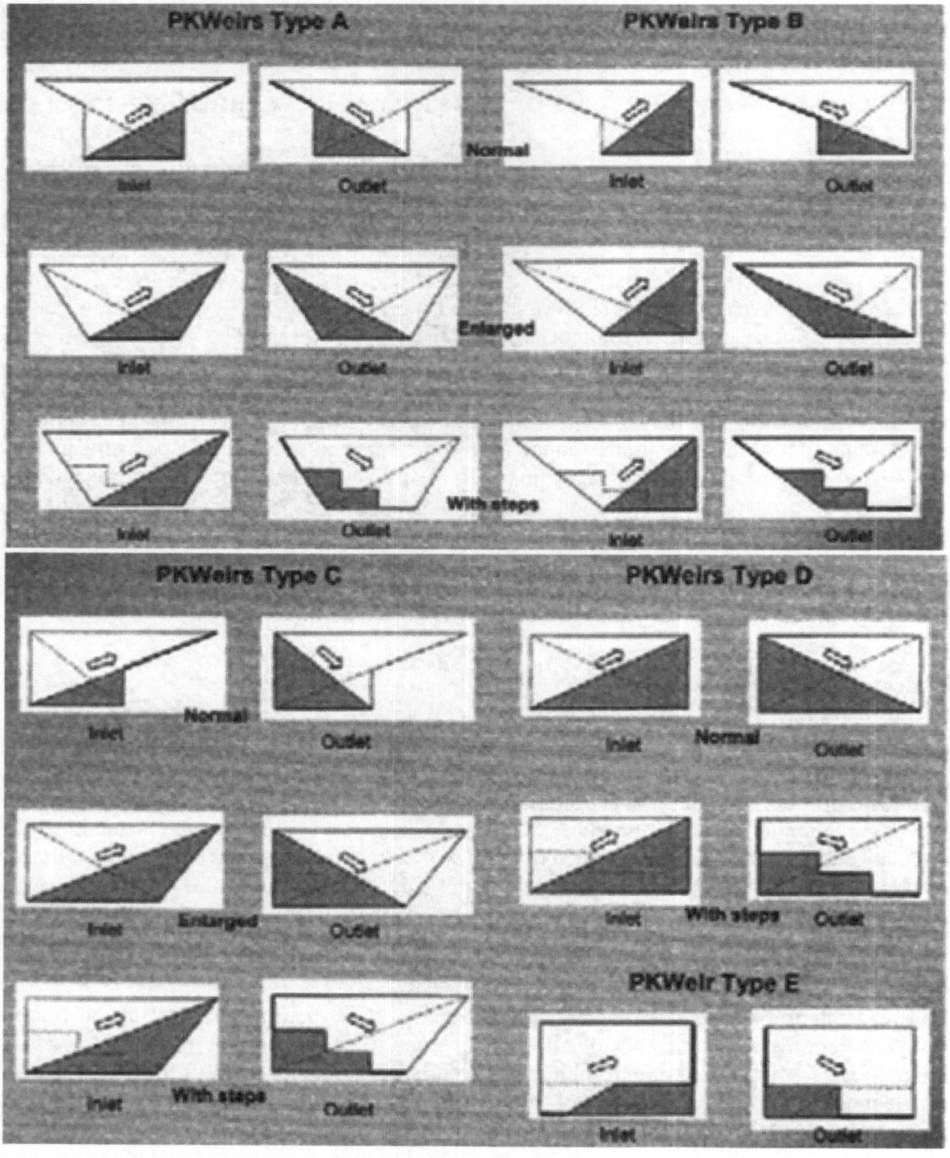

Figure 1. Different types of PKW.

1.2 *PKW utilization in recent projects (up to 2012)*

Up to 2012 the main applications of PKW in France by EDF concern the increase of the dam discharge capacity (dam rehabilitation). They are placed on the crest of gravity dams or on gravity walls. For the four first PKW (Goulours, Saint Marc, Etroit, Gloriettes), the maximum discharge is 660 m^3/s and the maximum head on the weir is 1.35 m.

In Vietnam several PKW were constructed but the most important is the Van Phong dam (Reynolds, 2011; Ho Ta Khanh et al., 2011b). This dam is long (475 m) but low (12 m). It includes a central part equipped with 10 radial gates and two PKW on each side. The max total discharge is 12,500 m^3/s and the max discharge on the two PKW is 8700 m^3/s for the Design flood.

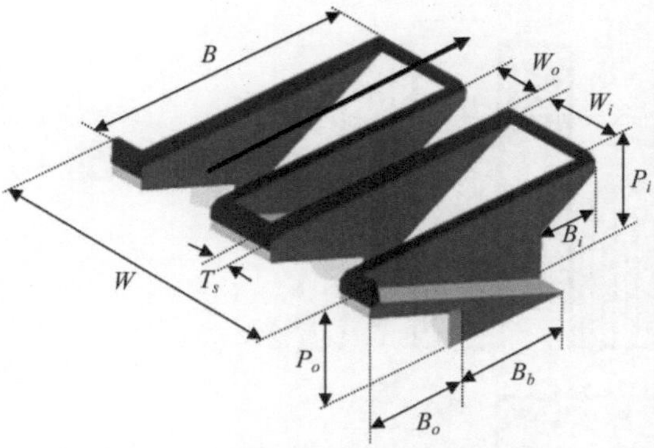

Figure 2. Main geometrical parameters of a PKW (from Pralong et al., 2011b).

Other projects are also under consideration in India (Das Singhal & Sharma, 2011; Sharma & Das Singhal, 2012).

1.3 *The future*

Many PKW are now under design or construction through the world: in France, the most important is the Malarce dam which can be visited during this symposium, in Vietnam, in India, in Burkina Faso and in South Africa, etc.

All these projects utilize what we can call "traditional PKW" but some new suggestions such as the PK morning glory (Cicéro et al., 2011) or the Fractal PKW (Laugier et al., 2011a) proposed in 2011, can lead in the future to interesting applications.

More future research concerning some issues linked to the utilization of PKW, such as evacuation of floating debris, dissipation of energy downstream PKW, etc. will be very useful.

2 THE PHYSICAL MODELS

2.1 *Geometrical parameters of the PKW*

The PKW are defined by many geometrical parameters. These parameters are more than 30 but 20 are generally enough for the most frequent cases (Fig. 2). It is then useful to have a naming convention for them. This work was carried out in 2010 by a team of EDF-CIH, HECE-ULg and LCH-EPFL (Pralong et al., 2011b). It is consequently recommended to use now this naming convention to compare the results of the different PKW.

If the numerous parameters complicated a bit the optimization, they allow a great flexibility in the design of the PKW as demonstrated by the existing PKW and the projects under design or construction.

2.2 *Rating curves of different types and sizes of PKW and their sensitivity with the variations of the main parameters*

As already indicated, the tests performed with the physical models up to 2006, coordinated by Hydrocoop, aim to define some "standard models" of PKW to facilitate preliminary designs, cost estimation and comparison with other solutions. For instance, the "standard model" for the Type A (Fig. 3) can be used for a new project. It is probably not the optimum but it is admitted that it is not very far from this optimum.

It is noteworthy that for this PKW, with $n = 5$, the specific discharge: $q_{SW} = 4.3\, P_m^{0.5} H_u$, with H_u the upstream head and P_m the height of the PKW at inlet and outlet apron intersection, is practically

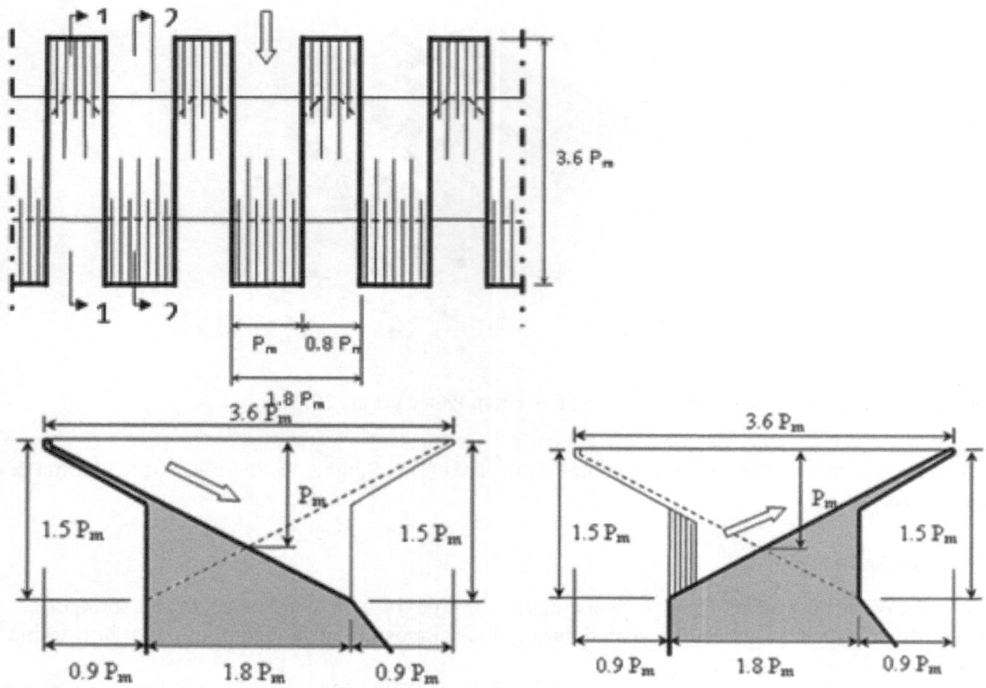

Figure 3. The "standard model" for PKW type A.

linear versus H_u. Knowing q_{sW} and H_u, the PKW can be completely defined by the value of P_m at a preliminary stage of the study. For a new project, a more precise rating curve can be measured later on a specific model taking into account the other constraints of the design and in particular the structural aspect.

The tests performed after 2006 by the different laboratories focused on the rating curves of the different types and sizes of the PKW with the extensive variations of all the parameters which characterize the PKW for free flows (Karelle, 2008; Machiels et al., 2009a,b, 2010a,b,c,d, 2011a,b,c,e, 2012a,b,c,d; Le Doucen et al., 2009; Crookston & Tullis, 2010; Erpicum et al., 2010a, 2012a,b; Anderson, 2011; Leite Ribeiro et al., 2011b; Laugier et al., 2011b; Pralong et al., 2011a) and later for submerged flows (Belaabed & Ouamane, 2011; Dabling & Tullis, 2012; Kabiri-Samani & Javaheri, 2012; Ho Ta Khanh et al., 2013). Some tests were carried out in the framework of R&D programs, but most of them were carried out for specific projects.

It is not possible in this paper to summarize all the interesting results provided by these tests, but the list of papers in the Reference allows getting a lot of information on this subject.

It is interesting to point out hereunder some main results:

- The main parameters of a PKW which influence the values of the rating curve are: $n = L/W$, H_u/P and the ratio of the widths of the inlet and outlet keys may have a significant influence according to the types of PKW.
- For the free flow, q_{sW} being the specific discharge of the PKW and q_c the specific discharge of the Creager weir with the same width W, the ratio $e = q_{sW}/q_c$ characterizes the "efficiency" of the PKW, e is a bit lower than n for the small values of H_u/P (for example with $n = 5$, $e > 3$), but decreases as the value of H_u/P increases. A PKW is generally interesting, if $0.4 < H_u/P < 1$ and rarely, if $1 < H_u/P < 2$, because even these heads are hydraulically efficient on the PKW, they might be uninteresting from an economic point of view as compared to the construction cost of an ogee crest.

Figure 4. Tests of flows on stepped spillway with Creager weir and PKW.

- For the submerged flows, the reduction of the specific discharges versus the coefficient of submersion H_d/H_u is significant only if $H_d/H_u > 0.6$ and the "sensitivity" (decrease of q_{sW} versus H_d/H_u) is less for the PKW compared with the Creager and the labyrinth weirs.

As the quantity of information on this subject is already important and that it will certainly increase a lot in the future, it was envisaged during the seminar "Labyrinth and Piano Key Weirs–2011" in Liège (Belgium) to create a PKW Database. This proposal, if raising some practical issues, can certainly be useful for the future PKW research and applications.

Several laboratories and authors have tried to get, for the "discharge coefficients" of the PKW (C_{dW} or C_{dL}), a mathematical formula of the main parameters by an analytical approach in the same way it was done for the labyrinth weirs. It is to note that opposite to the linear weirs, this "discharge coefficient" is not a constant but a function of the upstream head H_u (very often C_w varies as $H_u^{-0.5}$ and then q_{sW} is approximately a linear function of H_u). The utilization of such "universal formula" at a preliminary stage of the study may present some drawbacks if not correctly applied:

- The formula contains many coefficients adjusted empirically on a limited number of tests.
- The formula is valid only inside a certain domain.

2.3 *Dissipation of energy downstream PKW*

This topic was addressed for St-Marc and Gloriettes dams (Bieri et al., 2009; Leite Ribeiro et al., 2007a,b, 2011a) but only for small discharges. The specific discharges of a PKW are high compared with a Creager weir and the problem of the dissipation of energy downstream a PKW can be serious for the large floods.

Some tests were performed in Vietnam with a mobile bottom (Ho Ta Khanh et al., 2011a). For the comparison, a Creager weir and a PKW were successfully installed on the crest of the dam, and several types of spillway chutes (smooth, with regular steps, irregular steps, steps with baffles) were used (Fig. 4). The tests show a better aeration of the nappe downstream the PKW in all conditions, compared with a Creager weir, and consequently a lower risk of cavitation on the spillway chutes for high nappe depth. But the measures of the scour downstream the dam do not show a real advantage due to the better aeration of the nappe and do not allow drawing a clear indication concerning the dimension required for the stilling basin. Some further tests must be consequently performed for this issue with different methods to evaluate the residual energy of the flow in the stilling basin.

More recently, the issue of the dissipation of the energy of the flow downstream a PKW and a stepped spillway was also addressed by Erpicum et al. (2011b) and Silvestri et al. (2013) with physical model tests in the HECE-ULg laboratory. The residual energy in the stilling basin was determined by the calculation of the hydraulic jump in the stilling basin. The results obtained are "atypical" according to the authors and it seems that the length of the spillway chute and the uniform flow above it are the main factors which intervene in the result. The conclusion of this research is not very clear and further studies on this topic are needed.

3 THE NUMERICAL MODELS

Given the complexity of the flows and the great number of the geometrical parameters in a PKW, an optimization of the characteristics of the PKW and a study of the sensitivity of the parameters by mean of physical models will be very costly and time consuming. The use of numerical models appears to be very interesting and complementary to the physical models.

In 2010, Erpicum et al. (2010b, 2011a) presented a first simplified 1D-numerical model for the PKW. It is based on a 1D modeling of the inlet and outlet separately, with a single upstream reservoir and interactions between both flows by exchange of mass and momentum along the lateral crest. The comparison of the numerical results with several experimental data has shown the ability of the numerical model to predict with reasonable accuracy the release capacity of a PKW, whatever its geometry, on a significant range of the head in the reservoir (not too low or too high). This software is now available on internet and may be useful for the preliminary study of a PKW in the common cases.

A 3D numerical model has been developed at EDF on the CFD-Software Flow-3D®. This model was calibrated and validated – for free flows and submerged flows – on the existing physical models. Several geometrical parameters have been tested for a large range of flow rates. Sensitivity analysis was performed with more than several hundred of numerical simulation. Comparison with data available on EDF's PKW physical models have also led to good results with a mean relative deviation in specific flow rates not exceeding 5%. Recently, the research on submerged flows for different types of PKW – utilizing this numerical model by EDF coupled with physical models tested in Vietnam – has provided good results, with interesting views of the velocity magnitude and vectors not possible to see with the physical models (Ho Ta Khanh et al., 2013).

The two numerical models indicated here above shows that it is now possible to obtain – quickly and not very costly – good approximate specific flow rates for any new PKW designs. The utilization of mathematical formula for the "discharge coefficient" valid only in restricted domains and presenting discrepancies according to the authors must be used with discernment. For the preliminary design of new projects, prior utilization of the numerical models, the "standard models" or the method exposed in Machiels et al. (2011d) can be useful.

4 POSSIBILITY AND ADVANTAGES OF PKW UTILIZATION IN RECENT PROJECTS

4.1 *Utilization of PKW on gravity dams*

The main advantages of a PKW are:

- It is a free fix overflow spillway.
- Compared with a linear Creager spillway, it may for example treble (or more for the low nappe depths) the capacity of the spillway for the same width and nappe depth. It can consequently

Figure 5. Downstream view of St-Marc (left) and L'Etroit (right) dams with the original spillway and with the additional PKW.

increase the FSL and the capacity of the reservoir, or it can increase the capacity of the spillway, for the same MWL and dam crest level.

- It can be installed on the crest of a thick wall or a gravity dam thanks to its relative narrow footprint compared with a labyrinth weir.

For an existing gravity dam with a spillway on the crest, the installation of an additional PKW is an attractive solution to increase the total capacity of the spillway whenever the Design flood must be upgraded, the MWL not modified and its width limited for several possible reasons. However it is to note that the use of a PKW is not restricted to gravity dams but – as a complementary free overflow spillway – can also be installed on the side of an arch dam or an embankment dam.

The installations of PKW on some EDF's existing dams of different types (Fig. 5) are good examples of the solutions to meet the new hydrology requirements. The details concerning the study and construction of these PKW are described in the following documents (Barcouda et al., 2006; Leite Ribeiro et al., 2007a,b, 2009, 2011a,b; Bieri et al., 2099; Laugier et al., 2009, 2012; Cicéro et al., 2010; Dugué et al., 2011; Erpicum et al., 2011c, 2012c; Pinchard et al., 2011; Vermeulen et al., 2011).

It can be observed with these examples that for the existing dams, the additional PKWs do not correspond to the "standard model" (with $n = 5$ and symmetrical overhangs) for two reasons:

- The constraints due to the difference between the FSL and the MWL and the limited width for the new spillway are not compatible with this standard model. The value of n is generally higher than 5 (often about 7 or more) and the PKW not symmetrical.
- The specific local conditions of the existing structure which oblige to adapt the PKW to this structure (for example the particular shape of the PKW and its anchorage to the swelling concrete for the St-Marc dam).

For the new projects, such as in Vietnam, without strict constraints, the "standard model" was very useful as it allows sizing quickly a preliminary shape for the PKW. The following tests with the physical models have confirmed the approximate values adopted in this first stage of the study. Finally, the main features of the PKW were adjusted but without great modification.

Note also that with the EDF's PKW, the values of H_u/P are always small (<0.3) with high values for e, since the flood discharges are generally low and a high efficiency of the PKW is searched. For the new Vietnamese projects $H_u/P \approx 1$ because the specific discharge q_{sW} is generally high and e is lower, although this choice is economically justified compared with a linear weir project.

For an existing or planned gravity dam, it is always safer to foresee a – totally or partially – free overflow spillway in place of a totally gated spillway. For such a solution, the PKW, as a free overflow spillway, allows reducing the width of the structure and the nappe depth for a given

Figure 6. The Van Phong barrage in Vietnam under construction.

discharge compared with a traditional spillway (Ouamane & Lempérière, 2007; Lempérière, 2009; Lempérière et al., 2011; Ho Ta Khanh, 2011).

4.2 *Utilization of PKW for low barrages*

An attractive solution for low barrages on large rivers is to combine large radial gates with PKW placed on one or two sides of the structure. The advantages of this solution are:

- Improvement of the safety of the dam in case of no-opening or jamming of the gates thanks to the free overflow PKW.
- Reduction of maintenance of the gates.
- Possible reduction of investment costs in some developing countries.

As this kind of dam can be submerged by high floods, such as for the Van Phong barrage in Vietnam (Fig. 6), a study was undertaken between EDF-CIH (numerical models) and the HCMUT laboratory (physical models) concerning free overflow and submerged flow on PKW of types A and D (or E) and a rectangular labyrinth weir. The main result is that, hydraulically, it is necessary to distinguish the "efficiency" and the "sensitivity" of the type of weirs regarding the submerged flows. Finally, the choice between these types of weirs depends less on hydraulic issues than on structural issues (in particular the type of foundation) taking into account the economic and structural considerations (Ho Ta Khanh et al., 2013).

Some results of this study (preference for PKW type D or type E) are also valid for a PKW installed in a lateral spillway (example of an embankment dam) for which there is no constraint for the footprint in the upstream/downstream direction of the channel.

5 CONCLUSION: FURTHER R&D FOR A SAFER UTILIZATION OF PKW

As indicated in this paper, the R&D concerning the PKW has made good achievement during these last 15 years with the results obtained by several laboratories and the return of experience of constructed PKW. However, it seems that some topics, if they have been already addressed by some authors, require a deeper knowledge to allow the extension of this type of spillway worldwide.

5.1 *The risk of floating debris*

This topic has been already addressed by some simple tests on physical models (Goulours, etc.) but recently Pfister M. et al. (2013) have carried out more sophisticated tests with a distinction between individual trunks and rootstock and accumulative debris. It seems that they have provided rather favorable results, but this study needs to be developed as this issue can influence some present schemes located near inhabited areas or in large catchment areas covered with forests.

5.2 *The dissipation of energy downstream PKW*

As indicated in §2.3, it is certainly now one of the most important topic concerning the PKW as they are foreseen to be installed soon in some large catchment areas with high flood discharges.

5.3 *The methods of construction*

Up to now, all the PKW were constructed with reinforced concrete fabricated on the site with formworks and scaffoldings, but it can be envisaged some alternatives for particular situations such as:

- Small PKW, 1 to 3 m high, to be placed in a small construction site without a concrete batching plant: an alternative with prefabricated steel elements can be the solution.
- PKW on a large river, the PKW units being identical but numerous: the use of prefabricated reinforced concrete elements can be an interesting solution in particular if the construction has to take place in a dry season.

This issue must be discussed with the contractors at the bidding stage to find the best solution depending on the local conditions.

REFERENCES

Anderson, R.M. 2011, Piano Key Weir Head Discharge Relationships, *All Graduate Theses and Dissertations*, Utah State University.

Barcouda, M., Cazaillet, O., Cochet, P., Jones, B.A., Lacroix, S., Laugier, F., Odeyer, C. and Vigny, J.P. 2006, Cost effective increase in storage and safety of most dams using fusegates or P. K. Weirs, *in proceedings of 22ème CIGB/ICOLD Congress*, Barcelona.

Belaabed, F. and Ouamane, A. 2011, Contribution to the study of the Piano Key Weirs submerged by the downstream level, *Labyrinth and piano key weirs-PKW 2011*, CRC press, London, 89–95.

Bieri, M., Federspiel, M., Boillat, J., Houdant, B. and Delorme, F. 2009, Spillway capacity upgrade of Gloriettes Dam: Environmental integration and energy dissipation, *HYDRO 2009*, Lyon, France.

Blanc, P. and Lempérière, F. 2001, Labyrinth spillways have a promising future, *International Journal of Hydropower and Dams*, 8(4), 129–131.

Cicero, G.-M., Guene, C., Luck, M., Pinchard, T., Lochu, A. and Brousse, P.-H. 2010, Experimental optimization of a Piano Key Weir to increase the spillway capacity of the Malarce dam, *1st IAHR European Congress*, Edinburgh, United Kingdom.

Cicero, G.-M., Barcouda, M., Luck, M. & Vettori, E. 2011, Study of piano-key morning glory to increase the spillway capacity of the Bage dam, *Labyrinth and piano key weirs-PKW 2011*, CRC Press, London, 199–206.

Crookston, B. and Tullis, B. 2010, Hydraulic performance of labyrinth weirs, *in proceedings of International Junior Researcher and Engineer Workshop on Hydraulic Structures*, Edinburgh, UK.

Dabling, M.R. and Tullis, B.P. 2012, Piano Key Weir Submergence in Channel Applications, *Piano Key Weir for in-stream storage and dam safety (PKWISD-2012)*, New Delhi, India, 13–23.

Das Singhal, G. and Sharma, N. 2011, Rehabilitation of Sawara Kuddu Hydroelectric Project – Model studies of Piano Key Weir in India. *Labyrinth and piano key weirs-PKW 2011*, CRC Press, London, 241–250.

Dugué, V., Hachem, F., Boillat, J.L., Nagel, V., Roca, J.P. and Laugier, F. 2011, PK Weir and flap gate spillway for the Gage II Dam, *Labyrinth and piano key weirs-PKW 2011*, CRC Press, London, 35–42.

Erpicum, S., Machiels, O., Dewals, B.J., Archambeau, P. and Pirotton, M. 2010a, Contribution to the study of Piano Key Weir hydraulics, *IECS 2010*, Graz, Austria.

Erpicum, S., Machiels, O., Archambeau, P., Dewals, B.J. and Pirotton, M. 2010b, 1D numerical approach to model the flow over a Piano Key Weir (PKW), *SimHydro 2010*, Nice, France.

Erpicum, S., Machiels, O., Archambeau, P., Dewals, B. and Pirotton, M. 2011a, 1D numerical modeling of the flow over a Piano Key Weir, *Labyrinth and piano key weirs-PKW 2011*, CRC Press, London, 151–158.

Erpicum, S., Machiels, O., Archambeau, P., Dewals, B., Pirotton, M. and Daux, C. 2011b, Energy dissipation on a stepped spillway downstream of a Piano Key Weir – Experimental study, *Labyrinth and piano key weirs-PKW 2011*, CRC Press, London, 105–112.

Erpicum, S., Nagel, V. and Laugier, F. 2011c, Piano Key Weir design study at Raviege dam, *Labyrinth and piano key weirs-PKW 2011*, CRC Press, London, 43–50.

Erpicum, S., Machiels, O., Archambeau, P., Pirotton, M. and Dewals, B. 2012a, Hydraulic modelling of Piano Key Weirs: A composite approach, *Piano Key Weir for in-stream storage and dam safety (PKWISD-2012)*, New Delhi, India, 53–64.

Erpicum, S., Machiels, O., Dewals, B., Pirotton, M. and Archambeau, P. 2012b, Numerical and physical hydraulic modelling of Piano Key Weirs, *ASIA 2012 – 4th International Conference on Water Resources and Renewable Energy Development in Asia*, Chiang Mai, Thailand.

Erpicum, S., Machiels, O., Pirotton, M., Nagel, V. and Laugier, F. 2012c, Piano Key Weir (PKW) solution to upgrade Raviège Dam spillway (France), *24th ICOLD congress*, Kyoto, Japan.

Hien, T.C., Son, H.T. and Ho Ta Khanh, M. 2006, Results of some piano keys weir hydraulic model tests in Vietnam, *in proceedings of 22ème CIGB/ICOLD*, Barcelona.

Ho Ta Khanh, M., Hien, T.C. and Hai, N.T. 2011a, Main results of the P.K weir model tests in Vietnam (2004 to 2010), *Labyrinth and piano key weirs- PKW 2011*, CRC press, London, 191–198.

Ho Ta Khanh, M., Sy Quat, D. and Xuan Thuy, D. 2011b, P.K weirs under design and construction in Vietnam (2010), *Labyrinth and piano key weirs- PKW 2011*, CRC Press, London, 225–232.

Ho Ta Khanh, M., Hien, T.C. and Quat, D.S. 2012, Research and development of P.K. Weirs in Vietnam since 2004, *Piano Key Weir for in-stream storage and dam safety (PKWISD-2012)*, New Delhi, India, 91–96.

Ho Ta Khanh, M., Hien, T.C., Pralong, J. and Pinchard, T. 2013, Utilization of PK Weirs for Low Barrages, *AFRICA 2013*, Addis Ababa, Ethiopia, 16–18 April 2013.

Kabiri-Samani, A.R., Javaheri, A., 2012, Discharge coefficients for free and submerged flow over Piano Key weirs, *Journal of Hydraulic Research*, 50(1), 114–120.

Karelle, Y. 2008, Etude expérimentale du comportement hydraulique des déversoirs en touches de piano, *Master thesis*, ArGEnCo, University of Liege, Liege, Belgium.

Laugier, F. 2007, Design and construction of the first Piano Key Weir (PKW) spillway at the Goulours dam, *International Journal of Hydropower and Dams*, 14(5), 94–101.

Laugier, F., Lochu, A., Gille, C., Leite Ribeiro, M. and Boillat, J.-L. 2009, Design and construction of a labyrinth PKW spillway at Saint-Marc dam, France, *International Journal of Hydropower and Dams*, 16(5), 100–107.

Laugier, F., Pralong, J., Blancher, B. and Montarros, F. 2011a, Development of a new concept of Piano Key Weir spillway to increase low head hydraulic efficiency: Fractal PKW, *Labyrinth and piano key weirs-PKW 2011*, CRC press, London, 281–288.

Laugier, F., Pralong, J. and Blancher, B. 2011b, Influence of structural thickness of sidewalls on PKW spillway discharge capacity, *Labyrinth and piano key weirs-PKW 2011*, CRC Press, London, 159–165.

Laugier, F., Vermeulen, J. and Pralong, J. 2012, Achievement of New Innovative Labyrinth Piano Key Weir Spillways (PKW), *in proceedings of Piano Key Weir for in-stream storage and dam safety (PKWISD-2012)*, New Delhi, India, 25–42.

Le Doucen, O., Leite Ribeiro, M., Boillat, J.-L., Schleiss, A. and Laugier, F. 2009, Etude paramétrique de la capacité des PK-Weirs, *Modèles physiques hydrauliques-outils indispensables du XXIe siècle*, SHF, Lyon.

Leite Ribeiro, M., Albalat, C., Boillat, J.-L., Schleiss, A.J. and Laugier, F. 2007a, Rehabilitation of St-Marc dam. Experimental optimization of a piano key weir, *32nd IAHR Congress*, Venice, Italy.

Leite Ribeiro, M., Boillat, J., Kantoush, S., Albalat, C., Laugier, F. and Lochu, A. 2007b, Rehabilitation of St-Marc dam–Model studies for the spillways, *HYDRO 2007*, Granada, Spain.

Leite Ribeiro, M., Bieri, M., Boillat, J.-L., Schleiss, A.J., Delorme, F. and Laugier, F. 2009, Hydraulic capacity improvement of existing spillways – Design of Piano Key Weirs, *23rd CIGB/ICOLD Congress*, Brasilia.

Leite Ribeiro, M., Boillat, J.-L., Schleiss, A.J., Laugier, F. 2011a, Coupled spillway devices and energy dissipation system at St-Marc Dam (France), *Labyrinth and piano key weirs-PKW 2011*, CRC press, London, 113–121.

Leite Ribeiro, M., Boillat, J.L. and Schleiss, A.J. 2011b, Experimental parametric study for hydraulic design of PKWs, *Labyrinth and piano key weirs-PKW* 2011, CRC press, London, 183–190.

Lempérière, F. and Ouamane, A. 2003, The piano keys weir: a new cost-effective solution for spillways, *International Journal of Hydropower and Dams*, 10(5), 144–149.

Lempérière, F. 2009, New Labyrinth weirs triple the spillways discharge – Data for an easy design of P.K.Weir. Available online on www.hydrocoop.org.

Lempérière, F., Vigny, J.P. and Ouamane, A. 2011, General comments on Labyrinths and Piano Key Weirs: The past and present, *Labyrinth and piano key weirs-PKW 2011*, CRC press, London, 17–24.

Machiels, O., Erpicum, S., Archambeau, P., Dewals, B.J. and Pirotton, M. 2009a, Large scale experimental study of piano key weirs, *33rd IAHR Congress*, ASCE, Vancouver, Canada.

Machiels, O., Erpicum, S., Archambeau, P., Dewals, B.J. and Pirotton, M. 2009b, Analyse expérimentale du fonctionnement hydraulique des déversoirs en touches de piano, *Colloque CFBR-SHF: "Dimensionnement et fonctionnement des évacuateurs de crues"*, Paris, France.

Machiels, O., Erpicum, S., Archambeau, P., Dewals, B. and Pirotton, M. 2010a, Analyse expérimentale de l'influence des largeurs d'alvéoles sur la débitance des déversoirs en touches de piano, *La Houille Blanche* 2010(2), 22–28.

Machiels, O., Erpicum, S., Archambeau, P., Dewals, B.J. and Pirotton, M. 2010b, Experimental study of the hydraulic behavior of Piano Key Weirs, *17th IAHR-APD Congress*, Auckland, New Zeeland.

Machiels, O., Erpicum, S., Archambeau, P., Dewals, B.J. and Pirotton, M. 2010c, Piano Key Weirs, experimental study of an efficient solution for rehabilitation, *in proceedings of FRIAR*, Wessex Institute of Technology, Milano, Italia, 95–106.

Machiels, O., Erpicum, S., Archambeau, P., Dewals, B.J. and Pirotton, M. 2010d, Hydraulic behaviour of Piano Key Weirs: Experimental approach, *3rd International Junior Researcher and Engineer Workshop on Hydraulic Structures, IJREWHS'10*, Edinburgh, United Kingdom.

Machiels, O., Erpicum, S., Archambeau, P., Dewals, B. and Pirotton, M. 2011a, Influence of the alveoli slopes on the discharge capacity of Piano Key weirs, *34th IAHR congress*, Brisbane, Australia.

Machiels, O., Erpicum, S., Archambeau, P., Dewals, B. and Pirotton, M. 2011b, Varying the alveoli slopes of Piano Key Weirs: Interests and limitations, *6th International Conference on Dam Engineering*, Lisbon, Portugal.

Machiels, O., Erpicum, S., Archambeau, P., Dewals, B.J. and Pirotton, M. 2011c, Influence of Piano Key Weir height on its discharge capacity, *Labyrinth and piano key weirs-PKW 2011*, CRC Press, London, 59–66.

Machiels, O., Erpicum, S., Archambeau, P., Dewals, B.J. and Pirotton, M. 2011d, Piano Key Weir preliminary design method – Application to a new dam project, *Labyrinth and piano key weirs-PKW 2011*, CRC Press, London, 199–206.

Machiels, O., Erpicum, S., Dewals, B., Archambeau, P. and Pirotton, M. 2011e, Experimental observation of flow characteristics over a Piano Key Weir, *Journal of hydraulic research*, 49(3), 359–366.

Machiels, O., Erpicum, S., Archambeau, P., Dewals, B. and Pirotton, M. 2012a, On the Piano Key Weir hydraulics, *2nd IAHR Europe Congress*, Munich, Germany.

Machiels, O., Erpicum, S., Archambeau, P., Dewals, B. and Pirotton, M. 2012b, Influence of the relative alveoli widths on Piano Key Weirs efficiency for varied weir heights, *24th ICOLD congress*, Kyoto, Japan.

Machiels, O., Erpicum, S., Archambeau, P., Dewals, B. and Pirotton, M. 2012c, Influence of weir height and keys slope on PKW discharge, *4th IAHR International Symposium on Hydraulic Structures*, Porto, Portugal.

Machiels, O., Erpicum, S., Pirotton, M., Dewals, B. and Archambeau, P. 2012d, Experimental analysis of PKW hydraulic performance and geometric parameters optimum, *Piano Key Weir for in-stream storage and dam safety (PKWISD-2012)*, New Delhi, India, 97–114.

Ouamane, A. 2006, Hydraulic and costs data for various Labyrinth Weirs, *in proceedings of 22ème CIGB/ICOLD Congress*, Barcelona.

Ouamane, A. and Lempérière, F. 2006a, Nouvelle conception de déversoir pour l'accroissement de la capacité des retenues des barrages, *Colloque international sur la protection et la préservation des ressources en eau*, Blida, Algérie.

Ouamane, A. and Lempérière, F. 2006b, Design of a new economic shape of weir, *International Symposium on Dams in the Societies of the 21st Century*, Barcelona, Spain, 463–470.

Ouamane, A. and Lempérière, F. 2007, Increase of the safety of dams. Rehabilitation of weirs, *ICOLD 75th annual meeting*, St Petersburg.

Pfister, M., Tullis, B. and Schleiss, A. 2013, Effect of driftwood on hydraulic head of Piano Key Weirs, *Labyrinth and piano key weirs-PKW 2013*, CRC press, London.

Pinchard, T., Boutet, J.M. and Cicero, G.-M. 2011, Spillway capacity upgrade at Malarce dam: Design of an additional Piano Key Weir spillway, *Labyrinth and piano key weirs-PKW 2011*, CRC Press, London, 233–240.

Pralong, J., Montarros, F., Blancher, B. and Laugier, F. 2011a, A sensitivity analysis of Piano Key Weirs geometrical parameters based on 3D numerical modeling, *Labyrinth and piano key weirs-PKW 2011*, CRC Press, London, 133–139.

Pralong, J., Vermeulen, J., Blancher, B., Erpicum, S., Machiels, O., Pirotton, M., Boillat, J-L., Leite Ribeiro, M. and Schleiss, A.J. 2011b, A naming convention for the Piano Key Weirs geometrical parameters, *Labyrinth and piano key weirs-PKW 2011*, CRC press, London, 271–278.

Reynolds, P. 2011, PK weirs in Vietnam, *Journal of International Water Power & Dam Construction*, January 2011.

Sharma, N. and Das Singhal, G. 2012, Physical modelling of Piano Key Weir for Sawra Kuddu Hydro Electric Project in India, *Piano Key Weir for in-stream storage and dam safety (PKWISD-2012)*, New Delhi, India, 123–132.

Silvestri, A., Erpicum, S., Archambeau, P., Dewals, B. & Pirotton, M. 2013. Stepped spillway downstream of a Piano Key Weir – Critical length for uniform flow. In Bung, D. & Pagliara, S. (Eds), *IWLHS 2013*, Karlsruhe: Bundesanstalt für Wasserbau (BAW), 99–110.

Vermeulen, J., Laugier, F., Faramond, L. and Gille, C. 2011, Lessons learnt from design and construction of EDF first Piano Key Weirs, *Labyrinth and piano key weirs-PKW 2011*, CRC press, London, 215–224.

Labyrinth and Piano Key Weirs II – PKW 2013 – Erpicum et al. (eds)
© 2014 Taylor & Francis Group, London, ISBN 978-1-138-00085-8

Improvement of labyrinth weirs shape

A. Ouamane
Laboratory of Hydraulic Planning and Environment (LAHE), University of Biskra, Algeria

ABSTRACT: The labyrinth spillway is an effective means in order to increase the capacity of the dam spillway. The concept of the labyrinth is to change the plan form of the crest then to increase its actual length and permit the passage of a higher discharge. This type of weir is particularly suitable for sites where topographic conditions limit the extent of its width. In addition, the labyrinth weir is characterized by an economic cost of construction and a more important efficiency than the standard weir. However, the non-rectilinear shape of crest labyrinth weir creates a disturbed three-dimensional flow at each discontinuity of the weir axis. These flow conditions reduce the performance of labyrinth weir. It is possible to make some changes to the classical shape of labyrinth weir to improve its hydraulic performance and make it more economical. These changes concern the plan shape of crest labyrinth weir, the form of the upstream and downstream apron of the cells and the shape of the walls of the spillway.

1 INTRODUCTION

The labyrinth weir is often used when the width of the weir or the maximum head is limited. The cost of the labyrinth weir is relatively low compared to weirs with gates, this led to its use simultaneously to increase the storage and discharge capacity of the spillway.

Although the labyrinth weir has been the subject of several applications, some changes in classic shape can have a positive effect on the hydraulic and economic performance.

The geometry in the plane of the crest of the labyrinth weir is characterized by a discontinuous centerline. This thus results in a complex flow mode (Lux & Hinchliff, 1985), which is considered as three-dimensional.

The literature shows that the flow over a labyrinth weir is influenced by the angle formed at the point of discontinuity of the crest. The effect of the latter causes a reduction of the discharge coefficient which is amplified when the head on the weir increases.

The analysis carried out by Indlekofer & Rouve (1975) showed that the flow on the labyrinth weir is rather complicated with a three-dimensional flow. Thus, the flow over the weir is divided into two areas, one near the corner characterized by a disturbed three-dimensional flow area, followed by two-dimensional flow area. The area defined by the disturbed three-dimensional flow is characterized by a discharge coefficient lower than that of the area to be two-dimensional flow.

The importance of the reduction of discharge coefficient depending of the head on the weir and the importance of the angle of the corner. Thus, for a maximum head on the sill of weir, the disturbed flow affects the entire length of the crest, therefore, lower values of discharge coefficient are obtained.

The study conducted at the University of Biskra in 1998 (Unpublished report) for four models of labyrinth weirs of triangular shape and different angles, showed that the coefficient of discharge is minimal for the lowest values of the angle and increases with the increase thereof.

It has to be noted that it is possible to make some changes in the configuration of the labyrinth weir to reduce the flow disturbance and consequently improve the hydraulic efficiency.

Figure 1. Labyrinth weir model with rounded front wall.

Figure 2. Model labyrinth weir with flat front wall.

2 IMPROVING THE ENTRANCE SHAPE

The main parameter that promotes flow disturbance and therefore reduce the performance of the labyrinth weir, corresponds to the discontinuity of the weir crest. Thus, to reduce the effect of this discontinuity, the profiling of the front wall of the labyrinth weir may have a positive effect on the hydraulic efficiency.

The shape of labyrinth weir proposed by Blanc & Lempérière (2001) is one of the economic forms of weir. This configuration is essentially characterized by a rounded entrance which improves its hydraulic performance.

An experimental analysis of the hydraulic performance has concerned two models of labyrinth weirs with the same geometrical sizes and a trapezoidal shape, the first with a flat shape of the upstream wall and the second with a rounded shape.

The comparison of results obtained on two models of weir (Fig. 3) noted that the design of a rounded shaped entrance leads for an improving of the performance of about 10%. This can be explained on the one hand, by the profiled shape of entrance which facilitates the flow on each side of the front wall along the latter. On the other hand, the rounded shape eliminates the discontinuity points and thus reduces the effect of the disturbance at the top of the weir. These two effects of the profiled shape have a positive affect the hydraulic performance.

3 ADAPTATION OF LABYRINTH WEIR FOR EVACUATION OF LARGE DISCHARGE

During the design the labyrinth weir of Ute dam in New Mexico (USA) the USBR has encountered difficulties relating to the dimensions of the weir walls (Hinchliff & Houston, 1984). Structural

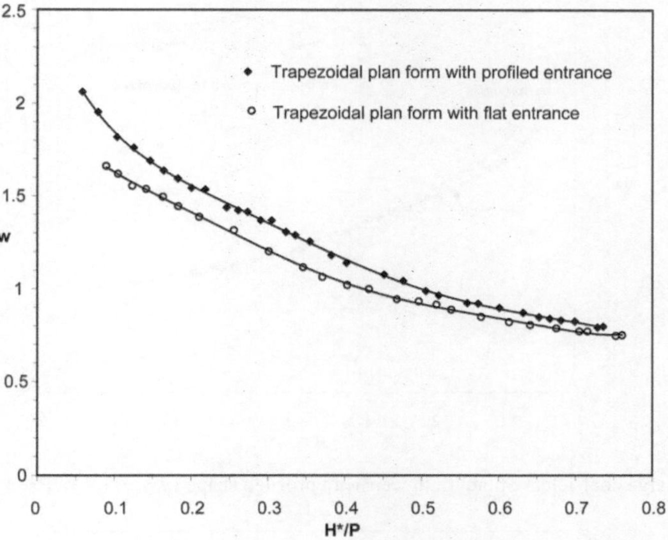

Figure 3. Flow coefficients according to the entrance shape of labyrinth weir L/W = 3.9 W/P = 1.

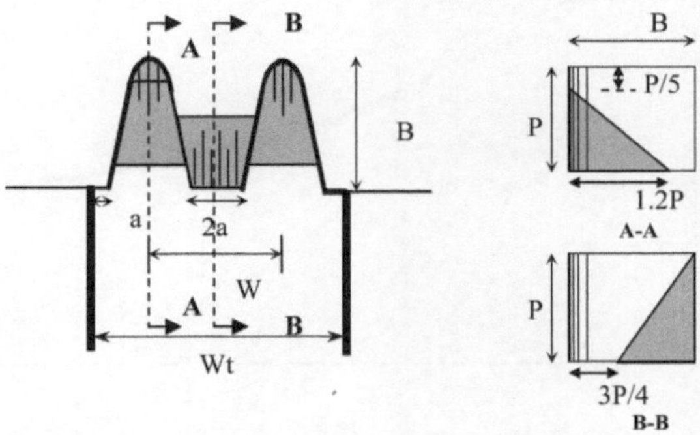

Figure 4. Schematic of labyrinth weir with entrance profiled and alveoli partially filled.

analysis of the labyrinth weir of Ute dam showed that the large hydrostatic heads caused by the height of the walls (P = 9 m) and the head on the sill of labyrinth (H = 5.8 m) required large quantities of reinforcement steel (03 tons/ml) (Lux III & Hinchliff, 1985). This makes the use of non-economic labyrinth weir for high discharges. As reported by the literature, the labyrinth weir is an economical solution as the specific discharge is less than 50 $m^3/s/m$. Beyond this value the labyrinth weir requires great height walls which implies a greater wall thickness and greater reinforcement. For higher specific discharge (50 $m^3/s/m$) the walls require an important quantity of steel (up to 200 $kg/m^3/s$) (Blanc & Lempérière, 2001).

A solution which helps to reduce the additional cost consists in filling the lower portion of the alveoli of the labyrinth by an ordinary concrete. This results in a reduction of the free height of walls, which allows to have hydrostatic pressure forces which act only moderately on the upper portion of the wall. This allows to have a smaller wall thicknesses and therefore a low reinforcement of the structure.

However, the partial filling of the alveoli can result in a reduction in the performance of labyrinth weir. To compensate for this loss of performance it would useful to choose a form of a labyrinth which has a high performance.

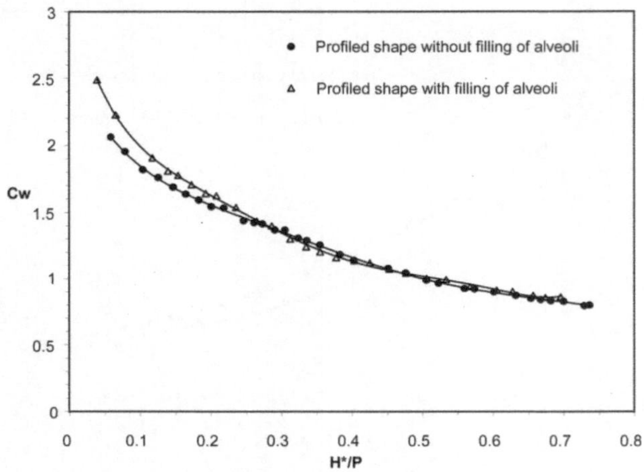

Figure 5. Discharge coefficient of labyrinth weir with profiled shape L/W = 3.9 W/P = 1.

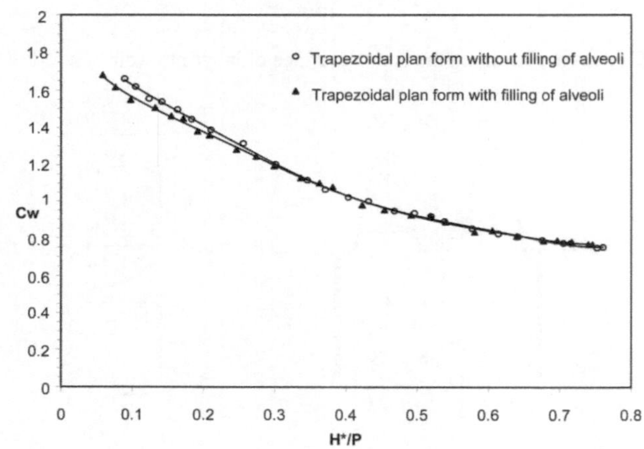

Figure 6. Discharge coefficient of labyrinth weir trapezoidal L/W = 3.9 W/P = 1.

To check the effect of partial filling of the alveoli on the hydraulic performance, two models of labyrinth weir are tested. The first with a classic trapezoidal plan form and the second with a profiled shape of entrance. The tests are carried out under conditions with and without filling of the alveoli.

The observation of the flow during the experiment showed that the flow on the weir with partially filled alveoli and rather airy and stable for low charge, which is not the case for weirs without filling alveoli. This justifies the offset between the two curves of discharge coefficient for the range of values of the relative head $H^*/P < 0.25$ (Fig. 5 and 6).

The comparison of curves discharge coefficient for both cases of functioning with and without filling (Fig. 5 and 6) shows that alveoli filling has no effect than for the low heads. For the range of values of $H^*/P > 2.5$, the two weirs have the same hydraulic performance. So, the partial filling of the alveoli has no effect on the performance of labyrinth weir and helps to avoid the problem of aeration for low heads on the weir.

As well, it's recommended for the greater specific discharges than $50\ m^3/s/ml$ to design the labyrinth weir with the alveoli partially filled in ordinary concrete this gives an economic structure while maintaining hydraulic performance similar to that of classical labyrinth weir.

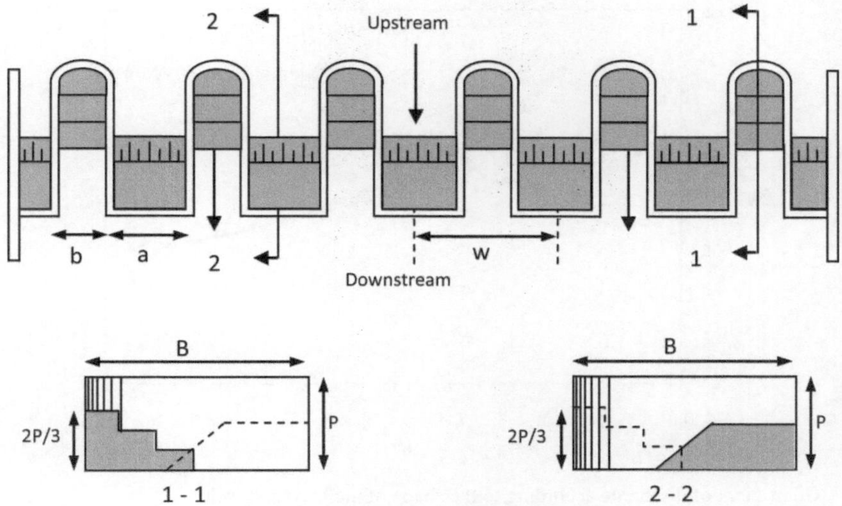

Figure 7. Rectangular labyrinth weir with shaped entrance and the alveoli partially filled with: a the width of upstream alveolus, b the width of downstream alveolus, B the length of the side wall and w the width of cycle of labyrinth.

4 IMPROVEMENT OF THE PLAN FORM

It is possible that other plan form can be effective as the trapezoidal and triangular shapes often used in existing dams. Thus, for practical, economic and hydraulic performance reasons, we may adopt a rectangular plan form with a profiled entrance and the apron of the downstream alveoli as a stairs and that of the upstream alveoli partially inclined. These structural particularities can improve the performance of labyrinth spillway and reduce the cost of construction (Fig. 7).

The choice of the rectangular shape allows vary the widths of the upstream and downstream cells to obtain a ratio of width which allows for better hydraulic efficiency and reduced cost (Ouamane & Ben Said, 2010).

The profiled shape of the entrance and the inclination of part of the apron of the upstream alveoli can improve the flow conditions at the entrance of the labyrinth; therefore the weir becomes more efficient. The partial filling of alveoli upstream and downstream of ordinary concrete can reduce the thickness of the walls and the quantity of steel, because the free part of the walls is reduced. The design of the apron of the downstream alveoli as stair step, also help to dispel some of the energy.

4.1 *Comparative study with the trapezoidal shape*

Often the labyrinth weir is designed with a trapezoidal shape in plan that is repeated periodically (Hinchliff & Houston, 1984). This is not justified; it is possible that the labyrinth weir take another geometric shape that can be as effective as the trapezoidal shape. The choice of a rectangular shape can be as effective of viewpoint, hydraulic and economic performance and is easy for construction, especially when it is used as a dam weir of low height in the rivers. The experiment of two models of labyrinth weirs of the same dimensions, successively, trapezoidal and rectangular rounded upstream showed that the rectangular shape can be as effective as the trapezoidal shape and even more effective for the relative heads lower than 0.5, corresponding to the practical range for the design of the labyrinth weirs (Fig. 8).

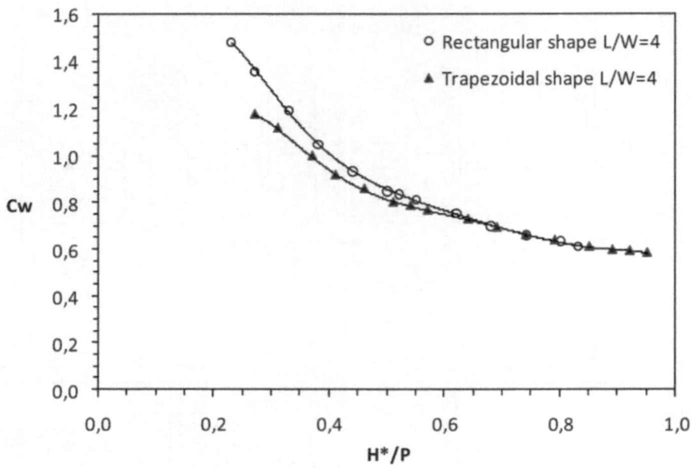

Figure 8. Coefficient of discharge according to the shape of the labyrinth weir.

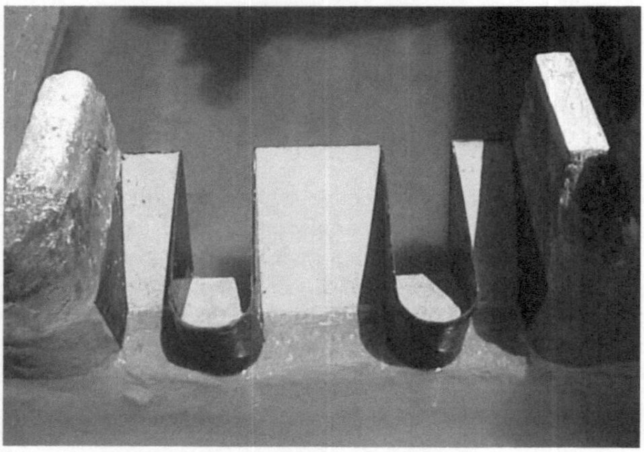

Figure 9. Model of the labyrinth weir with a downstream overhang.

5 IMPROVING THE SHAPE CROSS

Usually the labyrinth weirs are designed with vertical walls disposed frontally or laterally to the flow direction. This arrangement requires a substantial length of the side walls to achieve effective performance. This limits the application of the labyrinth weir and does not allow its installation on most crests of concrete dams. It is possible to reduce the length of the base of the side walls without affecting the length of the sill of labyrinth weir, therefore, maintain or even increase the efficiency of this type of weir. This is feasible by using a part of the length of the side walls in overhang upstream and downstream or only on one side. In this case the front walls become inclined and the upstream and downstream apron of alveoli can also be tilted.

This provision has been the subject of a test model labyrinth weir with downstream overhang and an inclined apron of the upstream and downstream alveoli. The results showed that this weir configuration provides a better performance which can be three times that the linear weir for medium heads on weir (Fig. 10).

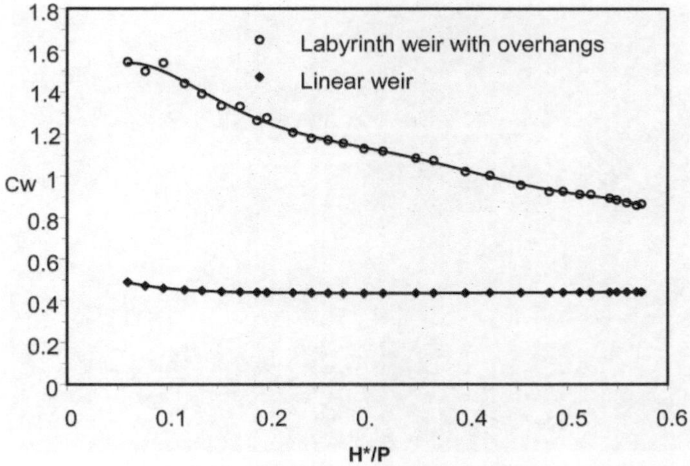

Figure 10. Curves of discharge coefficients of labyrinth weir with a downstream overhang and linear weir.

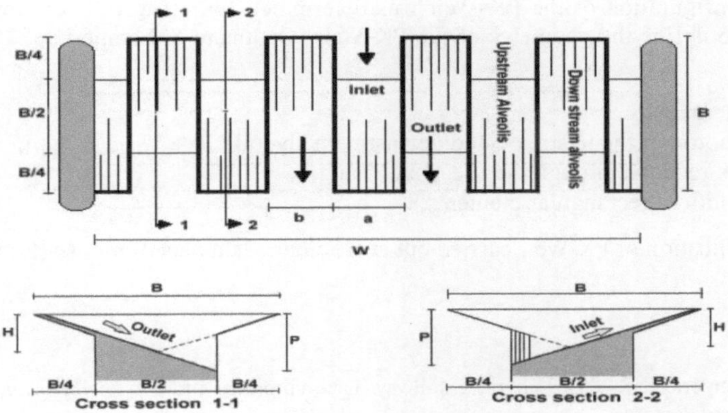

Figure 11. Piano Keys Weir model A.

6 CONCLUSION

So this shows that this type of weir is a similar alternative of the viewpoint hydraulic efficiency to that of the labyrinth weir with vertical walls. The disposition of the inclined apron between the side walls leads to reduction in the average height of the weir. Thus, the walls of the weir become less thick than the weir with horizontal apron due to the reduction of forces applied to the walls, which reduces the quantity of steel reinforcement. However, the volume of the concrete of the apron becomes larger. The use of inclined apron minimizes the effect of turbulence at the weir crest, producing a flow of the threads of liquid which are more or less regular.

The application of the previously mentioned improvements to the labyrinth weir can lead to a form of economic weir and of a better hydraulic performance. The piano keys weir (PK-Weir) brings all these advantages, it can be used for new dams or existing dams that require increase in the spillway capacity and/or storage capacity. It can be placed on small sections of existing or new concrete dams and allows discharging of specific flows up to 100 $m^3/s/m$ and multiply by at least three the discharge weir Creager (Lempérière & Ouamane, 2003).

Figure 12. The piano keys weir (PK-Weir) carried out in 2006 to the Goulours dam (France).

The final configuration of the PK-Weir was determined following tests on physical models, which permit to define the geometry of the PK-Weir (Ouamane & Lempérière, 2006), which is based on:

- A rectangular layout.
- An inclined apron of the upstream and downstream alveoli.
- A base length reduced through the use of overhang.
- A reduced width of rectangular elements.

The first application of PK Weir carried out at Goulours dam near Toulouse (France) (Fig. 12).

REFERENCES

Blanc, P., and Lempérière, F. 2001. Labyrinth spillways have a promising future, *The International Journal on Hydropower and dams*, 4.

Hinchliff, D.L., and Houston, K.L. 1984. Hydraulic Design and Application of Labyrinth Spillways, *Division of Research Engineering and Research Center, Bureau of Reclamation, Denver, Colorado.*

Lempérière, F., and Ouamane, A. 2003. The piano keys weir: a new cost-effective solution for spillways. *The international journal on Hydropower and Dams* 10(5).

Lux III, F., and Hinchliff, D.L. 1985. Design and Construction of Labyrinth Spillways, *15th Congress on Large Dams, ICOLD*, Vol. IV, Q59, R 15, pp. 249–274, Lausanne.

Ouamane, A., and Lempérière, F. 2006. Design of a new economic shape of weir, *International Symposium on Dams in the Societies of the XXI Century*. Barcelona.

Ouamane, A., and Ben Said, M. 2010. Economic conceptions of the labyrinth weir allowing the improvement of the management of floods and the increase of the useful capacity of the reservoir, *Rheinisch-Westfälische Technische Hochschule* Aachen, Germany, pp. 303–311. 158.

Labyrinth and Piano Key Weirs II – PKW 2013 – Erpicum et al. (eds)
© 2014 Taylor & Francis Group, London, ISBN 978-1-138-00085-8

Embankment dam spillways and energy dissipators

H. Chanson
The University of Queensland, School of Civil Engineering, Brisbane, Australia

ABSTRACT: For the last fifty years, the design floods of a number of embankment dams were re-evaluated and the revised spillway outflows are often larger than the original design discharges. Several embankment overtopping protection systems were developed for earthfill structures, and the applications range from river dykes to tsunami protections including embankment dams. Well-known designs include timber cribs, sheet-piles, riprap and gabions, reinforced earth, minimum energy loss (MEL) weirs, embankment overflow stepped spillways and the precast concrete block protection systems. In this review, several design techniques are reviewed and discussed based upon prototype experiences. A critical analysis of their performances highlights that a safe operation of embankment dam spillways and associated energy dissipators relies upon a sound design and a good quality of construction, suitable flow conditions, together with regular maintenance.

1 INTRODUCTION

Worldwide the design floods of numerous reservoirs were re-assessed and the revised discharge outflows are typically larger than the original design flow rate. The occurrence of such large flood events would result in dam overtopping, with catastrophic consequences in the case of embankment dams when an insufficient storage or spillway capacity is available. A number of overtopping protection systems were developed for embankment dams. These design techniques encompass timber cribs, sheet-piles, riprap and gabions, reinforced earth and concrete overtopping protection systems including minimum energy loss (MEL) weirs, embankment overflow stepped spillways, precast concrete block protection systems developed by the Russian engineers (ASCE 1994, Chanson 2009). In a broad sense an embankment is an earthfill structure designed to hold water. The definition includes river dykes, coastal barriers for storm surge and tsunami protections, as well as natural lakes and landslide dams (Fig. 1). Figure 1 presents some embankment dam structures: Figures 1a and 1b show some relatively large dams; Figures 1b to 1d illustrate some in-stream embankment structures. Figures 1b and 1d highlight the downstream energy dissipator operation during small flood events.

All embankment dams are potentially erodible when overtopped, unless an overtopping protection system is designed. During the last three centuries, a number of embankment structures failed worldwide (Smith 1971, Schnitter 1994). The most common causes of failures were dam overtopping and internal cracking. The former is linked with insufficient flood release capacity, and the latter results from a combination of poor understanding of geotechnical concepts, inappropriate construction standards and internal failure. Still today, embankment dam overtopping occurs during extreme rainfall events because of inadequate spillway capacity: e.g., Tous dam (Spain), Lake Ha! Ha! (Canada), Opuha dam (New Zealand), Glashüttte dam failure (Germany).

A number of embankment dam overtopping protection systems were developed during the last few decades. Herein several design techniques are presented and discussed, after a brief discussion of the embankment breaching process. The prototype experience gained during the past decades is analysed and discussed.

Figure 1. Photographs of embankment dam and spillways (a) Melton dam, Melton VIC (Australia) on 30 January 2000; (b) Chinchilla minimum energy loss (MEL) weir, Chinchilla QLD (Australia) in operation on 8 November 1997; (c) Timber crib structure: Greenup weir, Inglewood QLD (Australia) on 21 January 2009 (Courtesy of Damien Roman); (d) Sheet-pile concrete slab embankment: Joe Sippel weir, Murgon QLD (Australia) in operation on 5 March 2013.

2 EMBANKMENT BREACH DEVELOPMENT

The breaching process of an embankment dam is a relatively slow process, in comparison to the failure of a concrete arch dam. The latter may be a sudden, explosive failure, while an earthfill structure may be overtopped for sometimes before the breach develops progressively leading to the complete failure (Fig. 3). For example, the failure of the 100 m high Teton dam (USA) started around 11:00 am and the reservoir was drained by the evening; the breaching of the Zeyzoun dam (Syria) took more than three and half hours. In one case (Glashütte dam), witness reports indicated that the grass-lined downstream slope of the embankment was overtopped for more than 3 hours before the complete dam failure (Bornschein and Pohl 2003). Figure 2 presents some photographs of an embankment breaching experiment.

A number of studies on embankment breaching were conducted during the last fifteen years (Coleman et al. 2002, Rozov 2003, Chanson 2005, Dai et al. 2005, Hanson et al. 2005, Morris et al. 2007, ASCE/EWRI Task Committee on Dam/Levee Breaching 2011). Most experimental studies under carefully-controlled laboratory flow conditions, together with prototype observations, showed that the embankment breach starts with an initiation phase, followed by a rapid development of the breach, and then an enlargement of breach width once the breach invert reaches the main channel bed substrate. The visual observations highlighted a challenging similarity between the embankment breach process and the flow in a minimum energy loss weir inlet during the breach development (McKay 1970, Visser et al. 1990, Chanson 2004). This was nicely illustrated by two seminal laboratory studies of the breaching of non-cohesive embankment structures (Coleman et al. 2002, Rozov 2003). It is also seen in Figure 2. Figure 2a show the initial stages of the breach development, with a definition sketch in Figure 2b. Figure 3 shows some quantitative results based upon the re-analysis of data by Coleman et al. (2002).

A flow net analysis was performed with the laboratory data of Coleman et al. (2002). The flow cross-section areas were measured along the equipotential planes at different locations and times. Some typical results are shown in Figure 3a, with the breach cross-sectional shapes below the water line at several longitudinal sections for a given time t since the start of breach development. The results highlighted that the embankment breach flow was transcritical. Namely the flow was nearly critical between the inlet lip and throat, and the total head remained constant (Figs. 3b & 3c). The head losses occurred downstream of the throat when the flow streamlines diverged and some flow separation occurred at the lateral boundaries. Figure 3b and 3c show experimental results with the dimensionless total head H/H_1 as a function of the dimensionless centreline location, where H_1 is the upstream total head above downstream channel elevation and L is the embankment base length. Visual observations showed that the flow through the breach between the inlet lip to throat was very similar to the flow through a minimum energy loss (MEL) spillway inlet during the breach development. See, for example, a comparison Figure 2b and Figure 4.

The length of breach inlet, measured along the breach centreline between inlet lip and throat, was typically $L_{inlet}/B_{max} = 0.5$ to 0.6, where B_{max} is the free-surface width at the upper inlet lip (Fig. 2b). During the development of the breach, the outflow discharge satisfied the Bernoulli equation at the inlet lip:

$$Q = C_D \times B_{max} \times \sqrt{g \times \left(\frac{2}{3} \times H_1\right)^3} \quad (1)$$

where C_D = dimensionless discharge coefficient with $C_D \sim 0.6$ on average, and the breach dimensions increase with time, resulting in the hydrograph of the breach. That is, in Equation (1), both the breach free-surface width B_{max} and upstream total head H_1 are functions of time as well as embankment characteristics and reservoir size. For a semi-infinitely long reservoir, a re-analysis of embankment breach data showed that the inlet lip elevation z_{lip}, the inlet lip width B_{max} and the throat width B_{min} varied with time as:

$$\frac{z_{lip}}{H_1} = 1.08 \times \exp\left(-0.0013 \times t \times \sqrt{\frac{g}{H_1}}\right) \quad (2)$$

$$\frac{B_{max}}{H_1} = 2.73 \times 10^{-4} \times \left(t \times \sqrt{\frac{g}{H_1}}\right)^{1.4} \quad (3)$$

(a) Photographic observations: the first four shots were taken 5.2 s apart; the last two shots were taken respectively 50 s and 59 s after the first photograph

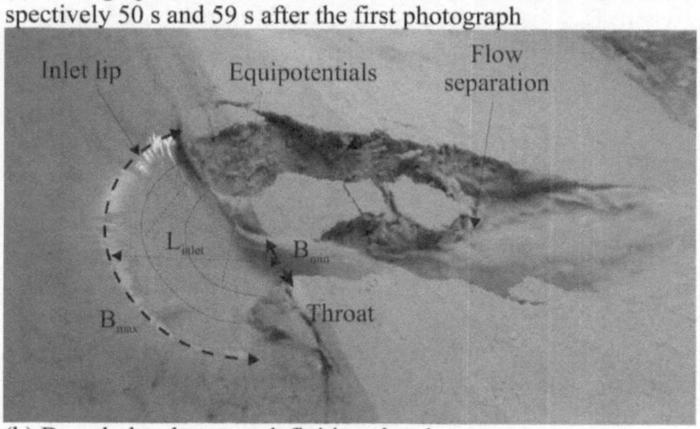

(b) Breach development definition sketch

Figure 2. Physical modelling of non-cohesive embankment overtopping and breaching at the University of Auckland – H = 0.3 m, L = 1.5 m, d_{50} = 0.3 mm, constant upstream head experiment – Flow direction from left to right with the reservoir on the right.

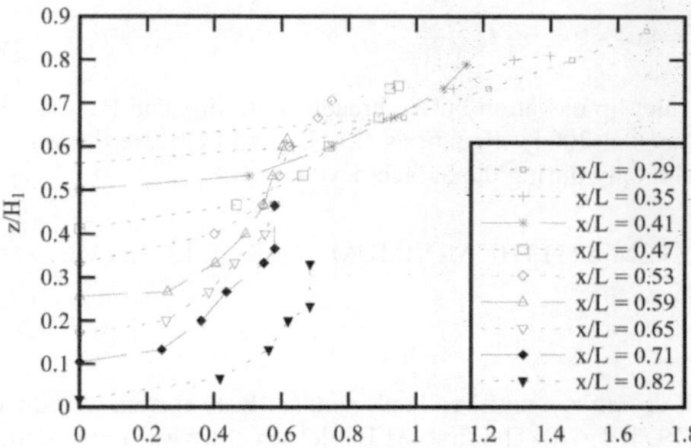

(a) Breach cross-sectional shape along equipotential planes below water line, t = 87 s, $Q_{breach} = 0.024$ m^3/s

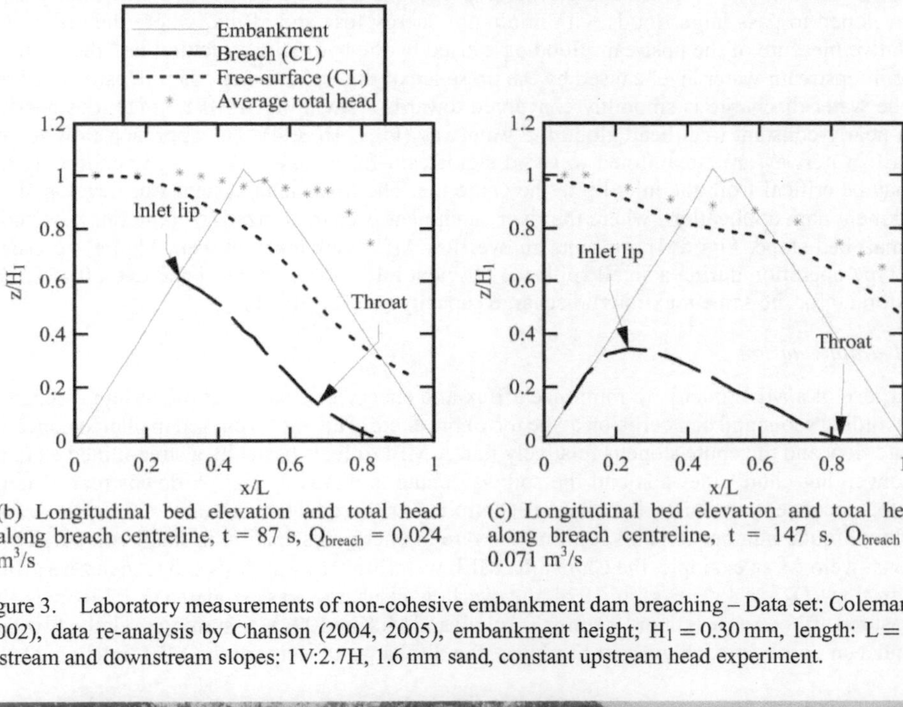

(b) Longitudinal bed elevation and total head along breach centreline, t = 87 s, $Q_{breach} = 0.024$ m^3/s

(c) Longitudinal bed elevation and total head along breach centreline, t = 147 s, $Q_{breach} = 0.071$ m^3/s

Figure 3. Laboratory measurements of non-cohesive embankment dam breaching – Data set: Coleman et al. (2002), data re-analysis by Chanson (2004, 2005), embankment height; $H_1 = 0.30$ mm, length: $L = 1.7$ m, upstream and downstream slopes: 1V:2.7H, 1.6 mm sand, constant upstream head experiment.

Figure 4. Minimum energy loss spillway inlet at Lake Kurwongbah (Brisbane QLD, Australia) in operation on 29 January 2013.

$$\frac{B_{min}}{H_1} = 4.01 \times 10^{-7} \times \left(t \times \sqrt{\frac{g}{H_1}} \right)^{2.3} \tag{4}$$

where g = gravity acceleration, z_{lip} = inlet lip elevation on the breach centreline and B_{min} = free-surface width at the breach throat (Chanson 2004). Equations (2), (3) and (4) were derived for cohesionless materials and they are valid only during the breach development.

3 EMBANKMENT SPILLWAY SYSTEMS: (1) THE MINIMUM ENERGY LOSS (MEL) INLET DESIGN

3.1 *Presentation*

The minimum energy loss (MEL) inlet design is a novel embankment spillway system introduced in Australia during the 1960s (McKay 1971, 1978). The first MEL inlet structure was the Redcliffe storm waterway system (1960); the structure is still in use and passed floods greater than its design flow ($Q_{des} = 25.8\ m^3/s$) without damage (McKay 1970, Chanson 2007). The MEL inlet design was developed to pass large floods with minimum energy loss and afflux, where the afflux is a quantitative measure of the upstream flooding caused by the hydraulic structure: i.e., the afflux is the rise in upstream water level caused by the presence of the embankment dam. Upstream of the inlet, the water discharge is smoothly converged towards a streamlined MEL chute, designed to yield a nearly-constant total head along the waterway (Figs. 1b & 4). The approach flow region and MEL waterway are streamlined to avoid significant form losses. At design conditions, the flow may be critical from the inlet lip to the chute toe. The MEL inlet system was developed for embankment dam applications where the river catchment is characterized by large rainfalls and a very small bed slope. Figure 1b presents an overflow MEL embankment weir. Figure 4 presents a prototype operation during a small spill: the efficient inlet design allowed and extra 0.457 m of water storage for the same maximum discharge capacity (McKay 1971).

3.2 *Hydraulic design*

The purpose of a MEL inlet is to minimize afflux and energy dissipation at the design discharge, while avoiding scour and bank erosion at the toe of the chute. The inlet is curved in plan to converge the chute flow and the chute slope is relatively flat. A MEL inlet is basically a streamlined channel with converging chute sidewalls and the spillway chute is relatively flat. A downstream energy dissipator is concentrated near the channel centreline at the downstream end. At the chute toe, the inflow Froude number remains low and the rate of energy dissipation is small compared to a traditional weir. As an example, the Chinchilla MEL weir (Fig. 1b) was designed to give zero afflux at design flow ($Q_{des} = 850\ m^3/s$); in 1974, the overflow discharge was estimated at 1,130 m^3/s and the measured afflux was less than 100 mm (Turnbull and McKay 1974). Assuming a relatively broad crest and a smooth approach without head loss, the discharge capacity of the MEL inlet equals:

$$Q = B_{max} \times \sqrt{g} \times \left(\frac{2}{3} \times (H_1 - z_{crest}) \right)^{3/2} \tag{5}$$

where $H_1 - z_{crest}$ = upstream head above spillway crest and B_{max} = crest width (see definition in Fig. 2b). A MEL spillway channel could be designed to achieve critical flow conditions at any position along the chute and, hence, to prevent the occurrence of a downstream hydraulic jump with high tailwater conditions. Assuming negligible energy loss along the inlet, the channel width B at any elevation $z - z_{crest}$ beneath the crest above the weir toe would satisfy:

$$B = B_{max} \times \left(\frac{H_{des} - z_{crest}}{H_{des} - z} \right)^{3/2} \tag{6}$$

where H_{des} = design upstream head. Equation (6) is only valid at design flow conditions. In practice, the variations of the tailwater elevations with discharge are important and a weak jump may take place at the inlet toe as seen in Figure 1b. The downstream conjugate depth is fixed by the tailwater conditions downstream of the hydraulic jump.

3.3 *Prototype experience*

The MEL spillway structures were designed with the concept of constant total head, hence zero afflux. The above pre-design calculations must be validated with some solid physical modelling, typically using 1:50 to 1:80 undistorted scale models with fixed bed.

The MEL overflow spillways may be built over earthfill structures and protected by concrete slabs. The construction costs must be minimum. The operations of a number of MEL spillways and weirs were documented, with a complement of field inspections and discussions with designers and operators (Chanson 2003, 2009). A number of MEL structures were observed to operate at design flow conditions and for floods larger than design. Inspections during and after flood events showed the sound operation together with little maintenance. The successful operation of several structures for over 40 years has highlighted some key operational considerations. Some improper approach flow conditions could affect adversely the spillway operation. MEL weirs are typically earthfill structures. An efficient drainage system must be installed underneath the chute slabs. A known issue is the overtopping risk during construction as for the Sandy Creek weir and Chinchilla weir (twice).

4 EMBANKMENT SPILLWAY SYSTEMS: (2) THE CONCRETE STEPPED SPILLWAY

4.1 *Presentation*

During the last three decades, a number of embankment dams were equipped with an overflow concrete stepped spillway (Chanson 2001, Gonzalez and Chanson 2007) (Figs. 1a, 1d & 5). Applications included both primary and secondary spillway structures: Figure 1d shows a sheet-pile concrete slab embankment weir with an overflow spillway across the whole width of the river bed. Figures 1a and 5 illustrate embankment dams equipped with a secondary embankment overflow stepped spillway.

Figure 5. Choctaw 8A auxiliary spillway (USA) (Courtesy of Craig Savela and USDA).

Most modern stepped spillways consist of flat horizontal steps, although different step configurations may be considered (Andre et al. 2004, Gonzalez and Chanson 2008, Guenther et al. 2013). The preferred construction method is the placement of roller compacted concrete (RCC) overlays on the downstream embankment slope (Ditchey and Campbell 2000, Gonzalez and Chanson 2007). During the construction, the RCC is placed typically in a succession of 0.2 to 0.4 m thick overlays with a width greater than 2.5 m for proper hauling, spreading and compacting. The advantages of the RCC construction include the cost effectiveness and the short duration of construction. Exposed RCC is frequently used for secondary spillways with infrequent overflows. In harsh climatic conditions, or for a primary spillway, a conventional concrete protection layer may be installed. In all the cases, a drainage layer beneath the concrete overlays is essential to prevent uplift pressures. Its purpose is to relieve pore pressure at the interface between the embankment and concrete stepped spillway. The drainage layer may be complemented by a series drain holes formed through the RCC during placement. At the downstream end of the overflow, a cutoff wall must be built to prevent the undermining of the concrete system during discharge.

4.2 *Hydraulic design considerations*

An embankment stepped spillway is designed to operate in a skimming flow regime (Chanson 2001). During the design process, the constraints include the embankment height, embankment downstream slope and design discharge. The variable parameters comprise typically the type of crest shape, the chute width and possibly the step height, although the step height h is always selected as a multiple of the RCC overlay height (Gonzalez and Chanson 2007).

In a skimming flow, the upstream flow region is characterized by a developing boundary layer (Amador et al. 2006, Meireles and Matos 2009) (Fig. 6). When the outer edge of the boundary layer interacts with the free-surface, the turbulent shear stress becomes greater than the surface tension force per unit area resisting the interfacial breakup and free-surface aeration takes place (Ervine and Falvey 1987, Chanson 2009b). The characteristics at the inception point of free-surface aeration are:

$$\frac{L_I}{h \times \cos\theta} = 9.72 \times (\sin\theta)^{0.080} \times \left(\frac{q}{\sqrt{g \times \sin\theta \times (h \times \cos\theta)^3}}\right)^{0.71} \tag{7}$$

$$\frac{d_I}{h \times \cos\theta} = \frac{0.403}{(\sin\theta)^{0.04}} \times \left(\frac{q}{\sqrt{g \times \sin\theta \times (h \times \cos\theta)^3}}\right)^{0.59} \tag{8}$$

where q = discharge per unit width (q = Q/B), L_I = longitudinal distance from the chute crest, d_I = flow depth at the inception point, g = gravity acceleration and θ = angle between the pseudo-bottom formed by the step edges and the horizontal (Fig. 6).

When the spillway chute is long and the flow reach uniform equilibrium (i.e. normal flow conditions), the characteristic flow depth d equals:

$$d = \sqrt[3]{\frac{f_e \times q^2}{8 \times g \times \sin\theta}} \tag{9}$$

where f_e = Darcy friction factor estimated based upon experimental air-water flow friction factor data (Chanson et al. 2002, Chanson 2006). If the flow does not reach normal flow conditions before the downstream end of the spillway, the flow is gradually varied downstream of the inception point of air entrainment. Combining some well-documented experimental results together with theoretical

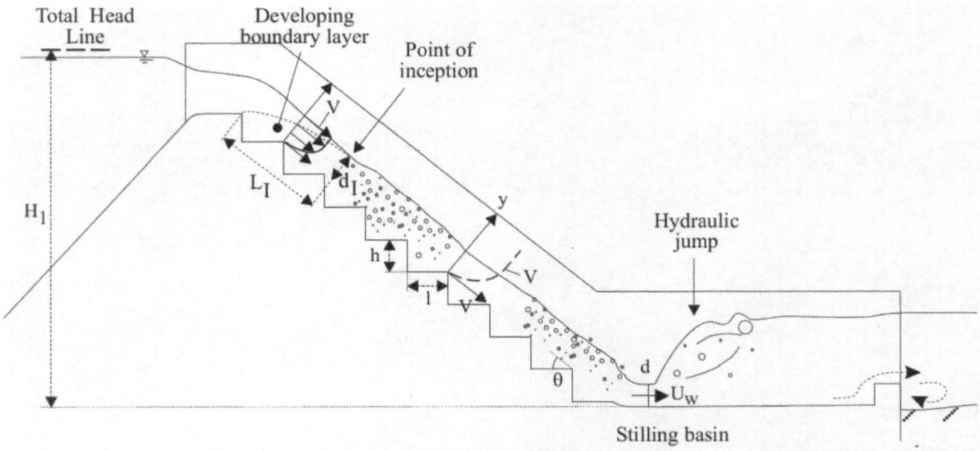

Figure 6. Definition of sketch of an embankment stepped spillway operation with downstream hydraulic jump stilling basin.

calculations, an empirical correlation was derived in terms of the downstream spillway velocity as a function of the upstream above crest and discharge (Gonzalez 2005):

$$\frac{U_w}{V_{max}} = 0.00105 \times \left(\frac{H_1}{\sqrt[3]{q^2 / g}} \right)^2 - 0.0634 \times \left(\frac{H_1}{\sqrt[3]{q^2 / g}} \right) + 1.202 \tag{10}$$

where H_1 = upstream total head above chute toe, d_c = critical depth, V_{max} = ideal flow velocity deduced from the Bernoulli principle, and U_w = downstream velocity. Such an approach may be used for pre-design calculations assuming a friction coefficient $f_e = 0.2$, although the method was only validated for moderate stepped spillway slopes ($15° < \theta < 25°$). These preliminary estimates must be checked with some solid physical modelling, based upon undistorted scale models with a scaling ratio no greater than 3:1 (Chanson and Gonzalez 2005, Felder and Chanson 2009).

For short stepped spillways and large discharges, the flow may not be fully-developed before the downstream end of the chute. That is, the chute length may be smaller than the distance between crest and inception point of free-surface aeration. A simple method was developed to predict the depth-averaged flow properties (Chanson 2001, Meireles and Matos 2009).

5 EMBANKMENT SPILLWAY SYSTEMS: (3) GABION STEPPED WEIR

A gabion is a basket filled with earth or stone for use in engineering, and the construction technique is extensively used for earth retaining structures and hydraulic structures. Its advantages are the stability, low cost, flexibility and porosity of the construction material. The gabion porosity in particular is important to prevent the build-up of uplift pressures. Modern box gabions consist of rockfill material enlaced by a basket or a mesh, shaped like a rectangular box. Typical gabion dimensions are heights of 0.5 to 1 m, with a width equal to the height and length-to-height ratio between 1.5 and 4. Longer gabions may be subdivided into cells by inserting mesh diaphragms to strengthen the box. The wire is normally made of soft steel with a zinc coating. The durability of gabion structures relies heavily upon the quality of mesh and wire. The gabion filling consists of loose or compacted rocks. The stone size must equal at least 1 to 1.5 times the mesh size but should not be larger than 2/3 of the minimum dimension of the gabion. The use of small-sized stone, typically 1.5 times the mesh size, permits a better adaptability of the gabions to deformation.

Figure 7. Gabion stepped spillway at Robina (Gold Coast QLD, Australia) on 2 April 1997, shortly after completion: $h \sim 0.5$ m, $h/l \sim 0.5$.

The dimensions of the gabion box and the design discharge are the two basic design parameters controlling the hydraulic operation of the stepped chute. The step height h is typically the gabion height, although it might be twice or three times the gabion height in some cases. The stepped chute slope ranges from 1V:4H to 1V:2H. For a gabion structure only, the choice of a steep slope with a skimming flow regime operation may reduce the number of gabions and the overall structure cost. For an embankment with gabion overtopping, a flat slope may be more appropriate for the stability requirements of the earthfill structure (Fig. 7). Figure 7 illustrates a gabion stepped weir. The design considerations for the stability of gabion weirs are generally the same as for any gravity structure. The calculations of structural stability involve checking the stability of the weir against overturning, sliding and uplift. Inclined (upward) gabion-stepped spillways may also be used (Peyras et al. 1991). Larger energy dissipation is achieved but their construction requires greater care. In comparison with concrete spillways, the flow above a stepped gabion chute is characterised by some interactions between the surface overflow and seepage flow, and a rougher surface of the gabion steps. The former aspect was discussed by Kells (1993, 1995), while Gonzalez et al. (2008) documented the effect of step roughness for a 1V:2.5H stepped chute.

The performances of gabion stepped weirs are often restricted by the gabion resistance to damage and their stability. Sediments and debris carried by the stream flow may affect and fracture the gabion mesh. With large-size debris, it is common practice to protect the step surfaces with timber, steel sheets, concrete facing or even reinforced concrete slab (Agostini et al. 1987, Peyras et al. 1991).

6 EMBANKMENT SPILLWAY SYSTEMS: (4) PRE-CAST CONCRETE BLOCK SPILLWAY

6.1 *Presentation*

The pre-cast concrete block spillway design was developed in Russia by late Professor Gordienko (Gordienko 1978, Pravdivets and Bramley 1989). The chute is made of individual blocks placed in an overlapping staircase fashion (Fig. 8). Figure 8 shows an experimental earth dam in Siberia equipped with a pre-cast concrete block spillway for primary flood release. An interesting feature is the flexibility of the channel bed allowing differential settlements of the earthfill embankment. Another feature is the fairly short construction time on site, while and stepped design contributes to some energy dissipation along the chute.

Figure 8. Precast concrete block spillway of Volymia dam (Siberia, Russia) (Courtesy of Prof. Y. Pravdivets) – This experimental structure (H = 20 m) operated for more than 15 years, the spillway discharging ice and water from early spring to late fall.

The Russian engineers developed a strong expertise in the design of concrete wedge blocks. This was supported by extensive testing. For large discharges, each block should be tied to adjacent blocks, possibly made of reinforced concrete. A step height-to-length ratio in the range 1:4 to 1:6 may ensure maximum stability of the blocks during the overtopping. Drains must be placed in areas of sub-atmospheric pressure to relieve uplift pressures.

6.2 *Practical considerations*

For any embankment dam, an uppermost basic criterion is the stability of the material. Seepage may occur in saturated embankment, resulting in uplift pressures which might damage or destroy the entire structure. In a typical precast block chute design, an adequate drainage is essential, and the blocks lay on a filter and erosion protection layer. The layer filters the seepage flow out of the subsoil and it protects as well the subsoil layer from erosion by flow in the drainage layer. The protection layer reduces drastically the uplift pressures acting on the concrete blocks. Usually a geotextile membrane is laid on the embankment before the placing of the layer, and another covers the protection layer before the installation of the blocks. Drains are typically placed in areas of sub-atmospheric pressures (e.g. vertical step face) to relieve uplift pressures. The location of drains must be appropriately selected to avoid reverse flow in the drains and dynamic pressures associated with hydraulic jumps at low flows. Grinchuk et al. (1977) recommended that the total area of the drainage holes should be 10–15% of the exposed step area, although Baker et al. (1994) suggested that an open area of 2 to 5% could be optimum. Laboratory model tests showed that the aspiration on the vertical step face increased with increasing downward slope of the steps (Frizell 1992).

The seepage flow in the embankment dam must be predicted accurately to make the appropriate provision for drainage and evacuation of seepage flow through the blocks. Note that the seepage may be influenced by the infiltration into the downstream slope caused by the spillway flow, in addition to the flow through the embankment.

7 HYDRAULIC DESIGN CONSIDERATIONS

For an embankment dam, the uppermost important design criterion is the stability of the earthfill embankment at any stage, including during droughts and flood events. The construction must be of good quality and the design must be sound and kept simple. Seepage may occur in saturated embankment yielding unacceptable uplift pressures; that is, an adequate drainage is essential. For all spillway designs, a filter and erosion protection layer is typically laid on the downstream embankment slope beneath the overflow protection.

The hydraulic design of embankment dam spillway is critical. Some key points include (a) the maximum discharge capacity estimate, (b) the downstream dissipation structure and (c) the high level of hydraulic expertise required. First the spillway capacity must be correctly estimated to prevent any overflow over the unprotected embankment section. At the downstream end of the spillway, the turbulent kinetic energy of the flow must be dissipated safely. The most common dissipation designs include the hydraulic jump stilling basin (Figs. 1b & 6) and a flip bucket to deflect the water away from the chute toe (Fig. 8). The experience has shown that the hydraulic design of embankment spillways require a high level of expertise in dam and hydraulic engineering.

The successful operation of many structures highlighted the importance of regular maintenance. Some basic down-to-earth considerations must be considered. There were accounts of vandalism in a few projects, including motor bikes riding up and down a precast block spillway, thus damaging the pre-cast concrete blocks, and locals stealing mesh of gabion structures to build local fences. Alternative embankment overtopping protection systems include timber cribs, sheet-piles, riprap, macro-roughness and reinforced earth (Chanson 2001, 2009).

7.1 *Energy dissipation*

A number of construction techniques have been used for embankment dam spillways. In all cases, the safe dissipation of the kinetic energy of the flood flow is critical. A common design is the stepped profile, which increases the rate of energy dissipation on the spillway chute, thus reducing the size of the downstream energy dissipator (Chanson 2001, Ohtsu et al. 2004). Importantly the turbulent kinetic energy dissipation above the steps must be carefully estimated, in particular for large discharges per unit width corresponding to the skimming flow regime. A characteristic feature of skimming flows is the high level of turbulence and free-surface aeration (Rajaratnam 1990, Peyras et al. 1991). The water flows downs the steps as a coherent free-stream skimming over the pseudo-bottom formed by the step edges. In the step cavities, the turbulent recirculation is maintained through the transmission of shear stress from the free-stream. At the free-surface, air is continuously trapped and released, and the resulting two-phase mixture interacts with the flow turbulence yielding some intricate air-water structure associated with complicated energy dissipation mechanisms (Chanson and Toombes 2002, Gonzalez and Chanson 2008).

Two energy dissipation designs are the hydraulic jump stilling basin and the flip bucket. The skip jump deflects the water away from the dam toe, but it is usually restricted to small structures. The stilling basin design is more common. The jump must be confined to a reinforced area, away from the unprotected embankment slope and natural river banks.

8 CONCLUSION

In recent years, a number of embankment overtopping protection systems were developed for earthfill dams, coastal barriers and river dykes. The overtopping protection systems include concrete stepped overtopping protection, minimum energy loss (MEL) spillway, gabion stepped spillways and precast concrete block protection systems. For embankments higher than 5 to 10 m, the concrete stepped spillway is a sound design technique well-suited to small to large discharges. The flow down the stepped cascade is characterised by some strong aeration, high turbulence of the flow and a significant rate of energy dissipation.

A number of embankment dam spillways have been in operation for three to four decades. The prototype experience provides valuable informations. Based upon past accident and failure forensic investigations, it is clearly understood that a safe operation relies upon a sound design and a good quality of construction, suitable flow conditions, together with regular maintenance. Ultimately there is no better proof of design soundness than successful prototype operation.

ACKNOWLEDGEMENTS

The author thanks all the individuals and organizations who provided him with relevant information. He acknowledges the invitation of the organizers to contribute to the workshop. The financial support of the Australian Research Council is acknowledged (Grants ARC DP0878922 & DP120100481).

REFERENCES

Agostini, R., Bizzarri, A., Masetti, M., and Papetti, A. 1987. Flexible Gabion and Reno Mattress Structures in River and Stream Training Works. Section One: Weirs. *Officine Maccaferri*, Bologna, Italy, 2nd edition.

Amador, A., Sanchez-Juny, M., and Dolz, J. 2006. Characterization of the Nonaerated Flow Region in a Stepped Spillway by PIV. *Jl of Fluids Engineering*, ASME, Vol. 128, No. 6, pp. 1266–1273.

Andre, S., Boillat, J.L., Schleiss, A.J., and Matos, J. 2004. Energy Dissipation and Hydrodynamic Forces of Aerated Flow over Macro-Roughness Linings for Overtopped Embankment Dams. *Proc. Intl Conf. on Hydraulics of Dams and River Structures*, Tehran, Iran, Balkema, The Netherlands, pp. 189–196.

ASCE Task Committee 1994. *Alternatives for Overtopping Protection of Dams*. ASCE, New York, USA, Task Committee on Overtopping Protection, 139 pages.

ASCE/EWRI Task Committee on Dam/Levee Breaching 2011. Earthen Embankment Breaching. *Jl of Hydraulic Engineering*, ASCE, Vol. 137, No. 12, pp. 1549–1564 DOI: 10.1061/ASCEHY.1943-7900.0000498.

Bornschein, A., and Pohl, R. 2003. Dam Break during the Flood in Saxon/Germay in August 2002. *Proc. 30th IAHR Biennial Congress*, Thessaloniki, Greece, J. Ganoulis and P. Prinos Ed., Vol. C2, pp. 229–236.

Chanson, H. 2001. *The Hydraulics of Stepped Chutes and Spillways*. Balkema, Lisse, The Netherlands, 418 pages.

Chanson, H. 2003. Minimum Energy Loss Structures in Australia : Historical Development and Experience. *Proc. 12th National Engineering Heritage Conference*, Institution of Engineers, Australia, Toowoomba Qld, Australia, N. Sheridan Ed., pp. 22–28.

Chanson, H. 2004. Overtopping Breaching of Noncohesive Homogeneous Embankments. Discussion. *Journal of Hydraulic Engineering*, ASCE, Vol. 130, No. 4, pp. 371–374.

Chanson, H. 2005. The 1786 Earthquake-Triggered Landslide Dam and Subsequent Dam-Break Flood on the Dadu River, Southwestern China. *Geomorphology*, Vol. 71, pp. 437–440 (DOI: 10.1016/j.geomorph. 2005.04.017).

Chanson, H. 2006. Hydraulics of Skimming Flows on Stepped Chutes: the Effects of Inflow Conditions? *Journal of Hydraulic Research*, IAHR, Vol. 44, No. 1, pp. 51–60.

Chanson, H. 2007. Hydraulic Performances of Minimum Energy Loss Culverts in Australia. *Jl of Performances of Constructed Facilities*, ASCE, Vol. 21, No. 4, pp. 264–272 (DOI: 10.1061/ASCE0887-3828200721: 4264).

Chanson, H. 2009. Embankment Overtopping Protections System and Earth Dam Spillways. in *Dams: Impact, Stability and Design*, Nova Science Publishers, Hauppauge NY, USA, Ed. W.P. Hayes and M.C. Barnes, Chapter 4, pp. 101–132.

Chanson, H. 2009b. Turbulent Air-water Flows in Hydraulic Structures: Dynamic Similarity and Scale Effects. *Environmental Fluid Mechanics*, Vol. 9, No. 2, pp. 125–142 (DOI: 10.1007/s10652-008-9078-3).

Chanson, H., and Gonzalez, C.A. 2005. Physical Modelling and Scale Effects of Air-Water Flows on Stepped Spillways. *Journal of Zhejiang University SCIENCE*, Vol. 6A, No. 3, March, pp. 243–250.

Chanson, H., and Toombes, L. 2002. Experimental Investigations of Air Entrainment in Transition and Skimming Flows down a Stepped Chute. *Canadian Journal of Civil Engineering*, Vol. 29, No. 1, pp. 145–156.

Chanson, H., Yasuda, Y., and Ohtsu, I. 2002. Flow Resistance in Skimming Flows and its Modelling. *Canadian Journal of Civil Engineering*, Vol. 29, No. 6, pp. 809–819.

Coleman, S.E., Andrews, D.P., Webby, M.G. 2002. Overtopping Breaching of Noncohesive Homogeneous Embankments. *Jl of Hydraulic Engineering*, ASCE, Vol. 128, No. 9, pp. 829–838.

Dai, F.C., Lee, C.F., Deng, J.H., and Tham, L.G. 2005. The 1786 Earthquake-Triggered Landslide Dam and Subsequent Dam-Break Flood on the Dadu River, Southwestern China. Geomorphology, Vol. 65, pp. 205–221.

Ditchey, E.J., and Campbell, D.B. 2000. Roller Compacted Concrete and Stepped Spillways. *Proc. Intl Workshop on Hydraulics of Stepped Spillways*, Zürich, Switzerland, H.E. Minor & W.H. Hager Editors, Balkema Publ., pp. 171–178.

Ervine, D.A., and Falvey, H.T. 1987. Behaviour of Turbulent Water Jets in the Atmosphere and in Plunge Pools. *Proc. Institution of Civil Engineers, London, Part 2*, Mar. 1987, 83, pp. 295–314.

Felder, S., and Chanson, H. 2009. Turbulence, Dynamic Similarity and Scale Effects in High-Velocity Free-Surface Flows above a Stepped Chute. *Experiments in Fluids*, Vol. 47, No. 1, pp. 1–18 (DOI: 10.1007/s00348-009-0628-3).

Frizell, K.H. 1992. Hydraulics of Stepped Spillways for RCC Dams and Dam Rehabilitations. *Proc. 3rd Specialty Conf. on Roller Compacted Concrete*, ASCE, San Diego CA, USA, pp. 423–439.

Gonzalez, C.A. 2005. An Experimental Study of Free-Surface Aeration on Embankment Stepped Chutes. *Ph.D. thesis*, Department of Civil Engineering, The University of Queensland, Brisbane, Australia, 240 pages.

Gonzalez, C.A., and Chanson, H. 2007. Hydraulic Design of Stepped Spillways and Downstream Energy Dissipators for Embankment Dams. *Dam Engineering*, Vol. 17, No. 4, pp. 223–244.

Gonzalez, C.A., and Chanson, H. 2008. Turbulence and Cavity Recirculation in Air-Water Skimming Flows on a Stepped Spillway. *Journal of Hydraulic Research*, IAHR, Vol. 46, No. 1, pp. 65–72,

Gonzalez, C.A., Takahashi, M., and Chanson, H. 2008. An Experimental Study of Effects of Step Roughness in Skimming Flows on Stepped Chutes. *Journal of Hydraulic Research*, IAHR, Vol. 46, No. Extra Issue 1, pp. 24–35.

Gordienko, P.I. 1978. Reinforced-Concrete-Earth Overflow Dams. *Dams & Spillways*, Collection of Works No. 61, Issue 2, MISI, Moscow, pp. 3–17 (in Russian).

Grinchuk, A.S., Pravdivets, Y.P., and Shekhtman, N.V. 1977. Test of Earth Slope Revetments Permitting Flow of Water at Large Specific Discharges. *Gidrotekhnicheskoe Stroitel'stvo*, No. 4, pp. 22–26 (in Russian). (Translated in *Hydrotechnical Construction*, 1978, Plenum Publ., pp. 367–373.)

Guenther, P., Felder, S., and Chanson, H. 2013. Flow aeration, cavity processes and energy dissipation on flat and pooled stepped spillways for embankments. *Environmental Fluid Mechanics*, Vol. 13 (DOI: 10.1007/s10652-013-9277-4). (in print)

Hanson, G.J., Cook, K R., and Hunt, S.L. 2005. Physical Modeling of Overtopping Erosion and Breach Formation of Cohesive Embankments. *Trans. ASABE*, Vol. 48, No. 5, pp. 1783–1794.

Kells, J.A. 1993. Spatially Varied Flow over Rockfill Embankments. *Canadian Journal of Civil Engineering*, Vol. 20, pp. 820–827.

Kells, J.A. 1995. Comparison of Energy Dissipation between Nappe and Skimming Flow Regimes on Stepped Chutes – Discussion. *Jl of Hydraulic Research*, IAHR, Vol. 33, No. 1, pp. 128–133.

McKay, G.R. 1970. Pavement Drainage. *Proc. 5th Australian Road Research Board Conference*, Vol. 5, Part 4, pp. 305–326.

McKay, G.R. 1971. Design of Minimum Energy Culverts. *Research Report*, Dept. of Civil Eng., Univ. of Queensland, Brisbane, Australia, 29 pages & 7 plates.

McKay, G.R. 1978. Design principles of Minimum Energy Waterways. *Proc. Workshop on Minimum Energy Design of Culvert and Bridge Waterways*, Australian Road Research Board, Melbourne, Australia, K.F. Porter Editor, Session 1, pp. 1–39.

Meireles, I., and Matos, J. 2009. Skimming Flow in the Nonaerated Region of Stepped Spillways over Embankment Dams. *Jl of Hydraulic Engineering*, ASCE, Vol. 135, No. 8, pp. 685–689.

Morris, M.W., Hassan, M., and Vaskin, K.A. 2007. Breach Formation: Field Test and Laboratory Experiments. *Journal of Hydraulic Research*, Vol. 45, Special Issue, pp. 9–17.

Ohtsu, I., Yasuda, Y., and Takahashi, M. 2004. Flow Characteristics of Skimming Flows in Stepped Channels. *Jl of Hydraulic Engineering*, ASCE, Vol. 130, No. 9, pp. 860–869.

Peyras, L., Royet, P., and Degoutte, G. 1991. Ecoulement et Dissipation sur les Déversoirs en Gradins de Gabions. *Journal La Houille Blanche*, No. 1, pp. 37–47 (in French).

Pravdivets, Y.P., and Bramley, M.E. 1989. Stepped Protection Blocks for Dam Spillways. *International Water Power and Dam Construction*, Vol. 41, No. 7, July, pp. 49–56.

Rajaratnam, N. 1990. Skimming Flow in Stepped Spillways. *Jl of Hydraulic Engineering*, ASCE, Vol. 116, No. 4, pp. 587–591.

Rozov, A.L. 2003. Modeling a Washout of Dams. *Jl of Hydraulic Research*, IAHR, Vol. 41, No. 6, pp. 565–577.

Schnitter, N.J. 1994. *A History of Dams: the Useful Pyramids*. Balkema Publ., Rotterdam, The Netherlands.

Smith, N. 1971. *A History of Dams*. The Chaucer Press, Peter Davies, London, UK.

Turnbull, J.D., and McKay, G.R. 1974. The Design and Construction of Chinchilla Weir – Condamine River Queensland. *Proc. 5th Australasian Conf. on Hydraulics and Fluid Mechanics*, Christchurch, New Zealand, Vol. II, pp. 1–8.

Visser, P.J., Vrijling, J.K., and Verhagen, H.J. 1990. A Field Experiment on Breach Growth in Sand-Dykes. *Proc. 22nd International Conference on Coastal Engineering*, Delft, Netherlands, B. Edge Editor, Vol. 2, pp. 2097–2100.

Yin, Y., Wang, F., and Sun, P. 2009. Landslide Hazards Triggered by the 2008 Wenchuan Earthquake, Sichuan, China. *Landslides*, Vol. 6, No. 2, pp. 139–152.

A. Physical and numerical modeling – Hydraulic capacity and scale effects

Labyrinth and Piano Key Weirs II – PKW 2013 – Erpicum et al. (eds)
© 2014 Taylor & Francis Group, London, ISBN 978-1-138-00085-8

Effects of the crest shape on the discharge efficiency of a type A Piano Key weir

G.M. Cicero & J.R. Delisle
EDF – National Hydraulics and Environment Laboratory, Chatou, France

ABSTRACT: The Piano Key Weir (PKW) is a type of labyrinth weir with efficient hydraulic performance especially at low heads. For ten years, many experiments conducted in channels have contributed to identify the primary parameters influencing hydraulic performance and some secondary parameters like the dam height, the noses under the outlet keys or the crest shape. EDF has launched a new experimental program including the study of the crest shape effect on the discharge capacity under free flow conditions. This article presents the crest shapes testing on a type A PKW with flat-topped, half-rounded and quarter-rounded crests. Some crest configurations were tested at two dam heights. The analyses confirmed that the crest shape influences the discharge capacity at low heads and that the half-rounded and the quarter-rounded shapes have better hydraulic performance than the flat-topped shape. The present study highlighted that the discharge efficiency was mainly influenced by the lateral crest, and that with the rounded shapes, the gain in discharge (relative to the flat-topped crest) increased with the dam height and could reach from 10% to 20% at low heads when the scale effects became negligible.

1 INTRODUCTION

The Piano Key Weir (PKW) is a type of labyrinth weir with efficient hydraulic performance especially at low heads. Since the first tests performed for Hydrocoop in 1998 at the National Hydraulics and Environment Laboratory (LNHE) in Chatou (Lemperiere et al. 2011), the experiments conducted in channels by Anderson (2011), Ho Ta Khanh et al. (2011), Ouamane (2011), Leite-Ribeiro et al. (2012), Kabiri-Samini & Javaheri (2012), Machiels (2011, 2012) contributed to improve the knowledge on the hydraulic behavior of PKWs under free-flow or submerged conditions. The primary parameters influencing hydraulic performance were identified such as the width ratio of the inlet and outlet keys, the developed length to width ratio, and the weir height. However, some secondary parameters like the dam height, the noses under the outlet keys, and the crest shape also have an influence on hydraulic performance. Furthermore, new correlations have been recently published to estimate the PKWs discharge capacity for the type A (Leite Ribeiro et al. 2012) or for general geometries (Kabiri-Samini & Javaheri, 2012, Machiels, 2012).

In 2011, Electricité De France (EDF) started a new experimental program at LNHE in Chatou to expand the knowledge on some geometric parameters effects such as the overhangs (Cicero et al. 2012), the crest shape, the dam height, and the noses under the outlet keys. This program includes the testing of a non rectangular PKW design with sidewalls angles, and the testing of all geometries under submerged conditions. This program will also provide a useful database to validate 1D or 3D numerical modeling and empirical correlations.

The present article is devoted to the analyses of the crest shape effects under free-flow conditions. These effects have been already studied on linear (Hager, 1985), labyrinth (Tullis et al. 1995) and PK weirs (Barcouda et al. 2006, Leite Ribeiro et al. 2007, Anderson, 2011). All these experiments revealed an influence of the crest shape especially at low heads and showed that the half-rounded and the quarter-rounded shapes should be preferred to the flat-topped shape. Thus, these shapes

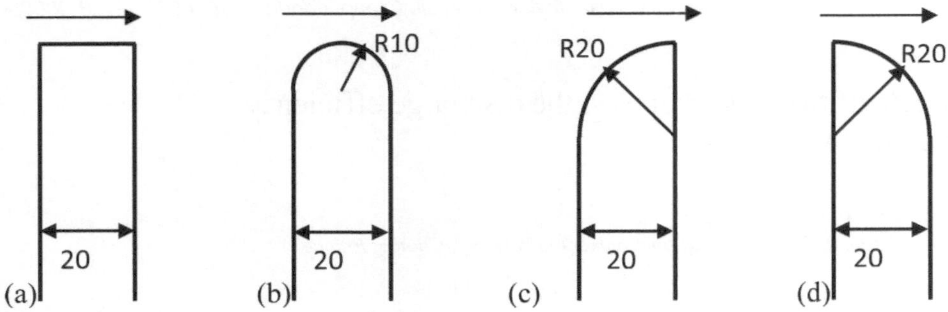

Figure 1. Shapes (dimensions in mm) of the lateral crest: flat-topped (a), half-rounded (b), upstream quarter-rounded (c) and downstream quarter-rounded (d).

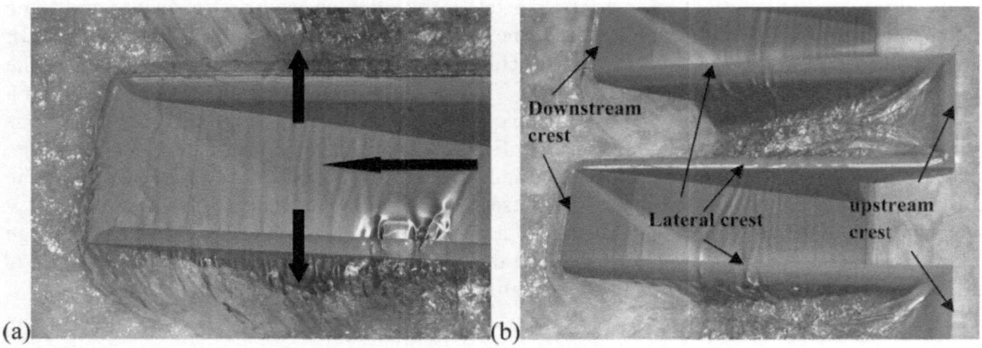

Figure 2. Lateral crest with the upstream quarter-rounded (a) and the downstream quarter rounded (b).

were tested on a type A PKW specifically designed to isolate and quantify the effects on the discharge capacity. The PKWs testing allowed isolating the effects of the lateral, the upstream and the downstream crest. Some of the crest shape configurations were tested at two dam heights. In the last part, this article presents a comparison of the measurements to the correlations published by Leite Ribeiro et al. (2012) and Machiels (2012). The standard notation as defined by Pralong et al. (2011) has been used herein.

2 PKW GEOMETRY AND CREST SHAPES TESTED

The objective of this study is to isolate the crest shape effects on the discharge efficiency of a type-A PKW, with symmetrical overhangs. The PKW geometry was the type A recommended by Lemperiere et al. (2011), (i.e. $W_i = W_o = 0.6P, B_b = 1.2P, B = 2.4P + 2T_s$), with a weir height $P =$ 0.222 m. The PKW width (W) was the total channel width to maximize the number of PKW units and to avoid side effects. The wall thickness ($T_s = 2$ cm) was chosen wide enough to enable various crest shapes testing. Considering the wall thickness recommended for prototype ($T_s \sim 35$ cm) by Vermeulen et al. (2011), the PKW scale model was around 1/20.

The PKW, fabricated in PVC, was specifically designed to allow testing various crest shapes of the inlet and the outlet keys. Four geometries were tested on the lateral walls (Fig. 1): a flat-topped shape, a half-rounded (R10) and a quarter-rounded one (R20). Since the latter is not symmetric it was tested with the rounded face in the upstream (Figs 1c, 2a) and in the downstream direction (Figs 1d, 2b). Also four geometries were tested on the upstream and the downstream crests (Fig. 3): a flat-topped, a half-rounded (R5), a rounded shape (R20) rather "horizontal", and a rounded shape (R20) rather "vertical".

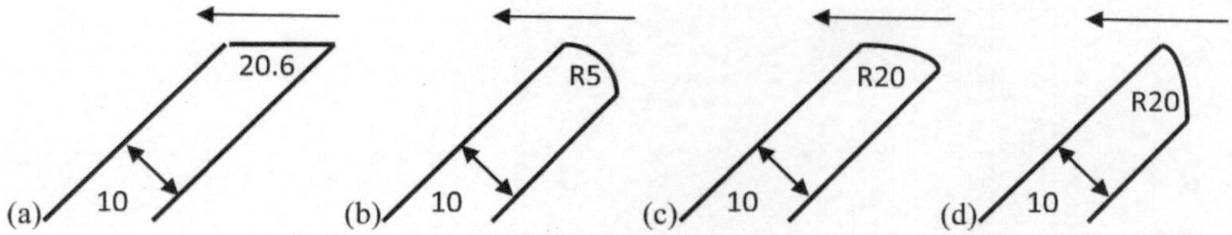

Figure 3. Shapes (dimensions in mm) of the upstream and downstream crests: flat-topped (a), half-rounded (b), "horizontal" rounded (c) and "vertical" rounded (d).

Table 1. Tests of the type A PKW with the various crest shapes and dam heights.

Test	dam height	upstream crest	downstream crest	lateral crest
1	0.6P	flat-topped	flat-topped	flat-topped
2	0.6P	half-rounded	half-rounded	half-rounded
3	0.6P	horizontal rounded	horizontal rounded	upstream quarter-rounded
4	0.6P	vertical rounded	vertical rounded	downstream quarter-rounded
5	0.6P	horizontal rounded	horizontal rounded	downstream quarter-rounded
6	0.6P	vertical rounded	vertical rounded	upstream quarter-rounded
7	0.6P	half-rounded	flat-topped	flat-topped
8	0.6P	flat-topped	half-rounded	flat-topped
9	0.6P	vertical rounded	flat-topped	flat-topped
10	0.6P	flat-topped	vertical rounded	flat-topped
11	1.6P	flat-topped	flat-topped	flat-topped
12	1.6P	half-rounded	half-rounded	half-rounded
13	1.6P	horizontal rounded	horizontal rounded	upstream quarter-rounded
14	1.6P	vertical rounded	vertical rounded	downstream quarter-rounded

3 EXPERIMENTAL RESULTS

3.1 *Testing apparatus*

The PKW testing was conducted in a 2 m wide by 1 m deep by 25 m long rectangular channel devoted to the discharge capacity measurements up to 500 l/s.

The weir discharge (Q) was measured by an electromagnetic flow-meter of +/−1% accuracy and the piezometric level (h) was measured at ~6 m upstream of the PKW with an accuracy of +/−0.18 mm. The discharge and the upstream level were recorded for a 3 min period at a 0.5 Hz frequency. As a result, the discharge and head measurements used for the analyses were the average and standard deviation of 90 temporal values.

The total upstream head (H) was calculated by adding the velocity head corresponding to the average cross-sectional velocity at the location of the piezometric level measurement.

3.2 *Experimental program*

The objective of the experimental program was to compare the hydraulic efficiency of the type A PKW with various crest shapes, under free-flow conditions. The PKW was installed on top of a platform in concrete without noses underneath the upstream overhangs. The platform height was increased to make the dam height vary from $P_d = 0.6P$ to $P_d = 1.6P$.

Table 1 details, for each test, the dam height, and the shapes tested on the lateral, the upstream, and the downstream crests. The first tests (1 to 6) were carried out at "low" dam height ($P_d = 0.6P$), with "uniform" crest shapes. The objective of tests 7 to 10 was to discriminate the effects of the lateral, upstream, and downstream crests, maintaining the flat-topped shape on the other crests.

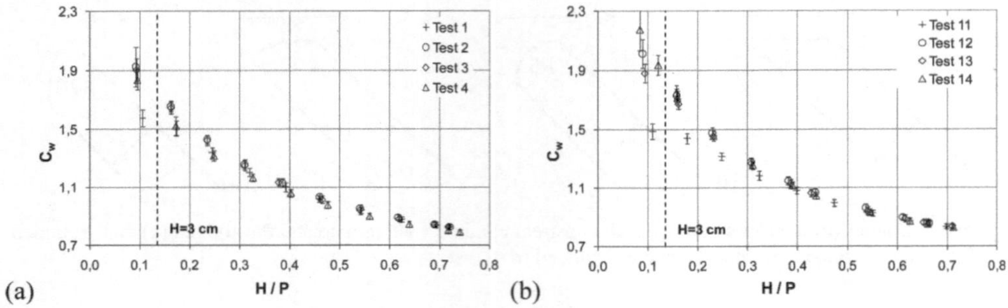

Figure 4. Crest shape effects on the experimental rating curves $C_w(H/P)$ for $P_d = 0.6P$ (a) and $1.6P$ (b).

Then, the dam height was increased ($P_d = 1.6P$) to examine the effect on the discharge efficiency of the PKW, with "uniform" crest shapes (tests 11 to 14).

3.3 *Experimental results*

For each test of Table 1, the rating curves $Q(H)$ were measured in 50 l/s increments from 50 l/s to 450 l/s, and we compared the non dimensional rating curves $C_w(H/P)$, when the discharge coefficient C_w was calculated by the classical formula:

$$C_W = \frac{Q}{W\sqrt{2g}H^{1.5}} \quad (1)$$

The measurement error on C_W was estimated by using the measurements accuracy of Q and h, and the standard deviations (σ_Q and σ_h) on the 90 temporal values by:

$$\frac{\Delta C_W}{C_W} = \frac{\Delta Q + \sigma_Q}{Q} + 1{,}5\frac{\Delta H}{H} \quad (2)$$

$$\Delta H = \Delta h + \sigma_h + \frac{Q^2}{gW^2(h+P+P_d)^2}\left(\frac{\Delta Q + \sigma_Q}{Q} + \frac{\Delta h + \sigma_h}{(h+P+P_d)}\right) \quad (3)$$

where $\Delta Q/Q = 0.01$ and $\Delta h = 0.18$ mm. For the various tests of Table 1, the average and minimal measurement uncertainty $\Delta C_w/C_w$ was 3% and 2% respectively.

The experimental curves $C_w(H/P)$ were well correlated ($R^2 > 0.997$) by polynomial interpolations of the 4th degree, which allow predicting the measurements with average errors lower than 1%. These functions were used to quantify the effects of the crest shape by comparing the discharge efficiency at the same upstream head. Since the measurement uncertainty $\Delta C_w/C_w$ is greater than 2%, the effects of the crest shape become significant if the absolute discharge ratio $\Delta Q/Q$ is greater than 4%.

Figure 4 compares the experimental rating curves $C_w(H/P)$ measured with a "uniform" crest shape, at both dam heights, with error bars showing the measurement accuracy on C_w. Figure 5 presents the corresponding gains in discharge relative to the flat-topped crest, i.e. the discharge ratios $\Delta Q/Q_{Test1}$ and $\Delta Q/Q_{Test11}$ for the tests performed at "low" and "high" dam height respectively. For both dam heights (Figs 4a, b), the hydraulic efficiency of the PKW was increased with the half-rounded crest (tests 2 and 12), and with the quarter-rounded shape associated to the "horizontal" rounded shape (tests 3 and 13).

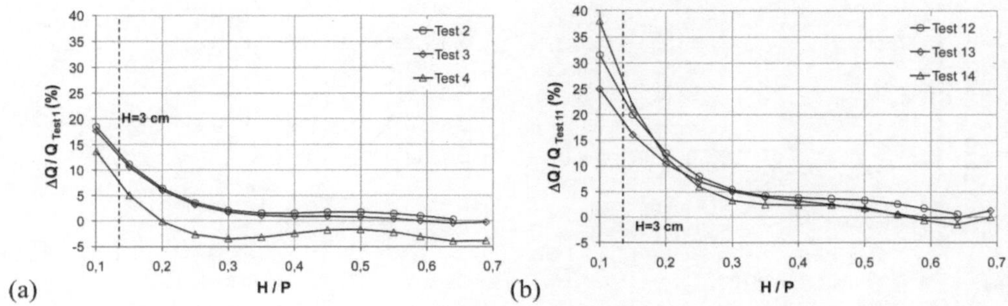

Figure 5. Crest shape effects on the gain in discharge relative to the PKW with the flat-topped crest for $P_d = 0.6P$ (a) and $P_d = 1.6P$ (b).

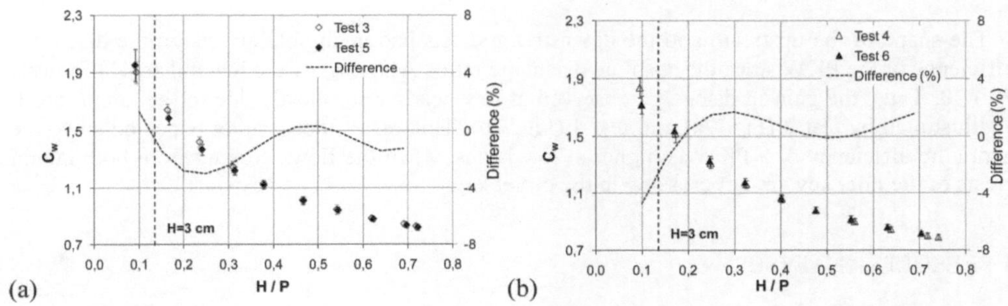

Figure 6. Effects on the rating curves $C_w(H/P)$, of the position (upstream or downstream) of the quarter-rounded shape on the lateral crest, associated to the "horizontal" (a) or the "vertical" (b) rounded shape on the upstream and the downstream crests.

Figures 4 shows that the PKW efficiency increased with the dam height for each crest configuration, whereas it was nearly unchanged with the flat-topped crest (tests 1 and 11). Thus, the gain in discharge (Figs 5a, b) increased with the dam height, by 5 to 20% at low heads, i.e. for $H/P < 0.2$.

The gain in discharge decreased rather strongly with the upstream head, so that the effects of the crest shape became negligible (<4%) if $H/P > 0.25$ for $P_d = 0.6P$ (Fig. 5a), and if $H/P > 0.35$ for $P_d = 1.6P$ (Fig. 5b). For the upstream heads higher than 3 cm, when the scale effects due to surface tension and viscosity became negligible, the PKW hydraulic efficiency could be increased by 10% (Fig. 5a) and 20% (Fig. 5b), with a half-rounded or a quarter-rounded instead of a flat-topped shape.

The tests 5 and 6 were performed to discriminate the effect of the position of the quarter-rounded shape on the lateral crest (Figs 1c, d). The rating curves $C_w(H/P)$ were compared respectively to the test 3 (Fig. 6a) and test 4 (Fig. 6b). As shown by Figure 6, the position of the rounded face had negligible effects on the rating curves (<4%), whenever the lateral crest was associated with the "horizontal" (tests 3 and 5) or the "vertical" (tests 4 and 6) rounded shapes on the upstream and downstream crest. However, the PKW was a little more efficient with the "horizontal" rounded shape than with the "vertical" rounded shape.

The present results agreed with Leite Ribeiro et al. (2006) experiments where the testing of a flat-topped crest, and a quarter-rounded crest in both positions of the rounded face, showed negligible effects on hydraulic performance.

The last tests (7 to 10) at "low" dam height were carried out to discriminate the effects of the upstream and the downstream crests, by replacing the flat-topped shape on one side only. Figures 7a and 7b show the discharge ratio $\Delta Q/Q_{Test1}$ measured with the half-rounded shape (tests 7 and 8) and the "vertical" rounded shape (tests 9 and 10).

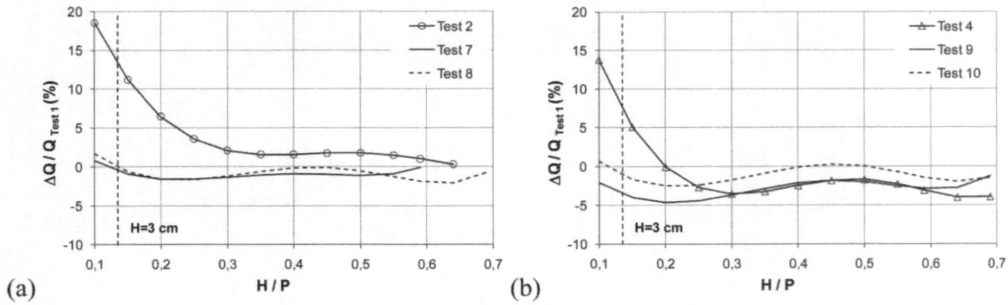

Figure 7. Effects on the gain in discharge relative to the flat-topped crest, of the lateral, upstream and downstream crests with the half-rounded shape (a), and with the "vertical" rounded shape (b).

The shape of the upstream and the downstream crests had negligible effects on the discharge efficiency of the PKW since the absolute discharge ratios $\Delta Q/Q_{Test1}$ were lower than 4% for tests 7 to 10. Thus, the gain in discharge observed at low heads were mostly due to the lateral crest, as illustrated by test 2 (Fig. 7a) and test 4 (Fig. 7b). This result looks rather expected since the hydraulic efficiency of a PKW is higher at low heads, when the flows coming from both lateral crests of the inlet key are not crossing in the outlet key.

4 RESULTS ANALYSES

4.1 *Discussion*

Anderson (2011) also tested the type A geometry recommended by Lemperiere, at the Utah Water Research Laboratory (UWRL), in a 93.4 cm wide by 61.0 cm deep by 7.4 m long rectangular flume. The PKW height was similar $P = 0.197$ m, but the ratio $W_i/W_o = 1.25$, and the wall thickness $T_s = 1.27$ cm were different from LNHE experiments. Both flat-topped and half-rounded crests were tested with parapet walls and noses under the upstream overhangs. For $0.1 < H/P < 0.6$, the gain in discharge of the rounded shape decreased from 22% to 4%.

To calculate the discharge coefficient of thick linear weirs, Hager (1987) proposed to correct the discharge coefficient of sharp-crested weirs by a factor C_s given by equations 4 and 5, for flat-topped and rounded crest respectively. Thus, the gain in discharge of the half-rounded shape relative to the flat-topped shape can be estimated by the ratio of both equations. Note that this estimation is only valid at very low heads, as long as the PKW can be considered acting as a linear weir. With the LNHE geometry ($T_s/P = 0.09$), the calculated value (~32%) for $H/P = 0.1$, was quite close to the gain in discharge measured at "high" dam height (Fig. 5b). With the UWRL geometry, which had a thinner crest ($T_s/P = 0.06$), the calculated value (~30%) was 8% higher than the measured one (22%).

$$C_s = 1 - \frac{2}{9\left(1 + \left(\frac{H}{T_s}\right)^4\right)} \tag{4}$$

$$C_s = \frac{\sqrt{3}}{2}\left(1 + \frac{\frac{22}{81}\left(\frac{H}{R}\right)^2}{1 + 0.5\left(\frac{H}{R}\right)^2}\right) \tag{5}$$

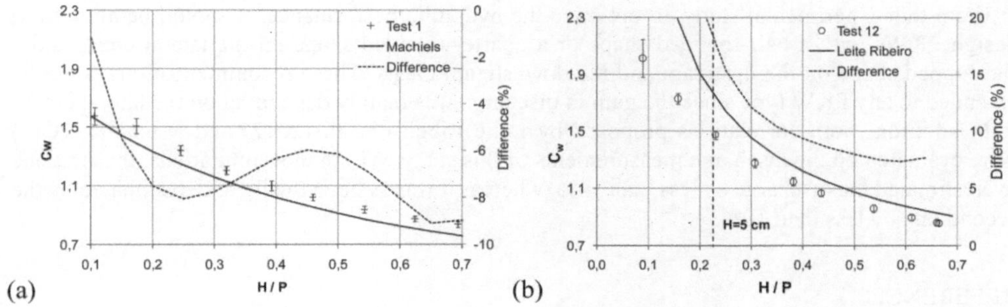

Figure 8. Rating curves $C_w(H/P)$, measured on test 1 and predicted by Machiels (2012) (a) – Rating curves $C_w(H/P)$, measured on test 12 and predicted by Leite Ribeiro et al. (2012) (b).

4.2 *Comparison to existing correlations*

Machiels (2012) developed a correlation for any type of PKWs, by testing PKWs with a flat-topped crest. The predictions of this correlation were compared (Fig. 8a) to the LNHE measurements with the flat-topped crest at "low" dam height (Test 1). The correlation underestimated the discharge capacity measured by less than 10% for all range of H/P.

Leite Ribeiro et al. (2012) developed a correlation for the type A, by testing PKWs with a half-rounded crest, and with noses under the upstream overhangs. The predictions of this correlation were compared (Fig. 8b) to the LNHE measurements with the half-rounded crest at "high" dam height (Test 12). Within its limitations ($H > 0.05$ m), the error was lower than 17% and even lower than 14% taking into account the beneficial effects of the noses. Actually, the UWRL experiments (Anderson, 2011) showed that the noses could increase the discharge by about 3% for $0.2 > H/P > 0.7$.

The correlation proposed by Kabiri-Samani & Javaheri (2012) was also evaluated on the LNHE experiments. Although some geometric parameters were slightly out of the range of validity, the discharge was highly overestimated (>26%) for all heads.

5 CONCLUSIONS

This experimental study, based on a type A PKW of the same width as the experimental channel, characterized the effects of the crest shapes on hydraulic performance under free-flow conditions. Some results already observed by Leite Ribeiro et al. (2007) and Anderson (2012) were confirmed:

- The discharge capacity was influenced by the crest shape mainly at low heads (for $H/P < 0.3$), when the jets coming from both the lateral crests of the inlet keys were not crossing.
- The half rounded and the quarter-rounded shapes should be preferred to the flat-topped shape because their discharge coefficient is higher at low heads.
- With the quarter rounded shape on the lateral crest, the discharge capacity was the same whatever the position of the rounded face (on the upstream or on the downstream crest).

The present study highlighted the following results:

- The discharge capacity was mainly influenced by the shape of the lateral crest, since the shape of the upstream and the downstream crest had negligible effects.
- For the crest configurations using a rounded shape, the gain in discharge (relative to the flat-topped crest) increased with the dam height.
- With the half-rounded and quarter-rounded crest, the gain in discharge reached 20% to 35% (according to the dam height) at very low heads, and 10% to 20% for $H > 3$ cm, when the scale effects became negligible.

From this experimental study, to optimize the hydraulic performance, it should be advised to design PKWs with a half-rounded shape or a quarter-rounded shape on the lateral crest, and a flat-topped shape on the upstream and the downstream crests. These recommendations could be extended to any PKW type since the gain in discharge was mainly depending on the lateral crest.

In addition, both correlations proposed by Leite Ribeiro et al. (2012) and Machiels (2012) were evaluated on the discharge measurements of this study. Within their limitations, the first one overestimated the discharge by less than 14%, whereas it was systematically underestimated by the second one by less than 10%.

REFERENCES

Anderson, R.M. 2011. Piano Key Weir Head Discharge Relationships. M.S. Thesis, Utah State University, Logan, Utah.

Barcouda, M., Cazailler, O., Cochet, P., Jones, B.A., Lacroix, S., Laugier, F., Odeyer, C. & Vigny, J.P. 2006. Cost effective increase in storage and safety of most dam using fusegates or P.K. weirs. *22ème congrès des grands barrages, Barcelona, Spain.*

Cicero, G.M., De Miranda D.A. & Luck, M. 2012. Assessment of the code Wolf 1D PKW for predicting the hydraulic behaviour of PK-Weirs. *Congrès SHF – 33èmes journées de l'hydraulique "Grands aménagements hydrauliques 2012", Paris, France.*

Hager, W.H. 1987. Lateral Outflow Over Side Weirs. *Journal of Hydraulic Engineering*, 113: 491–504.

Ho Ta Khanh, M., Chi Hien, T. & Thanh Hai, N. 2011. Main results of the P.K. weir model tests in Vietnam (2004 to 2010). *Proc. International Workshop on Labyrinths and Piano Key Weirs PKW 2011, CRC Press*, 191–198.

Kabiri-Samani, A. & Javaheri, A. 2012. Discharge coefficient for free and submerged flow over Piano Key Weirs. *Journal of Hydraulic Research*, 50(1): 114–120.

Leite Ribeiro, M., Boillat, J.L., Schleiss, A.J., Laugier, F. & Albalat, C. 2007. Rehabilitation of St-Marc dam-*Proceedings of the 32nd Congress of IAHR, Venice, Italy.*

Leite Ribeiro, M., Pfister, M., Schleiss, A.J. & Boillat, J.L. 2012. Hydraulic design of A-type Piano Key Weirs. *Journal of Hydraulic Research*, 50(4): 400–408.

Lemperiere, F., Vigny, J.P. & Ouamane, A. 2011. General comments on Piano Kew Weirs: The past and present. *Proc. International Workshop on Labyrinths and Piano Key Weirs PKW 2011, CRC Press*, 17–25.

Ouamane, A. 2011. Nine years of study of the Piano Key Weir in the university laboratory of Biskra "lessons and reflections". *Proc. International Workshop on Labyrinths and Piano Key Weirs PKW 2011, CRC Press*, 51–58.

Machiels, O., Erpicum, S., Dewals, B.J., Archambeau, P. & Pirotton, M. 2011. Experimental observation of flow characteristics over a Piano Key Weir. *Journal of Hydraulic Research*, 49(3): 359–366.

Machiels, O. 2012. Experimental study of the hydraulic behaviour of Piano Key Weirs. M.S. Thesis, Université de Liege, Liege, Belgium.

Pralong, J., Vermeulen, J., Blancher, B., Laugier, F., Erpicum, S., Machiels, O., Pirotton, M., Boillat, J.L., Leite Ribeiro, M. & Schleiss, A.J. 2011. A naming convention for the Piano Key Weirs geometrical parameters. *Proc. International Workshop on Labyrinths and Piano Key Weirs PKW 2011, CRC Press*, 271–278.

Tullis, J.P., Armanian, N. & Waldron, D. 1995. Design of labyrinth spillways. *Journal of Hydraulic Engineering*, 121(3): 247–255.

Vermeulen, J., Laugier, F., Faramond, L. & Gilles, C. 2011. Lessons learnt from design and construction of EDF first Piano Key Weirs. *Proc. International Workshop on Labyrinths and Piano Key Weirs PKW 2011, CRC Press*, 215–224.

Labyrinth and Piano Key Weirs II – PKW 2013 – Erpicum et al. (eds)
 ISBN 978-1-138-00085-8

Influence of geometrical parameters on PK-Weirs discharge with 3D numerical analysis

V. Lefebvre, J. Vermeulen & B. Blancher
Hydro Engineering Center, EDF-CIH, Le Bourget du Lac, France

ABSTRACT: Piano Key Weirs (PKW) are now well known in the hydraulic engineering sphere and their particular shape becomes familiar to everyone. However, the discharge capacity of a PKW is related to many parameters and is still complex to estimate. In order to have a better understanding of the flow behavior over such weirs, a numerical model has been developed at EDF-CIH on the CFD Software FLOW-3D®. The influence of several geometrical parameters has been tested on the PKW Model A of Lemperiere & Ouamane (2003) for different heads. This paper presents in-depth results of the sensitivity analysis concerning the total height of the PKW, the inlet and outlet key width, the use of noses or parapet walls and the inclination of the sidewalls. Some outcomes confirm the results of previous studies (e.g., PKW height), some are in disagreement (e.g., parapet walls influence), some are drawn from alternative approach (W_i and W_o are considered separately, not as a ratio) and some are original (noses and sidewalls vertical inclination). Numerical simulation allows performing many studies in a fairly short time and is promising for the future in terms of reliability on advanced stage for a PKW project.

1 INTRODUCTION

Many studies have been performed concerning the influence of different geometrical parameters on PKW discharge. The use of CFD-software is now well established at EDF – Hydro Engineering Center and numerical modeling is often conducted in parallel with physical model experimentations in order to compare the conclusions.

Some parameters are also rather difficult to study with physical models and FLOW-3D® is a good means to predict the flow behavior in these cases.

This paper focuses on these two vistas: strengthen our knowledge on important practical conclusions and reveal new outcomes regarding PKW design.

2 REFERENCE NUMERICAL MODEL CALIBRATION

The simulation results produced using FLOW-3D® have been compared to those of a reference geometry in terms of specific discharge capacity. The reference model chosen for this comparison is the first PKW design submitted by Lemperiere & Ouamane (2003). The numerical model has been validated on the experimental data and used for sensitivity analysis with a mean relative deviation in specific flow rates not exceeding 5% (Tab. 2).

The model represents half an inlet, half an outlet and a sidewall applying symmetric boundary conditions on the middle plane of each key. The solver uses first order momentum advection and the turbulence model is not activated since it has been proved that it does not have impact on the discharge capacity results. The used mesh is coarser than previously (Pralong et al., 2011), about 50,000 cells (minimum 20 cm size), but fine enough to adequately represent the solid geometry. The computation time is thus reduced to less than 30 minutes.

Table 1. PKW Model A geometrical parameters.

B m	B_i m	B_o m	P m	W_i m	W_o m	T_s m	W_u m	L_u m
12	3	3	4	2.4	2.4	0.02~0*	4.8	28.8

*Approximed to 0 for global dimensions.

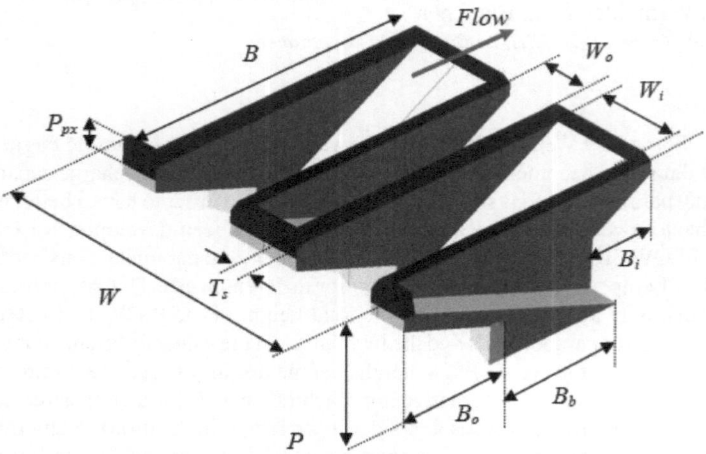

Figure 1. PKW main parameters (Pralong et al., 2011).

Table 2. Physical and numerical model specific discharges for reference PKW.

Upstream Head H m	Ratio H/P	Physical model q_{sW} $m^3/s/m$	Numerical model q_{sW} $m^3/s/m$	Deviation %
0.5	0.125	3.5	3.6	+3.3
1.0	0.250	8.2	7.9	−3.1
2.0	0.500	15.6	15.4	−1.1
4.0	1.000	28.7	29.2	+1.8

The following sensitivity analyses are drawn from numerical simulations using configurations similar to the reference one. Thus, the specific flow rates values are compared to those of the reference simulation on FLOW-3D® (not the experimental data from Lemperiere) for each parameter and upstream hydraulic heads equal to 0.5 m, 1 m, 2 m and 4 m.

The gain in specific discharge capacity is defined as: $(Q - Q_{ref})/Q_{ref}$.

3 NUMERICAL SENSITIVITY ANALYSES

3.1 *Total height of the PKW*

The height of the PKW is known as one of the parameters having a major influence on the discharge capacity of the structure. The main modifications induced by a variation of the total height are the inlet entrance section and the keys slopes.

On the PKW model A, this parameter is equal to $P = 4$ m. Different values are simulated in order to estimate the influence on the specific flow rate ($P = 2, 3, 4, 5, 6$ and 8 m).

These results confirm a clear conclusion: PKW discharge capacity increases when the total height increases. This trend is similar for all tested upstream heads. Such outcome was expected

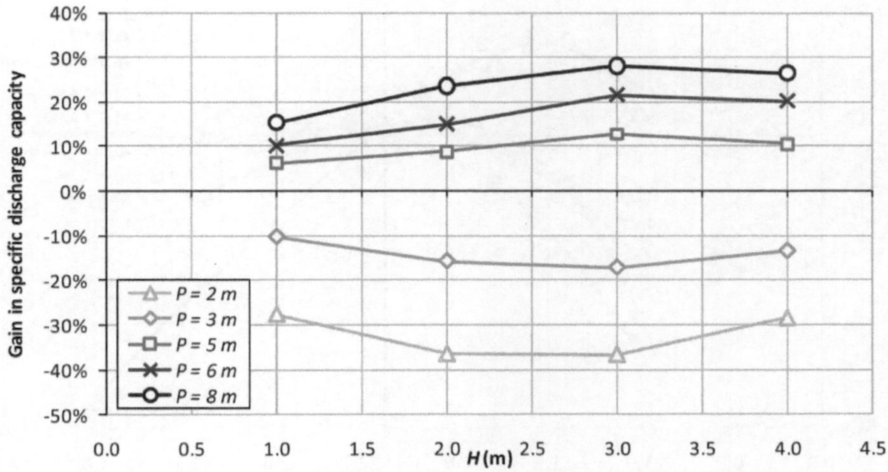

Figure 2. Specific discharge deviation versus upstream head for different PKW heights.

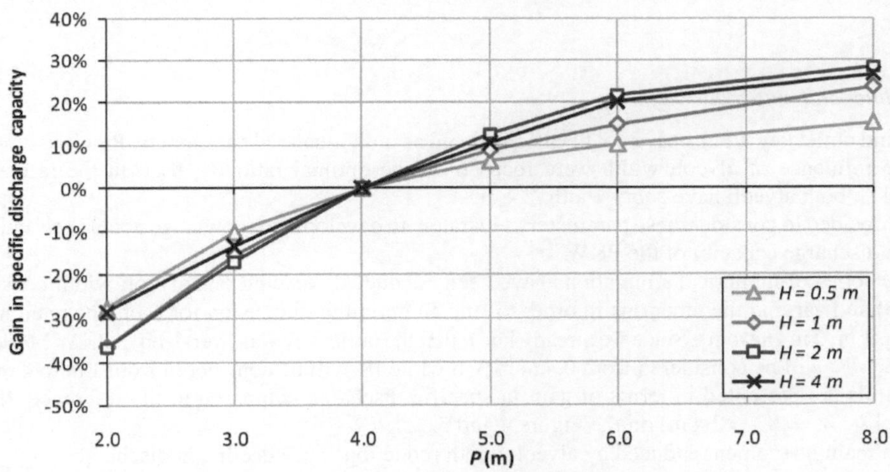

Figure 3. Specific discharge deviation versus PKW height for different upstream heads.

according to the previous studies (Machiels et al. 2011, Pralong et al. 2011) and the numerical simulation allows more detailed explanations:

- The inlet key section area is enlarged and the velocity is reduced, as well as the hydraulic head loss. Less energy is dissipated in this part;
- The outlet key slope is steeper when the PKW is higher and the flow evacuation is improved;
- Local submergence of the downstream unit is reduced.

Moreover, we see that the influence of a reduced P is more significant on the specific discharge for upstream heads of 1 m and 2 m. This is largely because the PKW Model A is more or less optimized for this range of heads and such modification to P triggers a bigger relative deviation in terms of discharge capacity.

Finally, even if this study confirms the manifest hydraulic interest of a higher PKW, this design remains subject to structural and economic validations under the overall constraints of a given project.

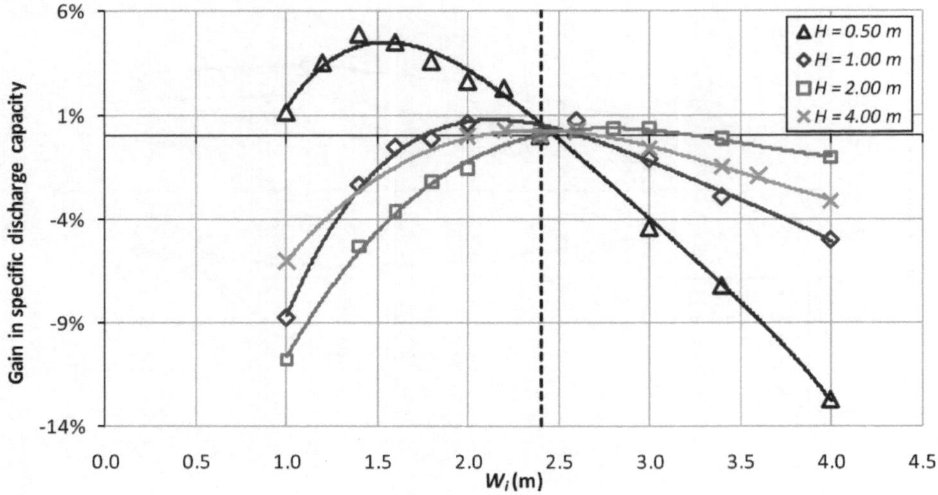

Figure 4. Specific discharge deviation versus inlet key width.

3.2 *Inlet and outlet key widths*

Inlet and outlet key widths have not been considered as individualized parameters. Previous studies on the influence of alveoli width were focused on the optimal ratio W_i/W_o. On the reference Model A, both alveoli have equal width, 2.4 m.

We decided to consider these parameters separately to develop our knowledge about their impact on the discharge capacity of the PKW.

Two series of numerical simulations have been conducted, keeping one alveoli width constant (2.4 m) and varying the other one in order to find an optimum dimension for each upstream head (0.5 m, 1 m, 2 m and 4 m). Since we already knew that the outlet key was oversized in PKW Model A scheme, W_o will be considered from 0.4 m to 3 m while W_i will be considered from 1 m to 4 m.

Results are presented in terms of gain in specific discharge compared to the reference PKW (Model A, $W_i = W_o = 2.4$ m) on the Figure 4 and 5.

Two main phenomena induced by alveoli width reduction cause decline in discharge:

- Lack of water supply in the inlet key if W_i is too small (except small H as L/W ratio increase balances this effect);
- Overfilling of the outlet key if W_o is too small.

Above these critical values, the specific discharge is reduced because the discharge gain is balanced by the geometrical expansion ($q_{sW} = Q/W$).

The results are drawn from discharge measures over each part of the PKW (inlet key crest, outlet key crest and sidewall crest) and we can observe different repartitions of the flow. Increasing the inlet (resp. outlet) width triggers a discharge gain on the inlet (resp. outlet) and the sidewall but a loss on the outlet (resp. inlet).

We can see that the optimal point for W_i and W_o is lower for small heads (up to 30% gain at $W_o = 1$ m for $H = 0.5$ m and 5% gain at $W_i = 1.5$ m for $H = 0.5$ m). The performances of the PKW are decreasing rapidly when W_i and W_o diverge from the optimal dimension, especially as the upstream head is small.

Combined optimization of W_i and W_o can lead to gain in specific discharge up to 75% for small heads (0.5 m).

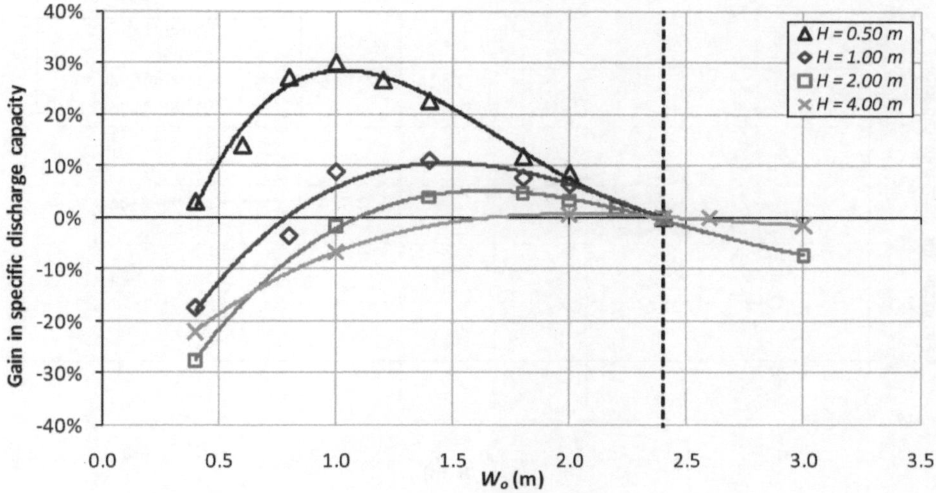

Figure 5. Specific discharge deviation versus outlet key width.

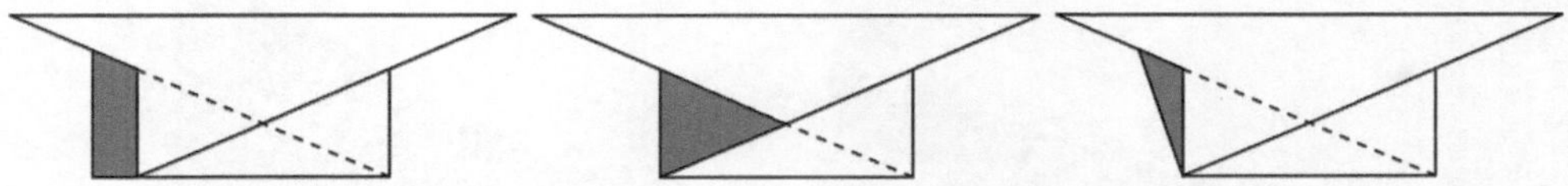

Figure 6. Triangular and circular (left), triangular extruded (center) and triangular variable (right) noses.

3.3 *Noses*

Noses are elements added under the upstream overhang to guide the flow into the inlet keys. It can be any type of hydrodynamic shape and is comparable to bridge piers.

On each PKW built, EDF has implemented noses as they are rather easy to construct but this is the first numerical study on the topic. Height different deflector models have been tested and four of them are presented here:

- 1 m triangular nose added upstream,
- 1 m half-circular nose added upstream,
- Triangular nose extruded in the outlet key base until geometrical cross of the slopes,
- Triangular nose added upstream with 1 m length at the top and 0 m at the base.

Figure 6 shows the position and the profile of the deflector models.

This graph (Fig. 7) confirms the limited impact of deflector on the discharge capacity of the PKW. Best gain is around 5% in terms of specific discharge.

We can see that the gain is less for small hydraulic head, probably because flow velocity is lower and the hydrodynamic profile of the structure has less impact on the fluid head loss.

The nose shape has a small influence on overall weir performance but we can still note that the triangular profile is better than the circular one. This result is consistent with the study concerning bridge piers (Ven Te Chow, 1959). Best effects are achieved with the triangular extruded deflector. Indeed, this one enlarge the inlet key cross section and contribute to a better supply of the weir opposing less resistance to the flow coming from the bottom.

It appears that the hydraulic impact of noses is positive but limited. Yet they can also be interesting in terms of structural stability because it adds weight on the upstream part of the PKW.

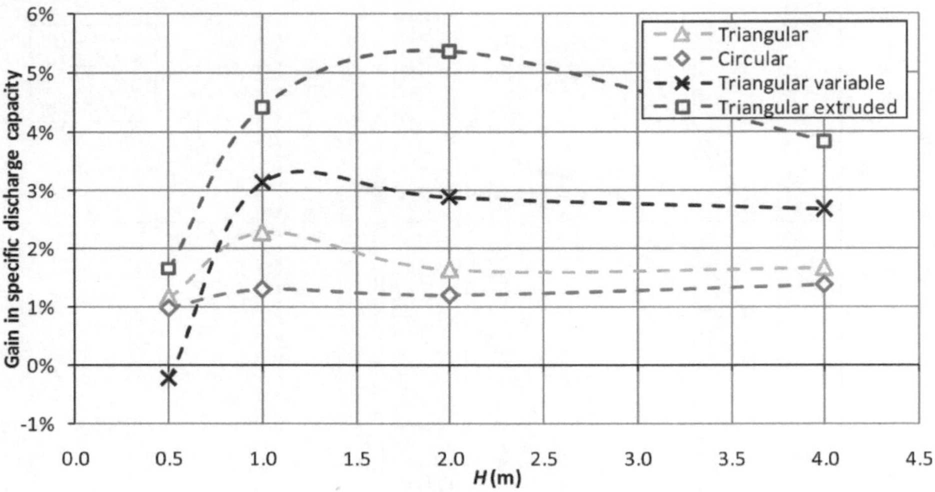

Figure 7. Specific discharge deviation versus upstream head for different deflectors.

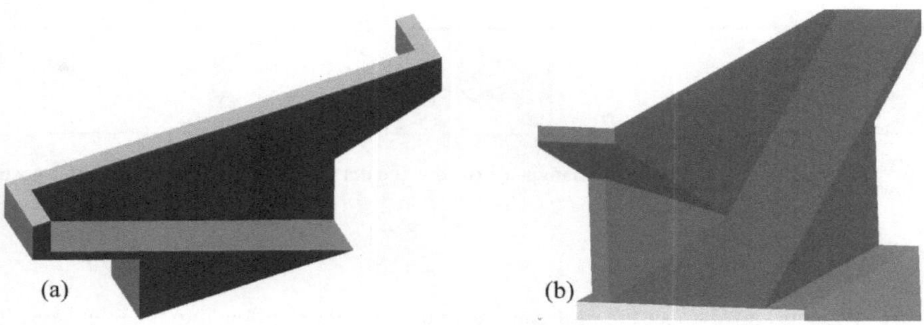

Figure 8. Parapet along the entire crest (a) and triangular extruded deflector (b).

3.4 *Parapet*

Parapet walls have been studied many times before (Pralong et al. 2011, Machiels et al. 2013), since they are considered as effective and easy to build structures. It consists in vertical walls placed on the crest of the PKW either to heighten the structure or to modify the inlet and outlet keys slopes, keeping the global height *P* constant. We consider only the second case since we already know that the influence of the parameter *P* on the discharge capacity is major.

Previous studies focused on the influence of parapet walls on each part of the PKW. The simulations were conducted with a parapet wall all along the unit crest and the specific discharge results reported separately for inlet, outlet and sidewall crests. Conclusion reveals that a parapet wall is beneficial on the outlet but not on the inlet keys.

New simulations using PKW models with parapet only at the inlet or outlet extremities show different outcomes. The prediction assuming that the outlet parapet would be very interesting is no longer right since it appears that such modifications permit a maximal gain around 5%.

3.5 *Inclination of the sidewalls*

According to the objective of inlet and outlet keys enlargement, making the sidewalls thinner is known as a major possibility of discharge capacity improvement. Yet, this kind of alteration is

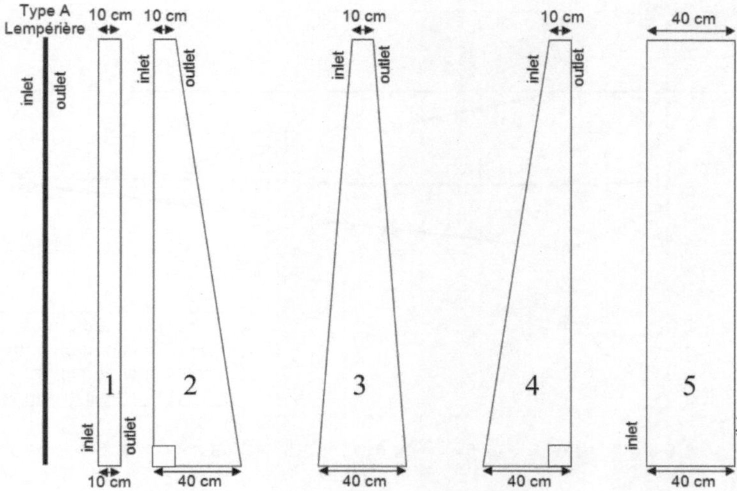

Figure 9. Type A PKW and studied sidewall profiles (not to scale).

often really hard to execute from a structural point of view because of the reinforcement limits. In fact there are less constraints at the top of the sidewall than at the bottom. An alternative solution consists in building sidewalls with a batter.

Different inclination configurations have been modeled and tested in FLOW-3D® (from left to right on Fig. 9):

- Vertical sidewall, 10 cm thick (1)
- Sidewall inclined in the outlet, 10 cm thick at the top and 40 cm thick at the base (2)
- Sidewall inclined on both sides, 10 cm thick at the top and 40 cm thick at the base (3)
- Sidewall inclined in the inlet, 10 cm thick at the top and 40 cm thick at the base (4)
- Vertical sidewall, 40 cm thick (5)

Figure 9 presents these profiles in a schematic; simulations are executed for upstream heads from 0.5 m to 4 m.

Since the PKW model A does not have structural sense and the configuration 1 is the closest of the PKW model A, it is taken as reference to compare the discharge capacity of other designs. Figure 10 shows the gain in terms of specific discharge due to the enlargement of the sidewall base compared to this reference.

Making the sidewalls thicker on the PKW crest will decrease its discharge capacity (according to Laugier et al. 2011) and Figure 10 confirms it. The 40 cm vertical sidewall (5) presents the largest loss for each hydraulic head. The objective is to reduce as far as possible this loss.

Considering these results, we could conclude that the best configuration is the sidewall inclined in the outlet (2) but simulations are using models with equal alveoli width (as model A PKW). It has been shown that $W_i/W_o > 1$ is a hydraulically more efficient ratio. The outlet key designed in the PKW model A is oversized compared to the inlet key, then reducing its width by placing the inclination of the sidewall on this side is just an artifact. Other simulations have been carried out with $W_i/W_o = 1.5$ and the optimal hydraulic configuration appears to be the symmetrical one (3).

Finally, setting the inclination on both sides (3) is the best configuration on the hydraulic point of view but such modifications of the concrete structure are subject to drastic constraints. It is easier to build formworks with at least one vertical side. Then the designer will have to consider all the other geometrical parameters to decide which key is too large and where to place the sidewall inclination.

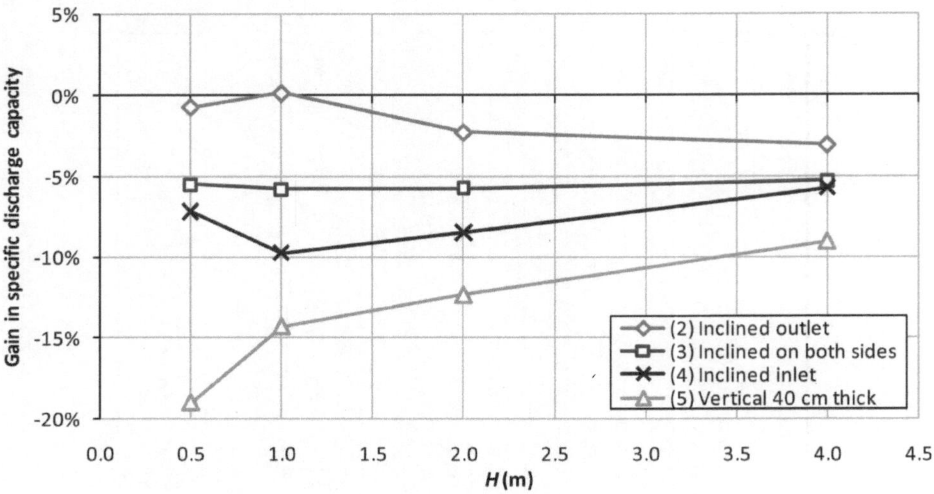

Figure 10. Specific discharge deviation versus upstream head for different sidewall inclinations.

4 CONCLUSION

The development of a new simulation model on the CFD software FLOW-3D® allows running different geometrical schemes of PKW in a fairly short time. The latest sensitivity analysis performed at EDF – Hydro Engineering Center confirms the importance of the PKW height and inlet/outlet width optimization to maximize the discharge capacity of the spillway. It also permits exploring the advantages and disadvantages of small structural devices on the hydraulic point of view such as noses, battered sidewall and parapet walls. Estimating the influence of this kind of equipment by physical model experimentations would have been longer, cumbersome, and more expensive.

This validation of the software on various cases is really promising for the future in terms of reliability on advanced stage for a PKW project.

REFERENCES

Laugier, F., Pralong, J. & Blancher, B. 2011. Influence of structural thickness of sidewalls on PKW spillway discharge capacity. *International Workshop on Labyrinth and Piano Key weirs*, Liège, Belgium.

Lempérière, F. & Ouamane, A. 2003. The piano key weir: a new cost-effective solutions for spillways. *Hydropower & Dams* (5): 144–149.

Machiels, O., Erpicum, S., Archambeau, P., Dewals, B. & Pirotton, M. 2011. Influence of the Piano Key Weir height on its discharge capacity. *International Workshop on Labyrinth and Piano Key weirs*, Liège, Belgium.

Machiels, O., Erpicum, S., Archambeau, P., Dewals, B. & Pirotton, M. 2013. Parapet Wall Effect on Piano Key Weir Efficiency. *Journal of Irrigation and Drainage Engineering*. 139(6), 506–511.

Pralong, J., Vermeulen, J., Blancher, B., Laugier, F., Machiels, O., Erpicum, S., Pirotton, M., Leite Ribeiro, M., Boillat, J.L. & Schleiss, A.J. 2011. A naming convention for the piano key weirs and geometrical parameters. *International Workshop on Labyrinth and Piano Key weirs*, Liège, Belgium.

Pralong, J., Montarros, F., Blancher, B. & Laugier, F. 2011. A sensitivity analysis of Piano Key Weirs geometrical parameters based on 3D numerical modeling. *International Workshop on Labyrinth and Piano Key weirs*, Liège, Belgium.

Ven Te Chow. 1959. Open-Channel Hydraulics. *McGraw-Hill Classic Textbook Reissue.*

Labyrinth and Piano Key Weirs II – PKW 2013 – Erpicum et al. (eds)

Staged labyrinth weirs and their effects on the outflow hydrograph

M.R. Dabling & B.P. Tullis
Utah Water Research Laboratory, Department of Civil and Environmental Engineering, Utah State University, Logan, USA

ABSTRACT: Nonlinear weirs (e.g., piano key and labyrinth) often replace less hydraulically efficient linear weirs (e.g., sharp- or ogee-crested) in existing spillway channels to increase spillway discharge capacity; a nonlinear weir can usually pass a larger design flood at a given upstream head relative to linear weirs of the same width. Using a hydraulically more efficient spillway means that flood routing through the reservoir will occur with less reservoir level fluctuations and the maximum base-flow pool elevation of the reservoir can be increased (more water storage). The increased discharge efficiency of the nonlinear weir will also cause the outflow hydrograph to compress temporally and the peak outflow discharge rate will increase, potentially increasing flood risk and hazard downstream. Utilizing staged labyrinth weirs (i.e., weir sections with different crest elevations) can confine base flows to a subset of the weir, alter the reservoir flood-flow storage characteristics, and/or satisfy outflow hydrograph discharge requirements. In this study, the flow characteristics of a staged labyrinth weir with one cycle lowered (laboratory-scale) were evaluated. As an illustration of use, flood-flow hydrographs were numerically routed through a sample reservoir featuring a spillway utilizing a standard (nonstaged) labyrinth weir, a staged labyrinth weir, and an ogee-crested weir. The resulting outflow hydrographs quantified and discussed. Results indicate that staged labyrinth weirs can be used to alter the outflow hydrograph and reduce the risk of downstream flooding.

1 INTRODUCTION

1.1 *Labyrinth weirs*

Dams are critical infrastructure components that provide water supply (municipal, agricultural, industrial), flood control, hydropower, navigation, and recreation. Aging infrastructure, changes in land use, and higher peak flow predictions for extreme flood events often require upgrading or rehabilitating existing spillways. The replacement of linear weirs with nonlinear weirs (ogee-crested weirs with labyrinth weirs in many cases) is often considered as a method to increase spillway discharge capacity without increasing the existing spillway channel width.

Geometric variables for labyrinth weir design are shown in Fig. 1, where P is the weir height, α is the sidewall angle, A is the apex width, w is the cycle width, W is the channel width, R_{crest} is the radius of the upstream crest wall, and l_c is the sidewall centerline length. The geometry of a labyrinth weir can significantly increase the weir crest length within a fixed-width channel relative to a linear weir. Because a weir's discharge is proportional to its length, this can increase discharge capacity by 3 to 4 times (Tullis et al. 1995). Due to their hydraulic performance, labyrinth weirs have been of interest to practitioners and researchers for many years. Many notable labyrinth weir design publications that have focused on discharge performance including Hay and Taylor (1970), Darvas (1971), Hinchliff & Houston (1984), Lux & Hinchliff (1985), Magalhães & Lorena (1989), Tullis et al. (1995), Melo et al. (2002), Falvey (2003), Tullis et al. (2007), Crookston (2010), Crookston & Tullis (2012a, b, 2013a, b), and Dabling et al. (2013). Labyrinth weirs are often used to increase spillway capacity and manage upstream flooding. However, increasing the discharge efficiency of the flow control structure can have implications on downstream flood routing as well. As the

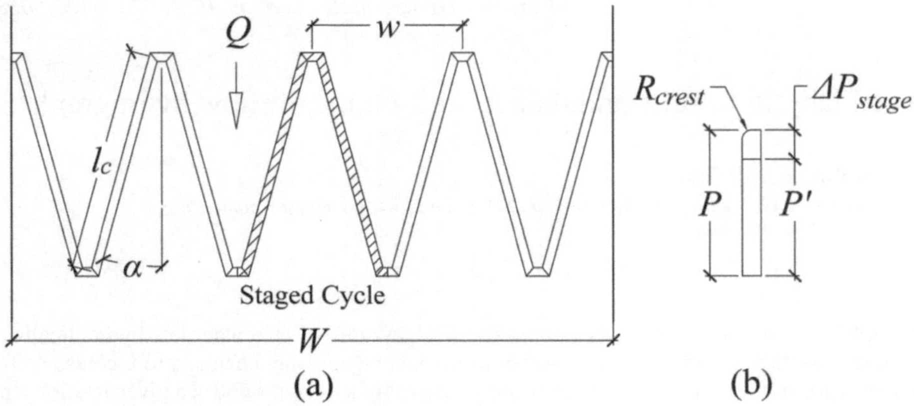

Figure 1. Labyrinth weir design parameters in plan (a) and elevation (b).

discharge efficiency of the outlet structure increases, the reservoir retention volume decreases and the peak outflow increases. The downstream conveyance system must be able to accommodate the increased discharge or flooding may result. This study looked at the influence of the reservoir flow control structure head-discharge characteristics on the outflow hydrograph.

In addition to spillway head-discharge requirements for the design storm (e.g., passing the full PMF), an upgraded higher flow capacity spillway design may also be required to match pre-development peak outflows for more frequent storm events. For example, a new spillway may be required to pass an enlarged PMF but the 25- and 100-year post-development peak outflows must be less than or equal to those of the existing spillway.

Different spillway types installed in parallel (e.g., gated and ungated spillways) can be an effective means of meeting a prescribed outflow hydrograph requirement during flood routing events. Another method would be to design a labyrinth weir that incorporates multiple crest elevations – creating a staged labyrinth weir.

1.2 *Staged Labyrinth Weirs*

In the literature and in practice, the terms 'notch' and 'stage' are often used interchangeably. For weirs with segments at different crest elevations, 'stage' refers to a continuous subsection of labyrinth weir with a common crest elevation. This study uses a labyrinth weir with two crest elevations; subsequently the term 'high stage' and 'low stage' are used. A variety of staged weir spillway types [e.g., ogee spillways (Tom Miller Dam, Austin, TX, USA), and labyrinth spillways (Lake Townsend Dam, Greensboro, NC, USA)] have been designed and built featuring segments with different crest elevations. Geometric variables specific to staged labyrinth weirs are presented in Fig. 1 where P' is the height of the low stage and ΔP_{stage} is the difference between P and P'.

When designing a staged labyrinth weir, the lower stage would likely be set at the normal pool elevation and convey base flows and runoff from relatively frequent storm events (e.g., 2-year, 10-year, etc.). The higher stage would provide additional required discharge capacity during more extreme flood events. Confining base flow and smaller storm event outflows to staged sections can also be beneficial by limiting algal or other biological growth that can develop on wet surfaces, reducing weir maintenance costs.

1.3 *Research objectives*

As previously noted, multiple hydraulic design methods have been published for labyrinth weirs; however, minimal studies have focused on their effects on the downstream hydrograph. One objective of this study was to investigate the hydraulic performance of a 4-cycle staged and non-staged

Table 1. Laboratory model geometries.

Weir Geometry		Description of Weir	Stage/Notch Geometry	
			l_{stage}	ΔP_{stage}
$\alpha = 15°$	$P = 152.4$ mm	Staged Labyrinth	995.7 mm	0.2P
$N = 4$	$w = 305.9$ mm	Non-staged Labyrinth	N/A	N/A
$L = 3982.8$ mm	$t_w = 18.4$ mm			
$L_{c-cycle} = 995.7$ mm				

labyrinth weir using laboratory-scale models. Using the head-discharge data obtained, a theoretical reservoir and inflow flood hydrograph were created with an appropriately sized spillway. The inflow hydrograph was routed through the reservoir featuring an ogee-crested linear weir, a labyrinth weir, and a staged labyrinth weir as the outlet control structure in order to evaluate the effects of spillway efficiency on the downstream hydrograph.

2 STAGED LABYRINTH WEIR DISCHARGE CHARACTERISTICS

2.1 *Experimental setup*

Investigations were conducted in a rectangular flume (1.2-m × 14.6-m × 1.0-m deep) at the Utah Water Research Laboratory (UWRL) at Utah State University. A 4-cycle ($N = 4$), 15° sidewall angle ($\alpha = 15°$) labyrinth weir with a quarter-round crest shape [$R_{crest} = 1/2$ the wall thickness (t_w)] was tested with and without a lowered section of crest (staged and non-staged). When staged, one full cycle centered at the centerline of the channel was lowered by 20% of P. Both model labyrinth weirs featured the same total weir centerline length (L) and a quarter-round crest shape. The geometries of both models are summarized in Table 1.

2.2 *Head-discharge relationship*

Eq. 1, a form of the standard weir equation (Henderson 1966), was selected to quantify the head-discharge relationship of the tested physical models and to calculate discharge coefficients for varying flow conditions.

$$Q = \frac{2}{3} C_d L \sqrt{2g} H_u^{3/2} \tag{1}$$

In Eq. (1), Q is the weir discharge; C_d is a dimensionless discharge coefficient that varies with weir type, geometry, crest shape, and flow conditions; L is the weir crest length; g is the gravitational acceleration constant; and H_u is the free-flow (non-submerged) upstream total head relative to the weir crest elevation (see Fig. 2). H_u was used rather than the piezometric head (h_u) in an effort to better account for the effects of approach flow velocity. Using H_u also allows the data to be applied when the approach flow velocity is zero (i.e. a reservoir application). Upstream depth measurements were made 6.5P (approx. 1 m) upstream of the weir using a stilling well equipped with a point gage ($\pm$0.15 mm). H_u was then computed as h_u plus the velocity head at the measurement location ($H_u = h_u + V^2/2g$).

Due to multiple crest elevations for a single weir, particular attention must be given to the definition of h_u for a staged labyrinth. In this study, head-discharge data were collected for the staged weir under two different scenarios: (a) discharges isolated to the lower stage segment and (b) discharges where the entire weir length was engaged. When flow was limited to the low stage, the upstream piezometric head for the low stage (h'_u) was measured relative to the lower crest elevation and low stage weir height (P') with the total head for the low stage (H'_u) computed as h'_u plus the velocity head (see Fig. 2a). When flow was conveyed over the entire labyrinth weir, h_u was

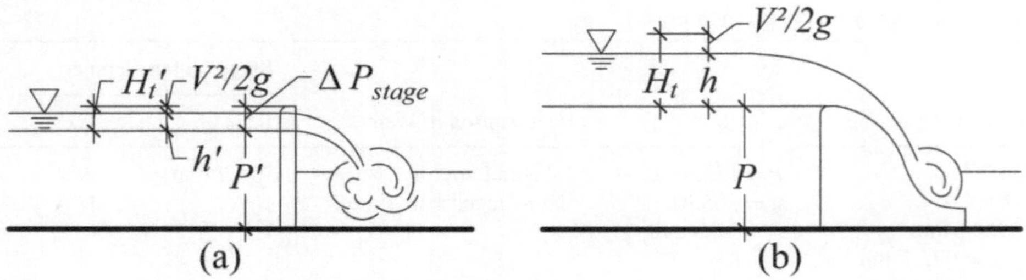

Figure 2. Flow parameters over a staged (a) and ogee crested (b) weir.

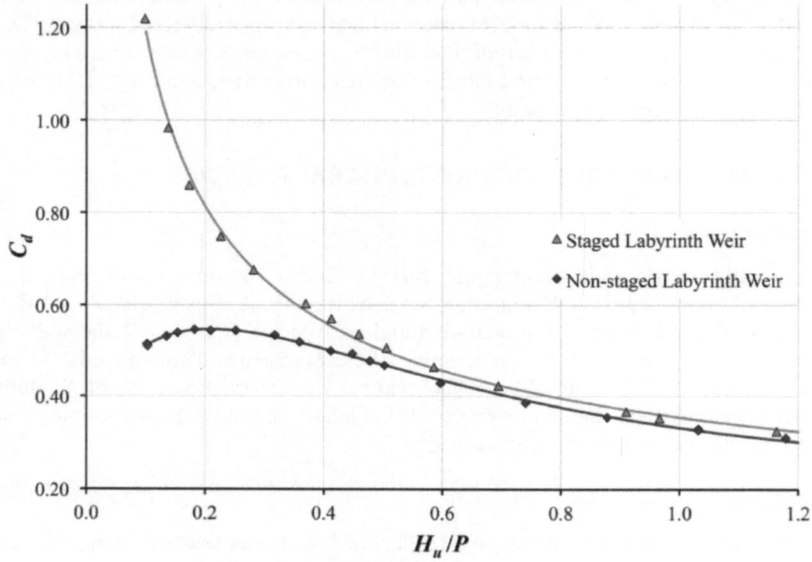

Figure 3. C_d vs. H_u/P for tested staged and non-staged labyrinth weirs.

measured relative to the high stage to compare the staged weir to the traditional labyrinth with a single crest elevation (see Fig. 2b).

With the entire labyrinth weir crest engaged (flow passing over both stages), approximately 15 to 30 head-discharge data points were collected (for the staged and non-staged labyrinth weir configurations) for dimensionless upstream heads ranging from $0.1 < H_u/P < 2.0$. C_d values for the entire spillway were computed for each measured flow condition and are presented in Fig. 3 along with C_d values for the standard labyrinth weir.

As a result of applying the higher stage crest reference to the lower stage, the experimentally determined 'composite' C_d values of the entire spillway (Fig. 3) are greater at small H_u/P values than typical for C_d values for traditional labyrinth weirs (the staged labyrinth weir produced values of $C_d > 1.0$).

3 DISCHARGE HYDROGRAPH MODELING

3.1 *Reservoir and weir characteristics*

A trapezoidal cross-section reservoir with a 16.1 m wide spillway was used as a model to route the inflow hydrograph. The reservoir was 1,500 m by 1,500 m with a normal pool water surface

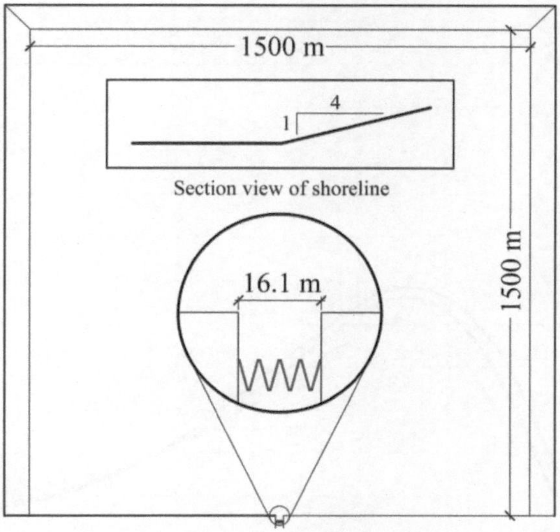

Figure 4. Model reservoir dimensions.

elevation of 200 m. Three sides of the reservoir had a slope of 1:4 (vertical:horizontal) to increase the area of the reservoir with increasing water depth, while the side that contained the spillway remained constant (see Fig. 4). The relatively basic shape of reservoir was selected to maintain simplicity in the model.

Both labyrinth weirs modeled were geometrically similar to the laboratory models, but were scaled so that $P = 4$ m. The ogee-crested weir maintained the same channel width, weir height, and crest elevation. The crest coefficient for the ogee-crested weir was consistent with USBR guidelines (USBR, 1987).

3.2 *Flood routing analysis*

An inflow storm hydrograph was routed through the model reservoir using the storage indication method (Walski et al., 2007) with a time step of 0.1 hours. The model inflow hydrograph had a 24-hour duration with a peak discharge of 300 m^3/s, as shown in Figure 5. The initial water level in the reservoir and crest elevations for the ogee and non-staged labyrinth weirs were set at an elevation of 200 m. For the staged labyrinth weir, the low stage was placed at 200 m to maintain the same normal pool elevation, with the high stage placed at 200.8 m. The storage (S), inflow (I), outflow (O), and reservoir elevation at each time step were calculated using Eq. (1) and Eq. (2). When the flow over the staged weir was limited to the lower stage, L in Eq. (1) was limited to the length of the lower stage.

$$\left(\frac{2S_{n+1}}{\Delta t}+O_{n+1}\right)=\left(\frac{2S_n}{\Delta t}-O_n\right)+\left(I_n+I_{n+1}\right) \quad (2)$$

4 EXPERIMENTAL RESULTS AND DISCUSSION

4.1 *Outflow hydrograph*

Using the modeling technique described, outflow hydrographs were obtained for ogee crest, labyrinth, and staged labyrinth weirs (see Fig. 5). With increased spillway efficiency (higher

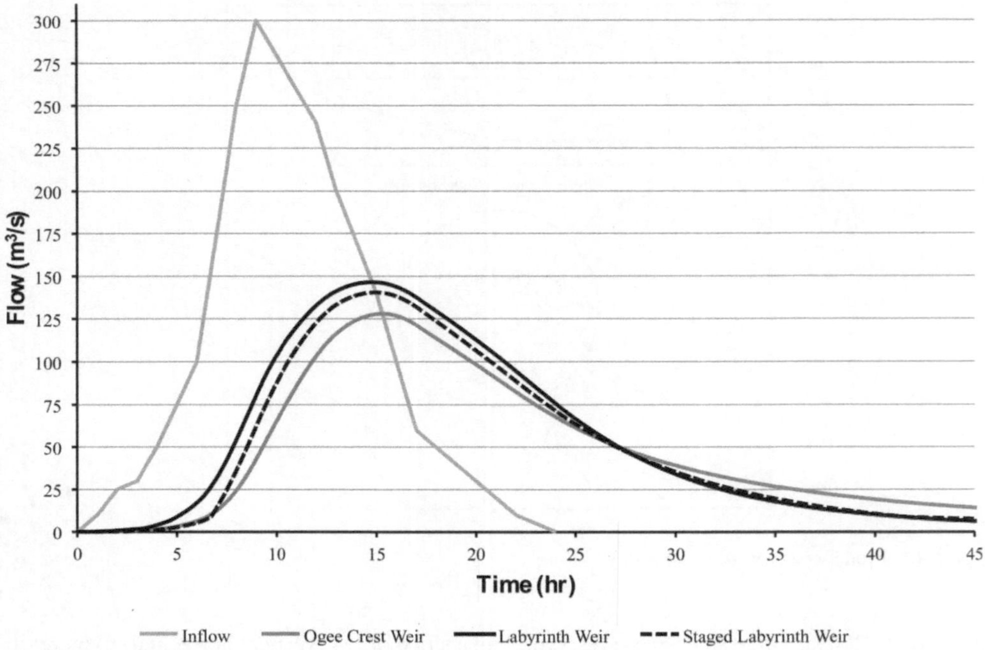

Figure 5. Outflow hydrograph model prediction.

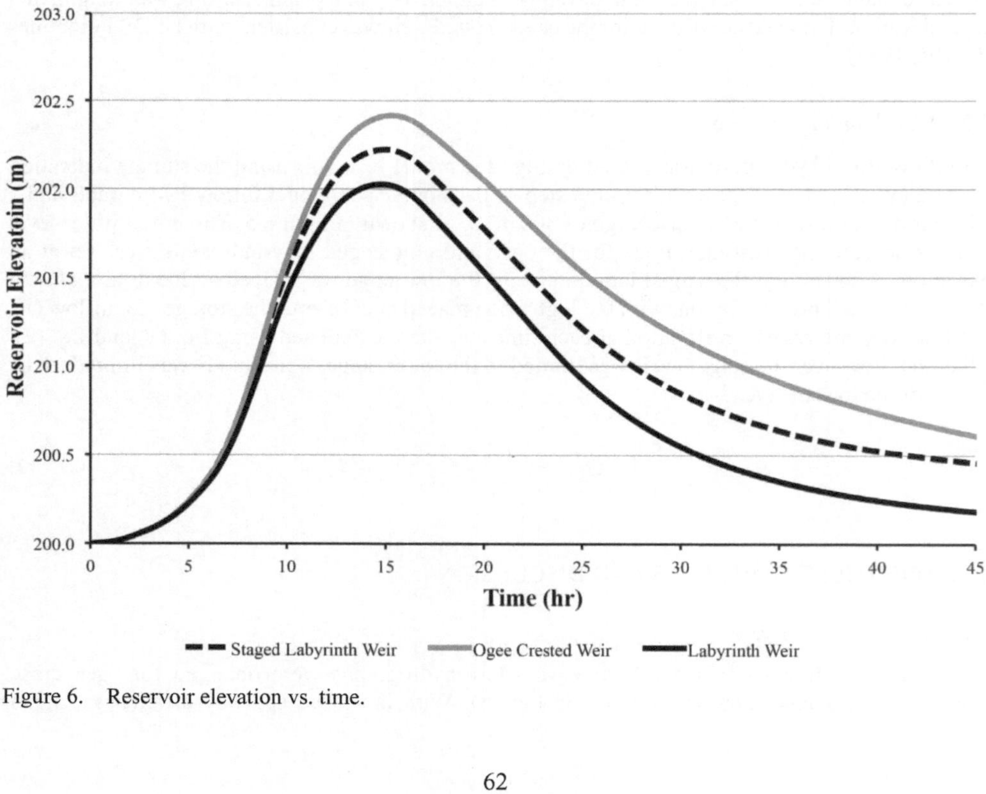

Figure 6. Reservoir elevation vs. time.

discharge for a given reservoir elevation), the outflow hydrograph occurred over a compressed timeframe, and the maximum outflow peak increased. This could create adverse effects downstream of the structure, as a larger flow over a shorter period can increase the flooding potential of the downstream channel. The model predicted the peak discharge of the standard labyrinth weir to be approximately 6 m^3/s (+4.3%) greater than the staged labyrinth. This is because the lower stage was placed at the normal pool elevation, requiring a 0.8 m increase in reservoir elevation before the high stage began spilling.

As the weir discharge efficiency increases, the amount of time required to pass the storm event through the reservoir decreases. This allows the reservoir to achieve a steady-state condition much sooner after the storm; both labyrinth weirs required significantly less time than the ogee-crested weir to return to the base-flow condition ($Q = 0$ in this case).

4.2 *Reservoir elevation*

By calculating the change in reservoir storage, the water elevation for all three weirs was obtained and is plotted versus time in Figure 6. As expected, the maximum reservoir elevation during the storm event decreases as the hydraulic efficiency of the spillway increases. The standard labyrinth weir required a 0.39 m (−16.1%) lower water elevation to pass the storm than the ogee-crested weir, while the staged weir's maximum water elevation was 0.20 m (−8.1%) less than the ogee-crested weir. With respect to the reservoir, improving the discharge efficiency of the outflow control structure means that flood routing storage (difference between the reservoir volume at base flow and maximum allowable pool) can accommodate a larger flood event, relative to a less efficient structure. Alternatively, with a higher discharge efficiency outflow control structure, the base-flow reservoir storage can also be increased by raising the weir crest elevation without impacting dam safety and flood routing capacity, relative to a less efficient structure. If discharge capacity of an existing linear weir spillway were found to be less than required, replacing the linear weir with a more efficient labyrinth weir would increase the spillway capacity while maintaining the existing spillway channel width.

4.3 *Reservoir storage*

The reservoir storage required to pass the inflow hydrograph with each of the three weirs examined was calculated by summing the differences between inflow and outflow values until the time of maximum reservoir discharge. The ogee-crested weir required 15,842 m^3 of additional storage to pass the model storm event, with the staged labyrinth requiring 14,505 m^3 (−8.4%) and the standard labyrinth requiring 13,184 m^3 (−16.8%). These results indicate that less additional storage is required as the spillway efficiency increases.

5 CONCLUSIONS

Labyrinth weirs can decrease the maximum reservoir water elevation required to pass a storm event when replacing an ogee-crest linear weir. A staged labyrinth weir can decrease the required water level to a lesser extent, while providing the benefits of confining smaller storm events to a specific lowered section of crest and decreasing the peak flow during a large storm event. Laboratory-scale models of a standard labyrinth and staged labyrinth weir were constructed and used to obtain head-discharge relationships. Using these data, a theoretical model of a reservoir and inflow hydrograph was created, and a comparison of the downstream hydrograph using the storage indication method for an ogee-crested spillway, labyrinth spillway, and staged labyrinth spillway was presented.

The comparison results for the maximum water elevation, maximum discharge, and storage required to pass the storm event are summarized in Table 2 with percent difference values normalized to the ogee-crested weir data.

Table 2. Summary of data presented.

Description of Weir	H_u Max. (m)	Q Max. (m^3/s)	Reservoir Storage Required (m^3)
Ogee-Crested	2.42	128.1	15,841
Standard Labyrinth	2.03 (−16.1%)	146.5 (+14.4%)	13,184 (−16.8%)
Staged Labyrinth	2.22 (−8.1%)	140.5 (+9.7%)	14,505 (−8.4%)

The more efficient spillways (labyrinth and staged labyrinth) decreased the maximum water elevation and storage required in the reservoir, they increased the peak outflow and decreased the time required to pass the model storm. These data are specific to the model reservoir, weirs, and inflow hydrograph evaluated. They are presented to illustrate the potential effects of discharge efficiency and flood routing, which should always be considered when designing a spillway rehabilitation project to evaluate the potential for downstream flooding.

REFERENCES

Crookston, B.M. 2010. *Labyrinth Weirs*. Ph.D. dissertation, Utah State Univ., Logan, UT.

Crookston, B.M. and Tullis, B.P. 2012a. Arced labyrinth weirs. *ASCE J. of Hydr. Eng.*, 138(6) 555–562.

Crookston, B.M. and Tullis, B.P. 2012b. Discharge efficiency of reservoir-application-specific labyrinth weirs. *ASCE J. Irrig. Drain. Eng.*, 138(6) 564–568.

Crookston, B.M. and Tullis, B.P. 2013a. Hydraulic design and analysis of labyrinth weirs. Part 1: Discharge relationships. *ASCE J. Irrig. Drain. Engr.*, 139(5), 363–370.

Crookston, B.M. and Tullis, B.P. 2013b. Hydraulic design and analysis of labyrinth weirs. Part 2: Nappe aeration, instability, and vibration. *ASCE J. Irrig. Drain. Engr.*, 139(5), 371–377.

Dabling, M.R., Tullis, B.P. and Crookston, B.M. 2013. Staged labyrinth weir hydraulics. *ASCE J. of Irrig. Drain. Engr.*, (in press).

Darvas, L. 1971. Discussion of performance and design of labyrinth weirs. *ASCE J. of Hydr. Eng.*, 97(80): 1246–1251.

Falvey, H. 2003. *Hydraulic design of labyrinth weirs*. Reston, VA: ASCE.

Hay, N. and Taylor, G. 1970. Performance and design of labyrinth weirs. *ASCE J. of Hydr. Eng.*, 96(11), 2337–2357.

Henderson, F.M. 1966. *Open channel flow*. New York: MacMillan.

Hinchliff, D. and Houston, K. 1984. Hydraulic design and application of labyrinth spillways. *Proc. of 4th Annual USCOLD Lecture.*

Lux III, F. and Hinchliff, D. 1985. Design and construction of labyrinth spillways. *15th Congress ICOLD, Vol. IV, Q59-R15,* 249–274.

Magalhães, A. and Lorena, M. 1989. Hydraulic design of labyrinth weirs. *Report No. 736*, National Laboratory of Civil Engineering, Lisbon, Portugal.

Melo, J., Ramos, C., and Magalhães, A. 2002. Descarregadores com soleira em labirinto de um ciclo em canais convergentes. Determinação da capacidad de vazão. *Proc. 6° Congresso da Água,* (CD-ROM), Porto, Portugal. (in Portuguese).

Tullis, B.P., Young, J., and Chandler, M. 2007. Head-discharge relationships for submerged labyrinth weirs. *ASCE J. of Hydr. Eng.,* 133(3), 248–254.

Tullis, P., Amanian, N., and Waldron, D. 1995. Design of labyrinth weir spillways. *ASCE J. of Hydr. Eng.,* 121(3), 247–255.

USBR. 1987. *Design of small dams: a water resources technical publication, 3rd ed.* Washington, DC: United States Department of the Interior, Bureau of Reclamation.

Walski, T.M., Barnard, T.E., Durrans, S.R., Meadows, M.E., and Whitman, B.E. 2007. *Computer applications in hydraulic engineering.* Exton, PA: Bentley Institute Press.

Labyrinth and Piano Key Weirs II – PKW 2013 – Erpicum et al. (eds)
© 2014 Taylor & Francis Group, London, ISBN 978-1-138-00085-8

Escouloubre Piano Key weir: Prototype versus scale models

S. Erpicum, A. Silvestri, B. Dewals, P. Archambeau & M. Pirotton
Research group HECE, ArGEnCo Department, University of Liege, Liege, Belgium

M. Colombié & L. Faramond
Hydro Engineering Center, EDF-CIH, Le Bourget du Lac, France

ABSTRACT: A 1.77 m high piano key weir in operation on the Escouloubre balance basin enabled the observation of real size flows for discharges ranging from 1.2 to 10 m^3/s, corresponding to upstream heads ranging from 11 to 48 cm. These observations are depicted and compared to the results gained on three physical models of the weir, built and operated considering a Froude similitude and geometric scale factors of, respectively, 1:7, 1:15 and 1:25. The study highlights the scale effects which may affect piano key weirs scale model studies and provides criteria to limit them. In particular, the results analysis shows that the scale models succeed in representing the flow conditions on the weir when the model upstream head is higher than 60 mm, and succeed in predicting the weir discharge capacity when the model upstream head is higher than 30 mm.

1 INTRODUCTION

In the framework of rehabilitation works to the Escouloubre and Nentilla powerplants in the South of France, a Piano Key Weir (PKW) was built by Electricité de France (EDF) on the Escouloubre balance basin. The basin is located between the two powerplants and aims at equilibrating the water volumes depending on the operation of the plants. During the rehabilitation of the Nentilla penstock downstream, the discharges from the Escouloubre powerplant cannot be stored in the basin. The PKW was built to release these discharges to the Aude River, directly downstream of the basin.

During operation tests on the upstream powerplant, it was possible to observe four different constant discharges over the PKW. The results from these observations constitute the first available prototype database about real size PKW operation. They are depicted in this paper and compared to the results gained on three physical models of the weir, built and operated in the Laboratory of Engineering Hydraulics (HECE) of the University of Liège.

2 THE ESCOULOUBRE PIANO KEY WEIR

2.1 *Prototype*

The Escouloubre PKW is a concrete structure located in a notch in the concrete dam at the downstream extremity of the balance basin, close to the Nentilla penstock inlet (Fig. 1a). It is of type A (upstream and downstream overhangs) and counts for one central inlet key, two outlets and a specific alveolus without ramped floor on each abutment (Fig. 1b). The crest is chamfered on both sides of a 10 cm long flat surface. Triangular noses are located below the outlet overhangs. Downstream of the PKW, a vertical chute 4 m high disconnects the flow on the structure from the downstream water levels.

The dimensions of the structure are given in Table 1 according to the nomenclature proposed by Pralong et al. (2011). The main aspect ratios are $L/W = 4.64$, $P/W_u = 0.63$, $W_i/W_o = 1.44$, $B_o/B_i = 1$, $B/P = 2.88$ and $B_{i,o}/P = 0.68$.

Figure 1. Escouloubre PKW prototype: (a) location in the dam body and (b) zoomed upstream view.

Table 1. Main dimensions of the prototype and scale models.

	W	W_u	L_T	L_u	$P_i = P_o$	P_d	W_i	W_o	T_s	B	$B_i = B_o$
Prototype	5.11	2.8	21.91	13	1.77	4.53	1.3	.9	.3	5.1	1.2
1:7 model	.73	.4	3.13	1.857	.253	.647	.186	.129	.043	.729	.171
1:15 model	.341	.187	1.461	.867	.118	.302	.087	.06	.02	.34	.08
1:25 model	.204	.112	.876	.52	.071	.181	.052	.036	.012	.204	.048

*All dimensions are in m.

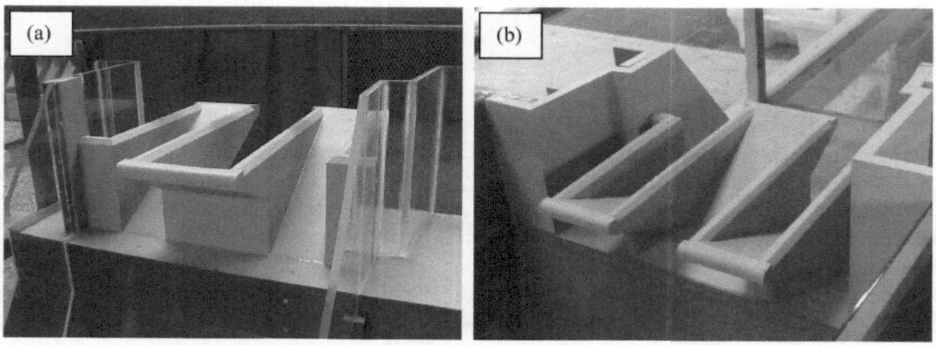

Figure 2. Escouloubre PKW scale models: (a) downstream view of the 1:7 model and (b) upstream view of the 1:15 model.

The Escouloubre PKW was designed to release a discharge of 10 m^3/s under a head of 0.5 m. The maximum H/P ratio is thus 0.28, which is small compared to usual PKW designs. The water level upstream of the PKW is measured using a limnimetric scale on the right bank of the basin. The discharge is estimated from the upstream turbine operating curve. The accuracy of these measurements is not known.

2.2 *Scale models*

Three scale models of the Escouloubre PKW were built with PVC considering geometric scale factors equal to 1:7, 1:15 and 1:25 respectively (Fig. 2). They were exact representations of the prototype PKW and abutment shapes (accuracy of 1 mm), including the dam height and the downstream chute. The dimensions of the models are given in Table 1.

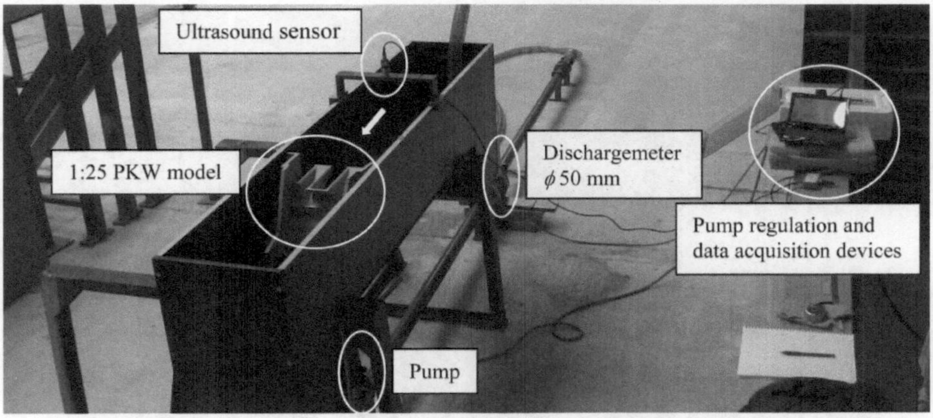

Figure 3. 1:25 scale model. General layout of the test facility and measurement devices position.

Table 2. Upstream head on the prototype and the scale models depending on the discharge.

		Discharge at the prototype scale m^3/s			
		1.2	2.8	5	10
Scale factor	Measurement accuracy at the prototype scale mm	Upstream head at the prototype scale *(at the models scale)* mm *(mm)*			
1:1	unknown	110 *(110)*	190 *(190)*	260 *(260)*	480 *(480)*
1:7	7	101 *(14)*	175 *(25)*	261 *(37)*	482 *(69)*
1:15	15	127 *(8)*	199 *(13)*	283 *(19)*	498 *(33)*
1:25	25	158 *(6)*	222 *(9)*	303 *(12)*	518 *(21)*

Upstream of the three models, a 8.4 m wide and 25 m long rectangular horizontal flume (prototype dimension) was considered. This flume is narrower than the prototype basin, the width of which is more than 20 m. Water was injected in the models through a permeable screen creating uniform flow velocity on the flume cross section. All models have identical approach flow conditions per discharge. These conditions differ slightly from the prototype ones.

Water levels upstream of the models were measured using an ultrasound probe (accuracy better than 1 mm) located in the center of the flume section, more than 10 m upstream of the weir (prototype dimension). Discharges were measured with an electromagnetic flowmeter (accuracy better than 0.5%) on the inlet pipe. The flowmeter diameter was 150 mm for discharges in the models higher than 5 l/s and 50 mm for the smaller discharges. A general view of the test facility for the 1:25 model is shown in Figure 3.

The models were operated considering a Froude similitude. The design upstream head was thus 71, 33 and 20 mm, respectively for model with 1:7, 1:15 and 1:25 scale factors. The design discharge ($10\ m^3/s$) was equivalent to 77.1, 11.5 and 3.2 l/s respectively.

3 RESULTS

Four discharges were observed on the prototype, i.e. 1.2, 2.8, 5 and $10\ m^3/s$. They have been reproduced and carefully observed on each of the three scale models. Additional discharges were tested on the scale models to draw the rating curve of the PKW. In the following analysis, it is

Figure 4. Flow over the prototype and the scale models for a discharge of 1.2 m^3/s.

assumed that discharges are correctly scaled and accurately measured. The differences between models and prototype are analyzed regarding the upstream heads.

The upstream head measured on the prototype and the scale models, depending on the discharge, is given in Table 2. The flows observed on the prototype and each scale model are illustrated in Figure 4 to 6 for three relevant discharges.

Figure 5. Flow over the prototype and the scale models for a discharge of 5 m^3/s.

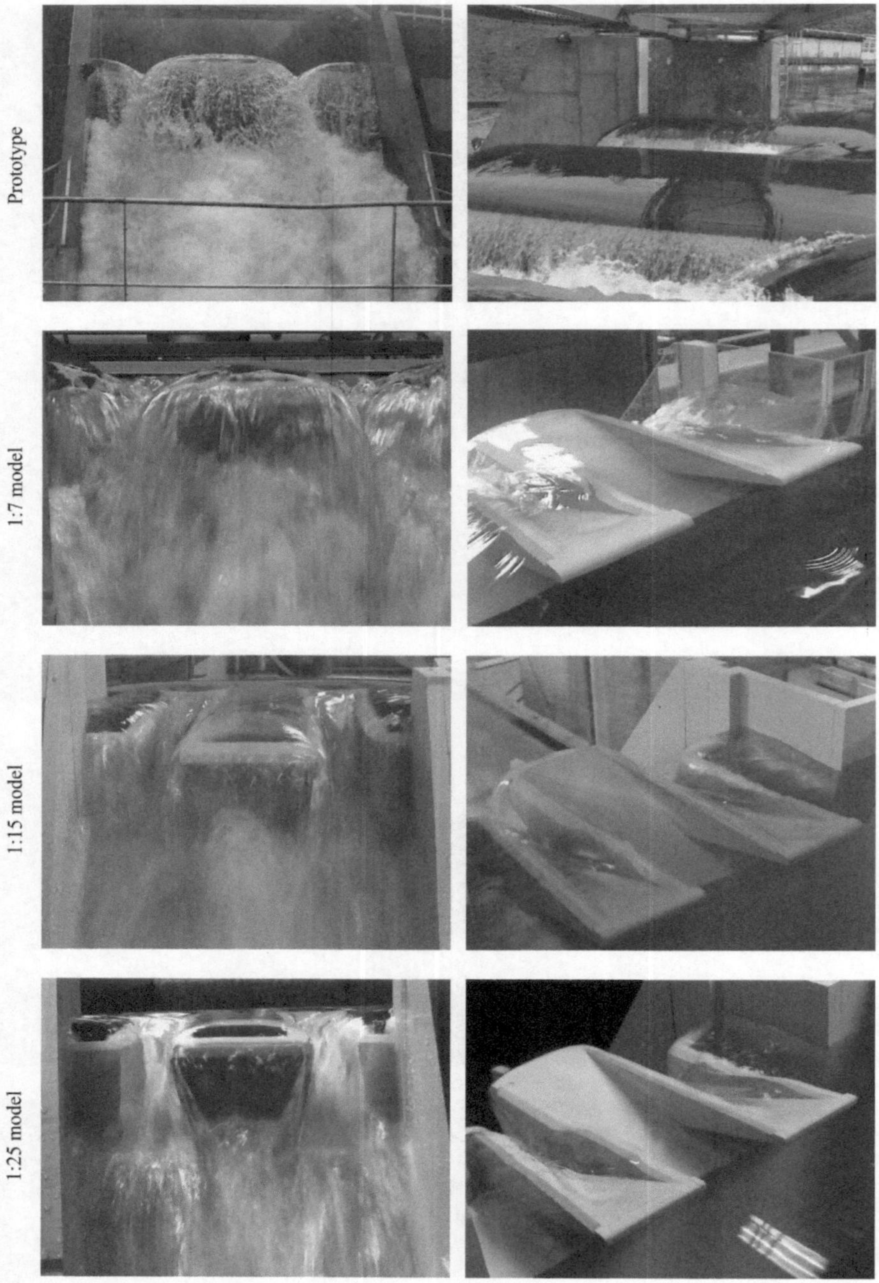

Figure 6. Flow over the prototype and the scale models for a discharge of 10 m^3/s.

4 DISCUSSION

4.1 *Flow characteristics*

Surface tension is not properly scaled considering the Froude similitude. Surface tension is negligible for most prototypes in hydraulic engineering, but is relevant in scale models considering small water depths (Heller, 2011). It affects the nappe shape over a weir as well as air entrainment by the flow. When evaluating the discharge capacity of an overflow weir, Bretschneider (1980) suggests a minimum upstream head of 20 mm to limit scale effects because of surface tension. Ettema (2000) proposes a limit of 25 mm to avoid scale effects on discharge capacity evaluation and of 60 mm to correctly model the nappe shape.

In the present study, the comparison of the flow conditions over the prototype and the scale models (Figs 4–6) highlights scale effects induced by surface tension depending on the scale factor.

Considering the 1:7 scale model, the flow conditions on the model are generally similar to what is observed on the prototype for all the tested discharges, except the smallest one. In the latter, the surface tension affects greatly the nappe shape. Indeed, the flow is attached to the side walls on the model. Air entrainment is also reduced compared to the prototype (Fig. 4). For a discharge of 5 m^3/s (Fig. 5), despite detached from the side wall, the jet downstream of the side crest is significantly closer to the side wall on the scale model than on the prototype. For a discharge of 10 m^3/s, the jet trajectory in the model is similar to the prototype one (Fig. 6). These observations are consistent with the upstream head value, which is lower than 60 mm for all discharges except the highest one (Table 2).

With the 1:15 scale model, the jet over the side crest reattaches the vertical wall for smallest discharges and detaches for the highest one. However, in the later, the jets downstream of the side crests are closer to the side walls on the scale model than on the prototype (Fig. 6). For the highest discharge, the upstream head on the scale model is only 33 mm (Table 2).

For the 1:25 scale model, the measured upstream head is always lower than 25 mm (Table 2). Flow over the side crest reattaches the vertical wall whatever the discharge, whereas such reattachment is never observed on the prototype. Very low aeration is observed downstream of the scale model compared to the prototype. With a 1:25 scale factor, the flow conditions on the model are never similar to those on the prototype for the tested range of upstream head.

Except for the highest scale factor and highest discharge, the scale models are not able to represent all the details of the flow conditions over the PKW. In particular, the smaller the scale factor, the worse the jet trajectory and air entrainment rate predictions. An upstream head at least equal to 60 mm is necessary to make the scale effects negligible on the scale models flow characteristics.

4.2 *Discharge capacity*

Figure 7 shows the weir rating curve data gained from the prototype and scale models measurements.

Considering an accuracy of 1 mm in the upstream head measurements, the data from the scale models are identical, i.e. the measurements for one model are included in the uncertainty intervals of model with higher scale factor and vice versa, when the measured upstream head is higher than 30 mm. In other words, the results from the 1:25 scale model match the results from both other models when the prototype head is higher than 75 cm and the results from the 1:15 scale model match the results from the 1:7 scale model when the prototype head is higher than 45 cm. Using this criterion, the 1:7 scale model succeeds in evaluating the discharge capacity of the prototype for prototype upstream head higher than 21 cm. This is confirmed by the good matching of the prototype and 1:7 scale model results in this range of head (Fig. 7).

For a model upstream head smaller than 30 mm, the scale model underestimates the discharge capacity of the PKW compared to the results from the other models with a higher scale factor. This is consistent with Bretschneider (1980) who mentions that a model weir provides a margin of safety when evaluating the discharge capacity. It is also consistent with Machiels et al. (2011) who show that the discharge capacity increases when the jet detaches from the crest. This is however contradictory with the results from Leite Ribeiro et al. (2009) who show that when the jet is attached to the lateral walls of the PKW, a suction effect increases the discharge compared to a detached jet.

In the present study, the head measured for small discharges on the 1:7 model is smaller than the prototype head, i.e. the 1:7 scale model overestimates the PKW discharge capacity for model upstream heads smaller than 30 mm. However, the uncertainty on the accuracy of the prototype measurements makes impossible a further analysis of these results.

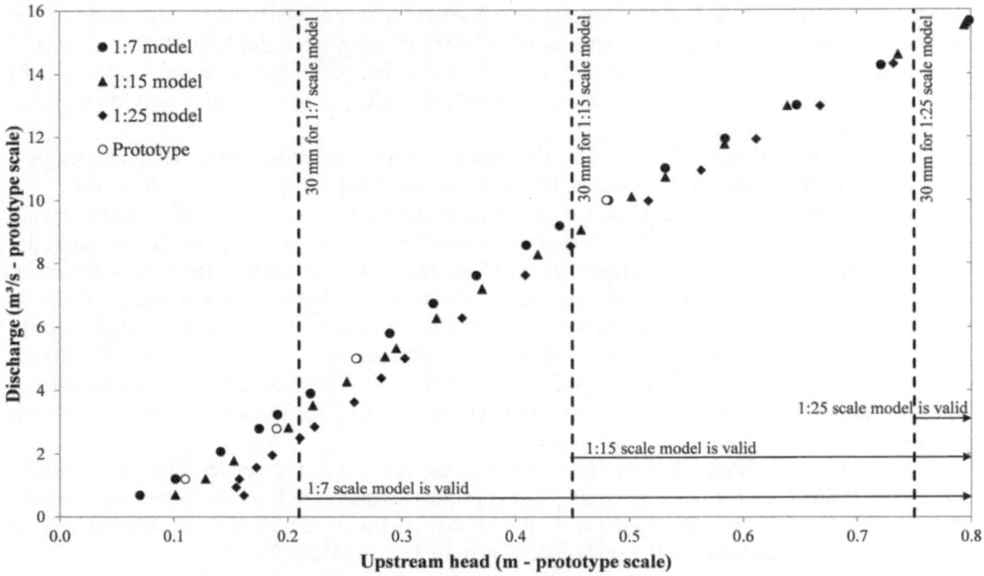

Figure 7. Prototype and scale models rating curve measurements and range of validity of scale models.

5 CONCLUSIONS

Three scale models of the Escouloubre PKW have been built with a scale factor equal to 1:7, 1:15 and 1:25 respectively and operated considering the Froude similitude. The comparison of the models and prototype flow characteristics and rating curve shows that the scale models succeed in representing the flow conditions on the weir when the model upstream head is higher than 60 mm, and succeed in predicting the weir discharge capacity when the model upstream head is higher than 30 mm. For smaller heads, surface tension induces non negligible scale effects which reduce air entrainment, move the jets closer to the walls with possible reattachment and decrease the weir discharge capacity.

REFERENCES

Bretschneider, H. 1980. Models of hydraulic structures: discharge conditions, energy dissipation, erosion. In Kobus H. (ed.), *Hydraulic modeling*: 183–204. Hamburg: Parey.

Ettema, R. (ed.) 2000. *Hydraulic modeling: concepts and practice*. ASCE Manuals and reports on engineering practices No. 97. Reston: ASCE.

Heller, V. 2011. Scale effects in physical hydraulic engineering models. *Journal of Hydraulic Research*, 49(3): 293–306.

Leite Ribeiro, M., Bieri, M., Boillat, J.-L., Schleiss, A., Delorme, F. & Laugier, F. 2009. Hydraulic capacity improvement of existing spillways – Design of piano key weirs. In *Proceedings of the 23rd ICOLD Congress (Q90, R43), Brasilià, 21–29 May 2009*. Paris: CIGB-ICOLD

Machiels, O., Erpicum, S., Dewals, B.J., Archambeau, P. & Pirotton, M. 2011. Experimental observation of flow characteristics over a Piano Key Weir. *Journal of Hydraulic Research*, 49(3): 359–366.

Pralong, J., Vermeulen, J., Blancher, B., Laugier, F., Erpicum, S., Machiels, O., Pirotton, M., Boillat, J.-L., Leite Ribeiro, M. & Schleiss, A. 2011. A naming convention for the Piano Key Weirs geometrical parameters. In Erpicum, S., Laugier, F., Boillat, J.-L., Pirotton, M., Reverchon, B. & Schleiss, A.J. (eds.), *Labyrinth and piano key weirs-PKW 2011*: 271–278. London: CRC Press.

Labyrinth and Piano Key Weirs II – PKW 2013 – Erpicum et al. (eds)

 ISBN 978-1-138-00085-8

Scale effects related to the rating curve of cylindrically crested Piano Key weirs

M. Pfister, E. Battisacco, G. De Cesare & A.J. Schleiss
Laboratory of Hydraulic Constructions (LCH), Ecole Polytechnique Fédérale de Lausanne (EPFL), Lausanne, Switzerland

ABSTRACT: Hydraulic scale effects on physical Piano Key weir (PKW) models were so far rarely discussed in literature, even though almost all prototypes are model-tested before construction. In parallel, physical weir models are generally known to include significant scale effects if operated under small heads, so that an up-scaling of the results derived from the latter is unreliable regarding the discharge-head function (rating curve). This comes from the fact that viscosity and surface tension of water are fluid properties which cannot be scaled simultaneously, whereas both affect the flow in models. Thus, scale effects occur particularly for small overflow heads. Literature mentions limiting heads to respect in the order of 0.03 to 0.05 m. Furthermore, a specific 'low-head behavior' regarding the transition from the clinging to the leaping nappe has been reported, which is different in a prototype and its scaled model. The latter is linked to the teapot effect and to air-water flow features. To derive general equations for the head-discharge relationship of PKWs, researchers thus have to exclude data which could be subjected to scale effects, without knowing the precise limit so far. However, the tests to be excluded are of specific interest to develop the aforementioned relationship, as PKWs are particularly efficient for these conditions. The paper discusses scale effects related to PKWs, based on an analogy considering cylindrical weirs.

1 INTRODUCTION

1.1 *Physical modeling*

Hydraulic modeling remains one of the principal engineering tools to design and optimize complex hydraulic processes, despite of the advancement in numerical simulations. One may suggest that almost all hydraulic structures and phenomena were physically modeled, given that most publications in the field refer to them. Nevertheless, physical modeling remains challenging, as model effects and scale effects have to be limited.

For free surface flows – as for instance at Piano Key weirs (PKWs) – the similitude according to Froude allows for a correct representation of the dominant forces, namely gravity and inertia. Beside these two forces, also the surface tension force and the viscosity force act on the fluid. The ratio of inertia to viscous forces gives the Reynolds number; the ratio of inertia to surface tension forces yields the Weber number. A true dynamic similarity requires achieving identical Froude, Reynolds and Weber numbers in both prototype and model. This is physically impossible when the same fluid (i.e. water) is used in both prototype and model. As a consequence, scale models based upon the Froude similitude may overestimate effects related to the fluid surface tension σ and the viscosity ν. This model-overestimation leads to retarded overflow under small weir heads, affecting the rating curve (i.e. the head-discharge relation). In order to allow for an adequate up-scaling of the model results to prototype dimensions, this overestimation of the surface tension and the viscosity has to be within acceptable limits. Such limits are typically respected when excluding model tests with heads below some 0.03 to 0.05 m.

Herein, a model family based on a numerical simulation is presented and compared with a analytical approach from literature, to validate such limiting heads for cylindrical weir crests. The latter are typical for PKWs, as they even increase their efficiency. This efficiency represents the main benefit of PKWs, spilling relatively high discharges Q under comparably small heads H. This comes from the non-linear nature of PKWs (folded back and forth in plan-view to make repeating cycles or keys) allowing for significantly longer developed weir lengths L than could be achieved using a traditional linear weir for a given spillway channel width W.

For small heads, the hydraulic effect of the edges and corners at the keys remains small, so that the flow features may be approximated with a linear weir of length L. Unfortunately, scale effects concerning the rating curve occur particularly for the "efficient" small heads, so that the accuracy of the physical model data is reduced for the most relevant operation regime of PKWs.

1.2 *Scale effects at free weir overflow*

As for reliable rating curves derived from scaled physical models, literature mentions minimal heads to respect. Otherwise, the effects of surface tension and viscosity influence the head-discharge relation, making impossible a correct up-scaling of the latter to prototype dimensions. Basically, two limitations can be identified, linked to (1) the onset of over-flow due to surface tension (Fig. 1a), and (2) flow affected by surface tension and viscosity (Fig. 1b).

The first limitation (1) may be illustrated by the following example, representing an extreme condition. It takes into account free flow over a cylindrical weir crest (discharge coefficient $C_d = 0.395$ from Castro-Orgaz (2012) for $H/R = 0.1$ and potential flow). Here, R = crest radius as shown in Fig. 2. Due to the fluid surface tension, an absolute minimal head of some $H_M = 0.004$ to 0.006 m is required for water to flow (Bollrich and Aigner 2000). In a scaled physical model, no discharge is flowing across the weir for this head (Fig. 1a). The related prototype, however, indicates a different behavior. There, the head H_P (subscript P for prototype) is equivalent to the model (subscript M) head H_M times the geometrical scale factor λ, which is $H_P = \lambda H_M = 0.300$ m, if $\lambda = 50$. With this H_P, a specific discharge of $q = 0.29\ \text{m}^2/\text{s}$ may be expected on prototype, if applying the Poleni equation. For a developed crest length of $L = 100$ m, an absolute discharge of almost $Q = 28.7\ \text{m}^3/\text{s}$ occurs, while the prediction from the model is $Q = 0.0\ \text{m}^3/\text{s}$.

Once the initial effect of the surface tension is overcome and water flows, then the fluid surface tension *and* the viscosity affects the rating curve (second limitation (2)). Both, surface tension and viscosity retard the flow, at least as long as no potential flow is achieved (with $\nu = \sigma = 0$). Small discharges require thus an over-proportional head to be conveyed, as compared to potential flow (Matthew 1991). Literature quantifies the minimum head necessary to avoid "relevant" scale effects related to the rating curve of weirs. The values depend of the particular weir crest type, nevertheless, typical limit heads H are:

- around 0.03 to 0.05 m following the data of Rehbock (1909) for sharp-crested weirs with aerated nappe
- 0.07 m following Kirschmer (1928) to reproduce the free nappe (flow separates from weir) on a cylindrical weir with a crest radius $R = 0.046$ m
- around 0.05 to 0.07 m following Dillmann (1933) for sharp-crested weirs
- 0.05 m for sharp-crested weirs (Sarginson 1972)
- 0.05 m under a crest-radius of more than 0.03 m for cylindrical weirs (Sarginson 1972)
- 0.02 m for standard ogee weirs referring to the rating curve, and 0.06 m to reproduce correct jet trajectories at sharp-crested weirs (Breitschneider 1978)
- 0.05 m for broad-crested weirs regarding the rating curve (Hager and Schwalt 1994)
- Some 0.025 m for determination of the rating curve and 0.06 m to correctly model the nappe shape (Ettema 2000).

Further minimal heads to limit the effects of surface tension and viscosity on the head-discharge relation are listed by Novak et al. (2010). As for Piano Key weirs, Erpicum et al. (2013) compared a prototype rating curve with these derived from a model family, including scale factors of 1:7,

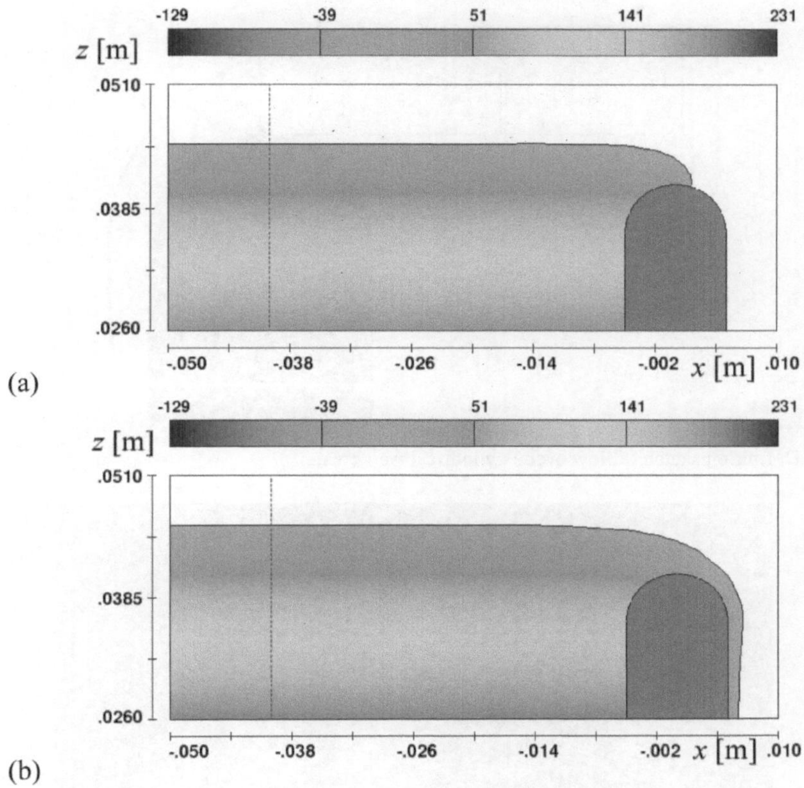

Figure 1. Water surface profiles and pressures (Pa) on a cylindrical weir of crest radius $R = 0.005$ m (from numerical simulation), for a head of $H =$ (a) 0.004 m for stagnant water (limitation 1), and (b) 0.005 m for flowing conditions with a small head (limitation 2).

1:15, and 1:25. The side wall thicknesses were $T_s = 0.300$ m (Fig. 2) at the prototype, 0.043 m in the 1:7 scaled model, 0.020 m in the 1:15 scaled model, and 0.012 m in the 1:25 scaled model. The authors conclude that model-heads smaller than 0.03 m underestimate the discharge capacity at PKWs, and that some 0.06 m are required to correctly reproduce the flow features in terms of nappe formation and jet geometry.

2 METHOD

Several numerical simulations were conducted, investigating the rating curve of a straight cylindrical weir under typical model and prototype dimensions. These are, for PKW prototypes near the crest around $T_s = 0.20$ to 0.35 m (Vermeulen et al. 2011), and in the physical model under geometrical scale factors of $20 \leq \lambda \leq 30$ in the range of $T_s = 0.005$ to 0.02 m.

As for cylindrical weirs, the effect of the absolute weir thickness $T_s = 2R$ on the rating curve was analytically derived by Matthews (1991) and Castro-Orgaz (2012), based on higher-order curved flow theory. This provides a second tool to investigate scale effects related to the wall thickness, and to compare them with the prediction of the numerical simulation. Of course, the analytical result is only valid for (transversally) linear weirs and for small relative heads H/R. As a first approximation, they can be applied to PKWs anyway, because the hydraulic effect of the edges and corners is small for minimal heads, as e.g. Fig. 3 indicates. It may be assumed that the contraction

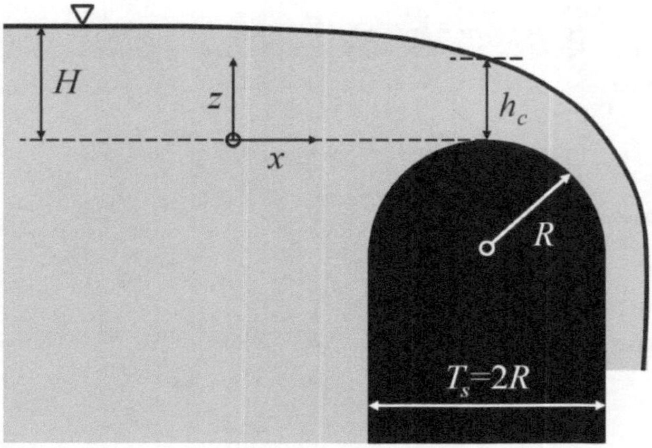

Figure 2. Definition sketch of flow over a cylindrical weir crest.

Figure 3. Flow features on sectional PKW model ($T_s = 2R = 0.02$ m, cylindrical crest) under a small head (Leite Ribeiro et al. 2012).

effect at edges compensates the widening at corners. Furthermore, particularly minimal heads are affected by scale effects, so that the herein derived limits may be a priori transposed from linear cylindrical weirs to PKWs considering the developed crest length as reference to derive the rating curve. For higher heads, the rating curve is influenced by the folded PKW crest, so that the effect on the discharge coefficient C_d reduces. These cases are, however, less important regarding scale effects.

The numerical simulations were performed using the Navier-Stokes solver *Flow-3D*, including the continuity and momentum equations based on a finite-volume approximation with a structured FAVOR grid. The flow region was subdivided into a mesh of fixed rectangular cells. The cylindrical weir was embedded by defining the fractional face areas and fractional volumes of the cells that are open to flow. The free water surface as interface between water and air was modeled with the VOF technique (Hirt and Nichols 1981).

Local average values of all dependent variables were defined within each cell, located at the cell centers, whereas the velocities were located at their faces (staggered grid arrangement). Most terms in the equations are evaluated using the current time-level values of the local variables explicitly. This produces a simple and efficient computational scheme but requires the use of a limited time-step size to maintain computationally stable and to provide accurate results. The pressure-velocity solver used the GMRES method.

The input parameters are (based on a Reynolds and Weber number investigation): gravity, viscous effects through the appropriate turbulence model, and surface tension effect. To take viscous effects into account, the two-equation k-ε model is used as the turbulence closure model. The surface tension model requires the static contact angle (chosen as 90° for water on smooth concrete wall between wetting and non-wetting condition), as well as the surface tension. Both viscosity and surface tension dependent on temperature: The simulations were performed isothermal at 20°C. For wall shear stress calculations, a surface roughness of 0.01 mm (equivalent to PVC as commonly used in physical models) was attributed to the crest for $0.005\text{ m} \leq R \leq 0.02\text{ m}$, and 1 mm (equivalent to concrete in prototype) for $R = 0.1$ and 0.2 m. Wall shear stresses at obstacle surfaces are modeled using the usual shear stress estimations by adding (to the molecular viscosity) terms taking into account the surface roughness and the local tangential velocity distribution perpendicular to the wall, in general in a sub-grid scale.

The computation was performed in 2D with a symmetry boundary on each side. The upstream boundary condition was a fixed water level corresponding to a given head. An outflow condition was imposed at the downstream, allowing incompressible fluid flowing out with a free surface. From experience, a minimum of five cells is required to represent any physical phenomena, so that a fine rectangular mesh was defined at the crest, with 0.5 mm high cells along z (Fig. 2), and with a x/z ratio of 2. The total number of cells is around 40,000 for the smallest R, and up to 1,800,000 for the largest. The approach flow section was sufficiently large and deep to avoid effects of the velocity and to achieve a horizontal water surface at the boundary.

3 EFFECT OF VISCOSITY AND SURFACE TENSION

3.1 *Discharge coefficient*

The developed PKW crest length L may be considered as characteristic length to derive the specific discharge, as least for small heads H, as

$$q = \frac{Q}{L} \tag{1}$$

The discharge coefficient C_d for a cylindrical weir crest follows then as

$$C_d = \frac{q}{\sqrt{2gH^3}} \tag{2}$$

Accordingly, C_d may be derived from the numerical simulations, based on the related prediction of H and q. As an alternative, and to validate the latter, analytical C_d values from literature were considered. Matthew (1991) derived C_d for cylindrical weir crests based on the second- and third-order equations of the plane potential flow taking into account the streamline curvature. Based on the latter work, Castro-Orgaz (2012) derived an equation including the relative head H/R (and R/H) as well as the absolute radius R, beside the fluid parameters, as

$$C_d = 0.385 \begin{bmatrix} 1 + 0.272\left(\dfrac{H}{R}\right) - 0.833\left(\dfrac{\sigma}{\rho g R^2}\right)\left(\dfrac{R}{H}\right) - 1.05\left(\dfrac{3}{g}\right)^{1/4}\sqrt{\nu}R^{-3/4}\left(\dfrac{R}{H}\right) \\ -0.045\left(\dfrac{H}{R}\right)^2 \end{bmatrix} \tag{3}$$

There, σ = surface tension, ν = kinematic viscosity, and ρ = density. Matthew (1991) limits his Eq. (3) to roughly $0.2 \leq H/R \leq 1.0$, while the numerical simulations imply a wider application range up to approximately $H/R \leq 3.0$. Note that Eq. (3) gives explicitly the effects of R, σ, ν, and ρ, i.e. scale effects (linked to a comparison of large "prototype" R and small "model" R) can be derived from the latter for an otherwise similar H/R (and R/H). The potential flow version of Eq. (3) with $\nu = \sigma = 0$ is accordingly (Castro-Orgaz 2012)

$$C_d = 0.385\left[1 + 0.272\left(\frac{H}{R}\right) - 0.045\left(\frac{H}{R}\right)^2\right] \tag{4}$$

Equation (4) gives the C_d values ignoring the effect of viscosity and surface tension. This is correct if describing flow with high heads with a marginal effect of the fluid parameters, thus without scale effects.

Figure 4 shows C_d values derived from the numerical simulations (Fig. 4a) and from Eq. (3) (Fig. 4b), as a function of the non-dimensional head H/R. Various crest radii are included, involving typical prototype and model values (R = 0.005, 0.01, 0.02, 0.1, 0.2, and 0.3 m). Note that the data for R = 0.3 m were provided by Castro-Orgaz (2012), measured on a physical model.

As for the limitation (1), one may see that $C_d = 0$ (no flow) occurs for small H/R combined with small R (Fig. 4a). For instance, a minimum of H/R = 0.8 to 0.9 is required to generate weir overflow for R = 0.005 m, whereas the latter occurs at $H/R < 0.1$ for R = 0.2 m. Expressed in absolute terms, the limitation (1) was observed at a constant head H = 0.004 to 0.005 m for all simulated tests, independent of R. This first limitation is less explicitly represented by Eq. (3) (Fig. 4b), which was developed for flowing water.

As for the limitation (2), it is visible in Fig. 4 that C_d is affected by R, for a constant ratio H/R. Small "model" R generate (for small H/R) C_d values which lay below those of "prototype" R (being close to potential flow according to Eq. (4)), so that the rating curve is incorrectly represented. This effect reduces, however, with increasing H/R, and is small for $H/R > 2$. Generally, the effect of viscosity and surface tension on C_d reduces with increasing relative head, with the consequence that the discharge coefficients collapse with Eq. (4) independently of R. For large relative heads, the (down-scaled) wall thickness of a cylindrically crested weir is thus not relevant, so that scale effects related to the rating curve disappear.

The numerical simulations and the analytical solution (Eq. 3) provide basically similar results, as the comparisons of Fig. 4a with Fig. 4b shows. The analytical result is slightly more conservative, i.e. indicates a marginally higher effect of R. Additionally, limitation (1) is not included in Eq. (3), as the latter presumes flowing water. Both methods asymptotically approach potential flow according to Eq. (4) for large H/R. Given the similitude of both methods, one may accept an extension of the application range of Eq. (3) up to $H/R = 3$ (as proposed before).

Minimal heads are to respect in order to avoid significant scale effects regarding the rating curve (as listed in the literature review of chapter 1.2). These heads may be derived from the Eqs. (3) and (4), and the numerical simulations. Therefore, the C_d value of a certain parameter set (including σ, ν, ρ, and R according to Eq. 3) is compared to that resulting for potential flow (Eq. 4). Two criteria were introduced to define the minimal heads: C_d following Eq. (3) or the numerical simulation is (1) 95%, and (2) 98% of C_d following Eq. (4). This provides the values H/R with negligible scale effects for each R. Multiplying the mentioned H/R with R results in the minimal head H to respect for each value R. As visible in Fig. 5a, the effect of σ, ν, and ρ is small if $H > 0.03$ m for the 98% criterion, and $H > 0.015$ m for the 95% criterion. This limits are applicable on models ($R < 0.02$ m) as well as on prototypes (R = 0.1 to 0.2 m)! Note that small $R < 0.005$ m are more sensitive, and require minimal heads exceeding the aforementioned values. A model wall thickness of R = 0.0025 m (T_s = 0.005 m), for instance, achieves a maximum precision of 80% regarding C_d as compared to the prototype (potential flow) if $H/R < 4$.

Figure 5b shows C_d values computed with Eqs. (1) and (2) as measured on various physical PKW models (cylindrical weir crests, R = 0.01 m) by Leite Ribeiro et al. (2012), zooming on small H/R.

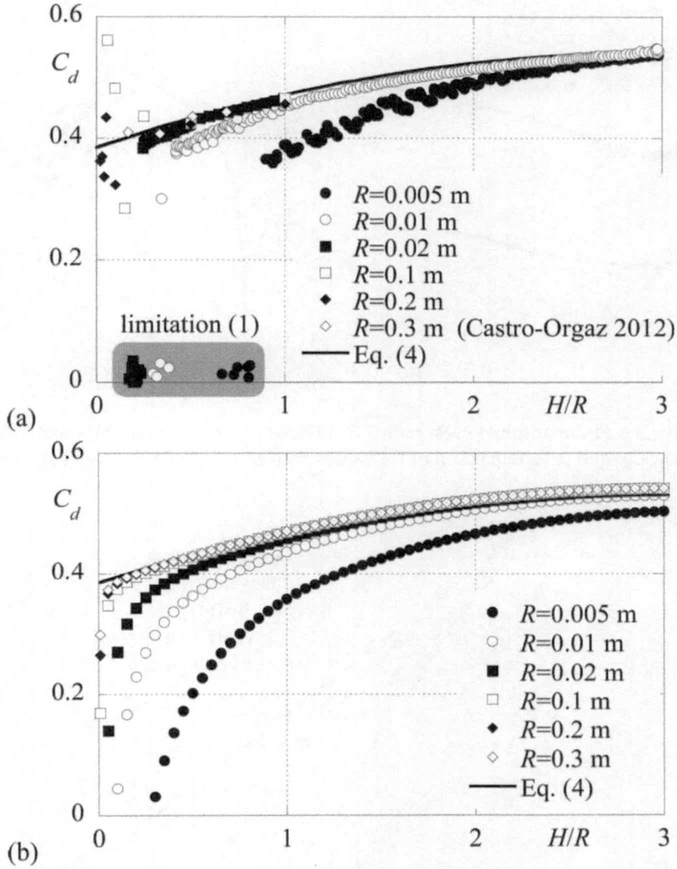

Figure 4. Discharge coefficient C_d versus H/R for different crest radii R, derived from (a) numerical simulations ($0.005\ \text{m} \leq R \leq 0.2\ \text{m}$) and physical modeling ($R = 0.3$ m), and (b) an analytical derivation resulting in Eq. (3) (Matthew 1991).

It is visible that the data tend to Eq. (3) for $H/R < 2$. The effect of the PKW cycles seems thus small for these relative heads, so that the overflow characteristic is close to that of a linear cylindrical weir. This indicates that the basic assumption, i.e. to consider a PKW under small relative heads as a linear cylindrical weir using L as reference, is appropriate.

3.2 *Crest pressure distribution*

Scale effects are not only observed related to the rating curve, but also to the behavior of the flow downstream of the crest. As shown by Erpicum et al. (2013), the prototype flow tends to separate from the crest if a certain discharge is exceeded, whereas the model flow then still clings on the structure (teapot effect). This affects the jet trajectories as well as the related disintegration and air entrainment (Pfister and Hager 2012, Pfister and Chanson 2012). The flow separation depends on the crest pressures, which are different in models as compared to prototypes.

Figure 6 shows an example of the vertical pressure profile in the flow at the crest center, derived from the aforementioned numerical simulations and under $H/R = 1$. The pressures are shown on the abscissa, normalization with the flow depth h_c at the crest (Fig. 2, subscript c). The ordinate gives the related elevation with $z/h_c = 1$ representing the water surface, and $z/h_c = 0$ the crest

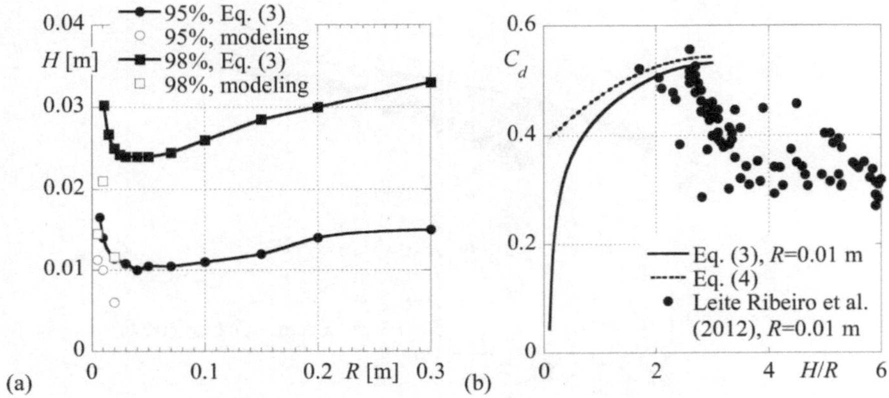

Figure 5. (a) Influence of the absolute crest radius R on head H for errors of 95% and 98% relative to the discharge coefficient C_d, and (b) comparison of C_d values derived from a PKW model and from Eq. (3), both for $R = 0.01$ m.

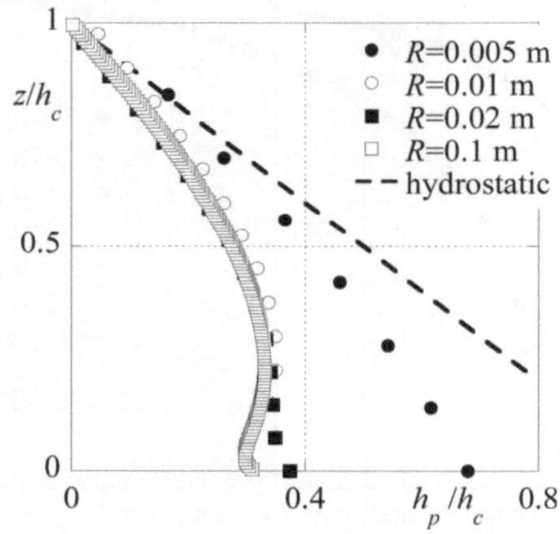

Figure 6. Vertical crest pressure h_p profiles (z/h_c versus h_p/h_c, with h_c as crest flow depth) for $H/R = 1$ under various crest radii R.

surface. It is visible that the pressures follow almost hydrostatic conditions near the surface, but then significantly reduce due to streamline curvature. Relevant is the difference of pressures on the crest ($z/h_c = 0$) for different R. Small values R (physical models) tend to larger pressures than large R (prototype), under otherwise identical conditions. Physical models with large scale factors ($R = 0.005$ m) tend thus to relatively higher crest pressures, retarding the separation of the flow from the weir surface. A precise limit is not derived herein, as not enough simulations were run in the request range.

4 CONCLUSIONS

Scale effects on weir flow occur in physical models for small heads, reducing the accuracy of the up-scaled head-discharge relation. As for PKWs, minimal heads to respect – in order to limit scale effects – were so far exclusively derived by Erpicum et al. (2013). They based their observation on a model family, comparing the related data with the flow features on an identical prototype.

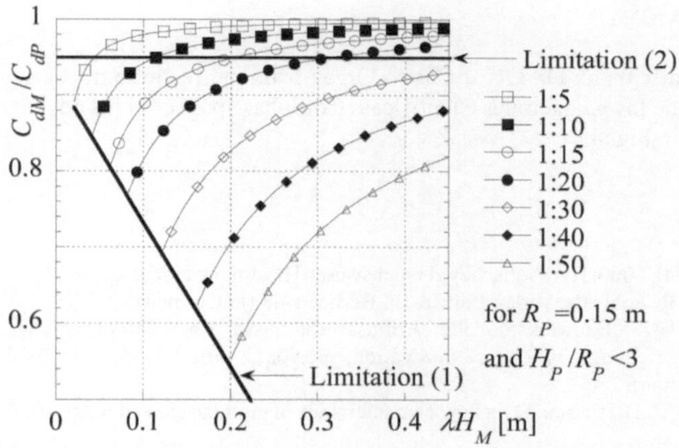

Figure 7. Relative error regarding the discharge coefficient ratio C_d prototype (subscript P) to C_d model (subscript M), for different scale factors λ.

Herein, numerical simulations and an analytical approach are used to derive the minimal heads on a cylindrically crested weir as typical on PKWs. The following conclusions are drawn:

- The flow features on PKWs are dominated by the crest shape for relative heads smaller than approximately $H/R < 2$ (Fig. 2). The effect of the folded crest is then small.
- For the aforementioned small heads, the developed crest length is relevant to derive the discharge coefficient.
- Onset of flow is observed for a absolute head of 0.004 to 0.005 m (limitation 1).
- Surface tension and viscosity affect the weir flow up to potential flow conditions (limitation 2). The related limiting head is $H = 0.03$ m for $0.005\text{ m} \leq R \leq 0.3$ m if assuming $C_{dM}/C_{dP} = 0.98$ as criterion, and $H = 0.015$ m for $C_{dM}/C_{dP} = 0.95$ (Fig. 5a). As for $R < 0.005$ m, more severe limits (i.e. larger limiting heads) have to be applied.
- Between limitation (1) and (2), physical models underestimate the discharge (i.e. the discharge coefficient) for a given head.
- Small R values (physical models) tend to generate overestimated crest pressures (as compared to large R on prototypes), so that flow separation from the profile (leaping nappe) is retarded.

Figure 7 summarizes the results by visualizing the limitations (1) and (2) for different scale factors λ, based on a typical prototype with $R = 0.15$ m ($T_s = 2R = 0.3$ m) and considering Eq. (3). If the related physical model is built for instance with $\lambda = 20$, then scale effects occur between limitations (1) equivalent to an up-scaled prototype head of $\lambda H_M = 0.08$ m and limitation (2) at $\lambda H_M = 0.31$ m. Exclusively prototype heads above 0.31 m are reliable (i.e. the model C_d values are at least 95% of those occurring under potential flow conditions) in terms of the rating curve. For an intermediate head of $\lambda H_M = 0.15$ m, the ratio $C_{dM}/C_{dP} = 0.88$, so that the model spills under these conditions only 88% of the prototype discharge.

Note that not only scale effects influence the precision of hydraulic model data. Additionally, the measurement error including head and discharge measurements is in the order of 1 to 2%, and the geometrical accuracy of the model is around 0.001 m. These errors are particularly significant, such as scale effects, under small heads. Nevertheless, one should not excessively question the accuracy of hydraulic model data, as the main input value to design a spillway, namely the discharge, results from hydrological analyses. These include typically higher uncertainties than the physical model data. A reliable hydraulic structure (e.g. a PKW) should operate as requested even if the effective boundary conditions (design discharge, concrete dimensions) are slightly different from design values.

ACKNOWLEDGMENTS

The authors cordially thank Dr. Oscar Castro-Orgaz from the Instituto de Agricultura Sostenible, Consejo Superior de Investigaciones Cientificas (Cordoba, Spain), for his advice and for providing the data included in Fig. 4a.

REFERENCES

Bollrich, G., Aigner, D. 2000. Hydraulisches Versuchswesen [Hydraulic modeling]. *Technische Hydromechanik* 4, H. Martin and R. Pohl eds. Verlag Bauwesen, Berlin. 1-48 (in German).

Breitschneider, H. 1978. Bauwerksmodelle: Abflussverhältnisse, Energieumwandlung, Erosion [Physical models: Discharge conditions, energy dissipation, erosion]. *Mitteilungsheft* 4, DVWW, H. Kobus, ed. 195–216 (in German).

Castro-Orgaz, O. 2012. Discussion Overflow characteristics of circular-crested weirs. *J. Hydraulic Res.* 50(2): 241–243.

Dillmann, O. 1933. Untersuchung an Überfällen [Investigation of weirs]. *Mitteilungen Heft* 7, TU München, D. Thoma ed. 26–52 (in German).

Erpicum, S., Silvestri, A., Dewals, B., Archambeau, P., Pirotton, M., Colombié, M., Faramond, L. 2013. Escouloubre PKW: Prototype versus scaled model. *Proc. Intl. Conf. Labyrinth and Piano Key Weirs* Chatou F. CRC Press, Boca Raton, FL.

Ettema, R. (ed). 2000. *Hydraulic Modeling, Concepts and Practice*. ASCE Manuals of Practice 97, Reston, US.

Hager, W.H., Schwalt, M. 1994. Broad-crested weir. *J. Irrigation and Drainage Engng* 120(1): 13–26.

Hirt, C.W., Nichols, B.D. 1981. Volume of fluid (VOF) method for the dynamics of free boundaries. *J. Computational Physics* 39(1), 201–225.

Kirschmer, O. 1928. Untersuchung der Überfallkoeffizienten für einige Wehre mit gerundeter Krone [Investigation of discharge coefficients of cylindrical weirs]. *Mitteilungen Heft* 2, TU München, D. Thoma ed. 5–28, TU München (in German).

Leite Ribeiro, M., Pfister, M., Schleiss, A.J., Boillat, J.-L. 2012a. Hydraulic design of A-type Piano Key weirs. *J. Hydraulic Res.* 50(4): 400–408.

Matthew, G.D. 1991. Higher order, one-dimensional equations of potential flow in open channels. *Proc. Instn Civ. Engrs, Part 2, 91, 187–201.* 138(8): 773–776.

Novak, R., Guinot, V., Jeffrey, A., Reeve, D.E. 2010. *Hydarulic modelling, an introduction*. Spon Press, London.

Pfister, M., Chanson, H. 2012. Discussion Scale effects in physical hydraulic engineering models. *J. Hydraulic Res.* 50(2): 244–246.

Pfister, M., Hager, W.H. 2012. Deflector-jets affected by pre-aerated approach flow. *J. Hydraulic Res.* 50(2): 181–191.

Rehbock, Th. 1909. Die Ausbildung der Überfälle beim Abfluss von Wasser über Wehre [Formation of overflow of water an weirs]. *Festschrift*, Technische Hochschule Fridericiana, Karlsruhe D (in German).

Sarginson, E.J. 1972. The influence of surface tension on weir flow, *J. Hydraulic Res.* 10(4): 431–446.

Vermeulen, J., Laugier, F., Faramond, L., Gille, C. 2011. Lessons learnt from design and construction of EDF first Piano Key weirs. *Proc. Intl. Conf. Labyrinth and Piano Key Weirs* Liège B, 215–224. CRC Press, Boca Raton FL.

B. Physical and numerical modeling – Submergence and energy dissipation

Labyrinth and Piano Key Weirs II – PKW 2013 – Erpicum et al. (eds)
© 2014 Taylor & Francis Group, London, ISBN 978-1-138-00085-8

Submerged flow regimes of Piano Key weir

F. Belaabed & A. Ouamane
Laboratory of Hydraulic planning and Environment, University of Biskra, Algeria

ABSTRACT: The piano keys weir can operate in two different flow conditions, in free flow and submerged flow by its use. It can be used to protect the dam against the extreme floods, to measure or control the flow in channels and regulate the flow in a river. As the free flow condition has been studied extensively, it was essential to study the PK-Weir in submerged conditions, basing primarily on theoretical analysis supported by experimental analysis and taking as a reference the linear weir. The obtained results showed a concordance between the theoretical development and experimental results.

1 INTRODUCTION

The spillway is regarded as one of the most important hydraulic infrastructures; this returns to its multifunction, it is often used as component of discharge control and spillway in the rivers. It can be designed in several forms depending on the characteristics required, namely, the dimensions and the profile of the cross-section, the shape of the indentation, the provision in relation to the direction of flow, and the upstream and downstream conditions of the flow.

Based on these last; there are two types of weirs according to the conditions of their operation. The first type corresponds to the free flow weir when water level the downstream does not influence on the flow to the upstream, as well, the aquifer is assumed to be free (Fig. 1-a).

The second type corresponds to the submerged weir when the water level downstream affects the water level upstream (Fig. 1-b).

The flow on the weirs in submerged conditions has been the subject of study in 1947 by Villemonte who has conducted an analysis theoretical and experimental for different forms of the sharp crested weirs. This study was based on simplifying assumptions in order to express the submerged flow by

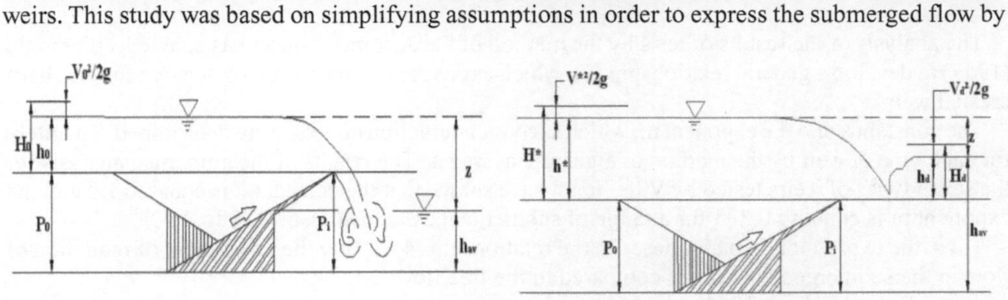

a- Free flow b- Submerged flow

Figure 1. Definition of hydraulic parameters for the weir under free and submerged flow conditions.

<u>Submerged flow</u>
– H^*: upstream total head; – h^*: upstream piezometric head;
– H_d: downstream total head; – h_d: downstream piezometric head.
<u>Free flow</u>
– H_0: upstream total head; – h_0: upstream piezometric head.

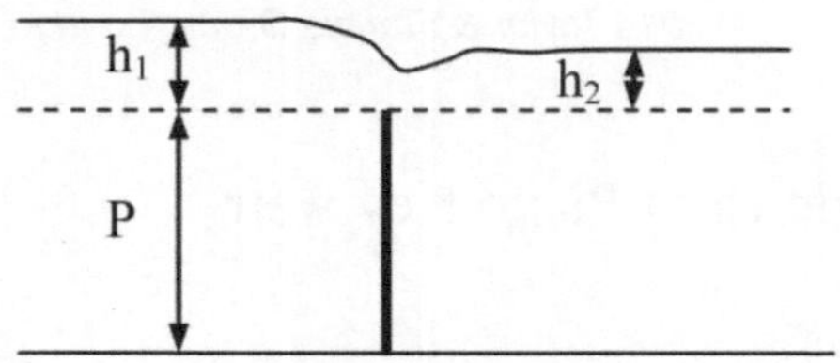

Figure 2. Description of submerged flow according Villemonte (1947).

a mathematical relationship. As well, it has been assumed that the flow on the submerged weir is composed of two parts, the first with a free flow under an effective load equal to the difference in the upstream and downstream levels, and the second part with a submerged orifice flow operating also under a load equal to the difference in the upstream and downstream levels. The second assumption, considers that the net discharge on the weir is the difference in the free flow discharge due to the upstream head h_1 minus the free flow discharge due to the downstream lead h_2 (Fig. 2).

$$Q = Q_1 - Q_2 \tag{1}$$

This last relationship can be written in the form:

$$\frac{Q}{Q_1} = 1 - \frac{Q_2}{Q_1} \tag{2}$$

The experimental tests carried out by Villemonte (1947) have shown that Q/Q_1 is related functionally to $(1 - Q_2/Q_1)$ but that the appropriate function is not the simple linear equation as presented in equation (2). However, it can be expressed by a relationship of the form:

$$\frac{Q}{Q_1} = f\left(1 - \frac{Q_2}{Q_1}\right) = k\left(1 - \frac{Q_2}{Q_1}\right)^m \tag{3}$$

By substituting in the equation (3) Q_1 and Q_2 by the universal discharge relationship $Q_i = C_i h_i^{ni}$, and considering the coefficients C_i and the exponent constant for a given Weir, the relationship (3) becomes:

$$\frac{Q}{Q_1} = k\left(1 - \frac{h_2}{h_1}\right)^m \tag{4}$$

The analysis of the results of tests by the method of the arithmetic mean has enabled Villemonte (1947) to develop a general relationship (5), which expresses the submerged discharge for the sharp crested weir.

The constant k and the exponent m, which account interaction effects, were determined separately for each type of weir by the method of algebraic averages. The results of the numerical analysis for the seven types of weirs tested by Villemonte have shown that the constant k is equal to 1.0 and the exponent m is equal to 0.385 for a range of submergence practice from 0.00 to 0.90.

Thus, the expression in parentheses in the relationship (4) is only the coefficient of reduction of flow in the submerged conditions compared to the free flow.

According to King (1954), this coefficient may be expressed also by the coefficient of submergence K.

$$K = \frac{Q_s}{Q_f} = \left[1 - \left(\frac{h_d}{h^*}\right)^n\right]^{0.385} \tag{5}$$

where K = coefficient of submergence; Q_s = discharge in submerged flow conditions; Q_f = discharge in free flow conditions; h_d = downstream total head; h^* = upstream total head; and n = exponent in free flow equation and numerator of submerged flow equation.

Table 1. Geometrical characteristics of the experimented model

Model of weir	n_0	n	L cm	W cm	W_0 cm	W_i cm	B cm	B_0 cm	B_i cm	B_b cm	$P_0 = P_i$ cm
PK-Weir model A	A2m	1.5	266	66	18	15	47	12	12	23	20
PK-Weir model A	A1	3	197	33	9.5	7	41	10.25	10.25	20.5	15

a- PK-Weir type A2m

b- PK-Weir type A1

Figure 3. Model of PK-Weir.

– Proportional weir: n = 1
– Parabolic weir: n = 2
– rectangular weir without contraction: n = 1.5
– triangular weir: n = 2.5

The K coefficient represents a reduction coefficient, which must be applied to the free flow discharge for obtain the discharge in the submerged conditions for a given weir. This coefficient depends on the type of weir and cannot be generalized.

The effect of the submergence on the labyrinth weir and PK-Weir has been studied by Tullis (2006), (2012) which determined experimentally the effect of submergence on the flow of these two types of weirs. The comparison of the obtained results on the two types of weirs has revealed that the PK-Weir seems to be relatively more efficient than the labyrinth weir for the low levels of submergence ($S < 0.75$). However, this is reversed for the highest levels of submergence. Tullis (2012) has pointed out that for the PK-Weir the module of submergence increases as the ratio H_0/P increases. The work which has been carried out by Ho Ta Khanh (2012) has shown that in a hydraulic point of view, the differences between the rectangular labyrinth weir and PK-Weir of types A and D are not very important, but with a slight advantage for the PK-Weir of type D for the strong floods.

Recently, an experimental and theoretical study has been conducted at the University of Biskra. This one represents continuity to that of Belâabed et al. (2011). This study has been based mainly on the theoretical analysis supported by an experimental investigation. As well, two models of PK-Weir have been tested to accomplish this work. The geometric characteristics of these models are listed in the Table 1.

2 EXPERIMENTAL PROGRAM

The experimental study conducted on the two models of PK-Weir was realized out in conditions of upstream approach with lateral contraction. Thus, the tests have been conducted in an experiment station of reduced hydraulic models made up of three channels of different sections. An upstream channel of 01 meter in width which simulates the flow in a river, followed by a center channel of 04 meters in width which simulates a reservoir and in its downstream end a channel with a 02 meters in width which simulates the discharge channel. The model of PK-Weir is installed at the exit of the center channel (reservoir), whose waters are discharged into the discharge channel (Fig. 6).

The experimental procedure has been as follows: after having fixed the flow which passes through the channels, we measure the level of water upstream in the free flow conditions. While holding the constant discharge, we are increasing the water level downstream of the PK-Weir by the installation

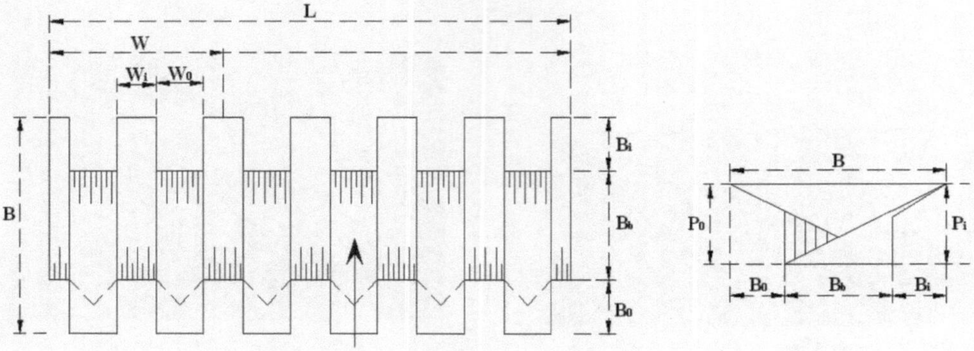

a- Plan view

b- Cross section

Figure 4. Geometrical configuration of the model A1.

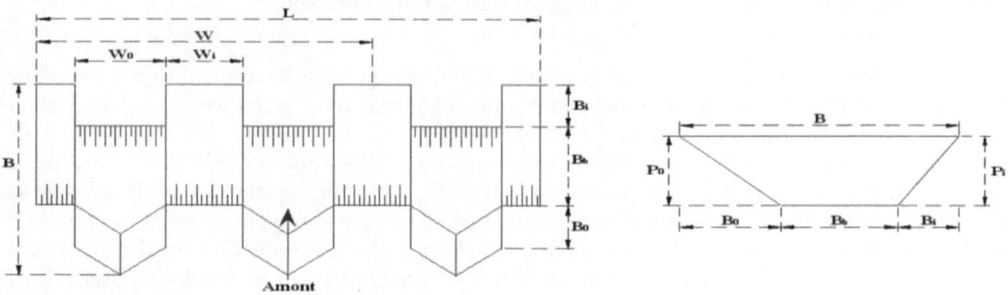

a- Plan view

b- Cross section

Figure 5. Geometrical configuration of the model A2m.

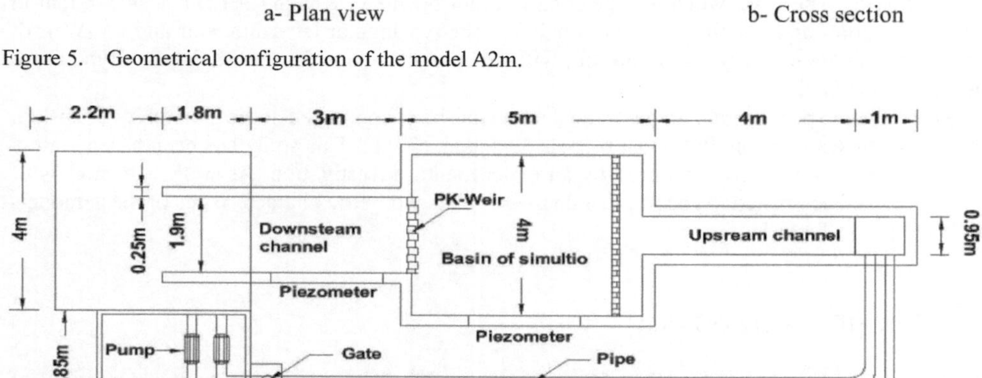

Figure 6. The experimental facility.

of a sill or the closure of a gate which is located at the downstream end of the reception channel. After stabilization of the flow we take the measurements of water depths upstream and downstream of the tested model. Once this operation is carried out we increase another time the height of the sill or the closure of the gate to increase the downstream level and we operate in the same way as previously.

3 EXPERIMENTAL RESULTS

The experimentation on reduced model is sanctioned by results that require an interpretation. In this study, these results are expressed either by the relative upstream head (H^*/H_0) according to the relative downstream head (H_d/H_0) or by the coefficient of submergence (Q_s/Q_f) depending on (H_d/H^*).

3.1 *Effect of the discharge on the submergence of the PK-Weir*

The model of PK-Weir type A2m which is characterized by the continuous overhangs to the base of the weir has been the subject of experimentation for eight values of fixed discharge. The results of these tests have been represented graphically in the form of $H^*/H_0 = f(H_d/H_0)$. From this graphical representation (Fig. 7) we can see that for the different values of discharge, the eight dimensionless curves are similar and confused, particularly for the curves which correspond to large values of flow. As well, we can say that in the case of the submerged flow, the variation of upstream level is done in a proportional manner in relation to the downstream level whatever the flow. This can be also noted for the results of the standards model of PK-Weir type A (Fig. 8).

3.2 *Comparison of the results obtained on the different models*

The geometry of the PK-Weir may take several configurations which can undoubtedly have effects on the flow. Thus, a comparison of the results obtained on the two tested models was made. It is to be noted that the geometry of the two models is different in particular, the form of the overhangs. The graphic representation of the relative upstream head according to the relative downstream head shows a discrepancy between the two curves of the order of 15% for the lower values of H_d/H_0

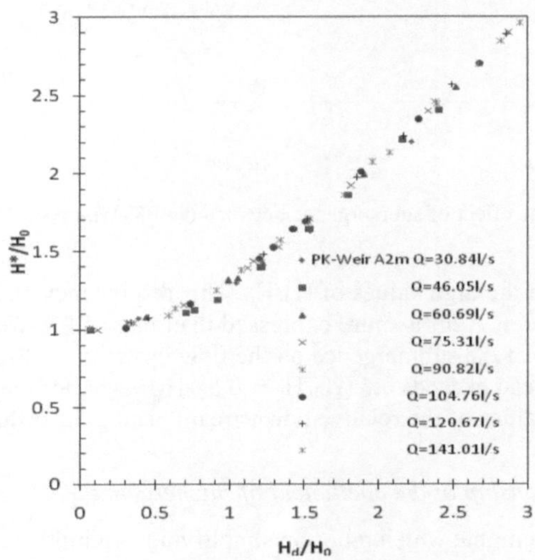

Figure 7. Relative upstream head according relative downstream head (model type A2m).

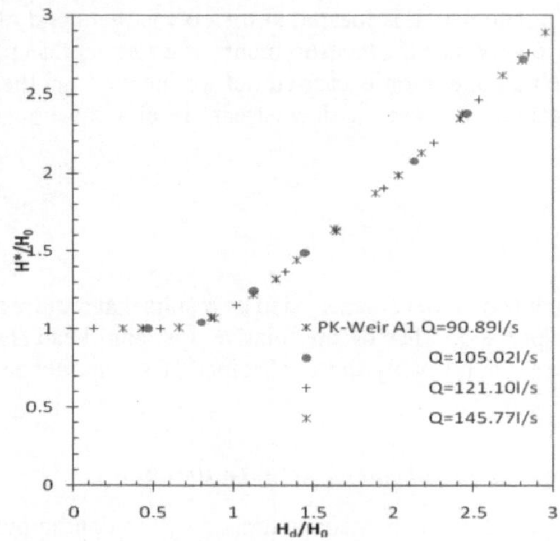

Figure 8. Relative upstream head according relative downstream head (model type A1).

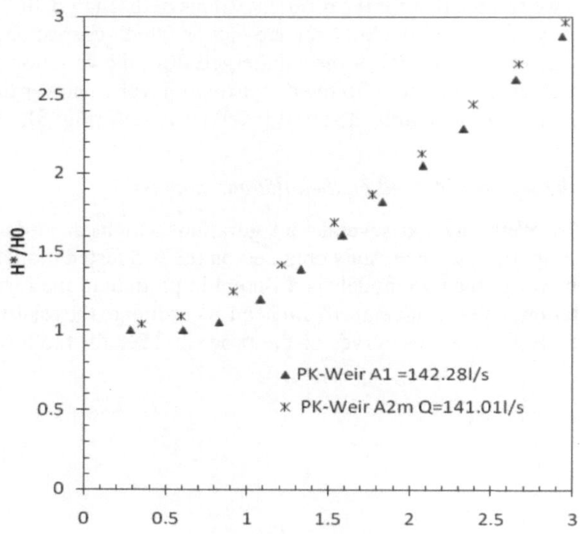

Figure 9. Comparison of the effect of submergence between the PK-Weir type A2m and A1.

and approximately 3% for the high values of H_d/H_0. This discrepancy shows that the effect of the submergence of the PK-Weir A2m is more expressed than that of PK-Weir A1 (Fig. 9). It is also to be noted that the effect of the submergence on the flow upstream of PK-Weir A2m began when the relative downstream head exceeds 0.5 ($H_d/H_0 > 0.5$). However, for the PK-Weir A1, this effect is observed only for the values of the relative downstream head greater than 0.8 ($H_d/H_0 > \sim 0.8$).

3.3 *Mathematical relationship of the coefficient of submergence*

Based on the study of Villemonte which relies on simplifying assumptions, a numerical analysis of the obtained experimental data has been carried out. This analysis had for objective to determine a mathematical relationship which expresses the coefficient of submergence of the PK-Weir A2m.

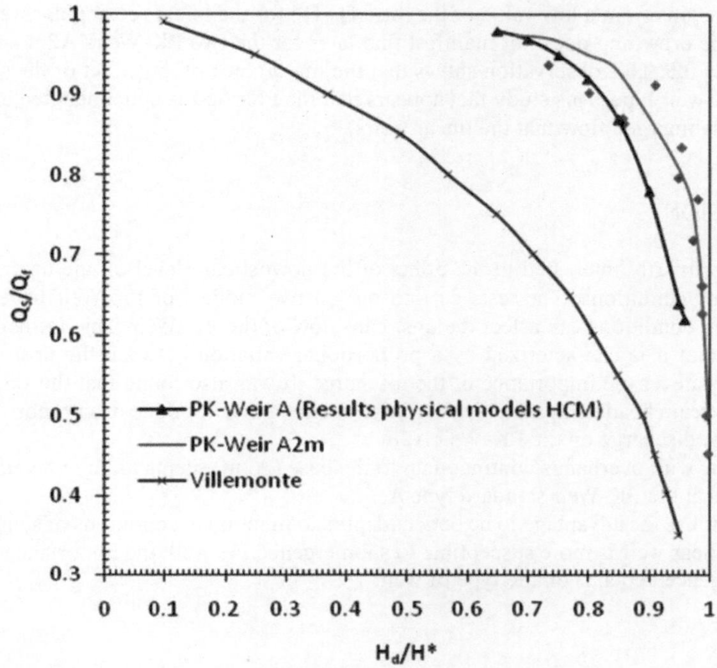

Figure 10. Discharge for submerged Weir.

This analysis has also concerned the PK-Weir type A (according to the data of Ho Ta Khanh, 2012). Consequently, the following relationships were obtained.

Rectangular Weir (Villemonte):

$$K = \frac{Q_s}{Q_f} = \left[1 - \left(\frac{h_d}{h^*}\right)^{1.5}\right]^{0.385} \tag{6}$$

PK-Weir type A2m:

$$K = \frac{Q_s}{Q_f} = 1.158\left[1 - \left(\frac{h_d}{h^*}\right)\right]^{0.126} \tag{7}$$

$0.68 \leq H_d/H^* \leq 0.99$

PK-Weir type A (Results of physical models HCM):

$$K = \frac{Q_s}{Q_f} = 1.289\left[1 - \left(\frac{h_d}{h^*}\right)\right]^{0.225} \tag{8}$$

$0.65 \leq H_d/H^* \leq 0.96$

The first relationship was developed by Villemonte for a linear rectangular weir without lateral contraction. While the last two relations (6) and (7) have been developed by a numerical analysis of the obtained data on the model A2m and the obtained data on the model A by Ho Ta Khanh (2012).

The graphical representation $Q_s/Q_f = f(H_d/H^*)$ (Fig. 10) for the three models of Weir (PK-Weir A2m, PK-Weir A (HCM) and the spillway rectangular (Villemonte) shows that the effect of the submergence begins from a low value of the ratio H_d/H^* for the linear rectangular weir. However, the effect of the drowning does not manifest that later for the two PK-Weirs A2m and A, for the values $H_d/H^* > 0.5$. This observation shows that the importance of the effect of the submergence depends on the weir type. This study fact appears that the PK-Weir is better adapted to function in conditions of submerged flow that the linear weirs.

4 CONCLUSION

This study is primarily interested in the effect of the downstream level on the upstream level in submerged flow conditions. The tests carried out on two models of PK-Weir have shown that the downstream conditions can affect the upstream flow of the PK-Weir. This influence has been demonstrated that it is characterized by a proportional variation between the upstream and the downstream whatever the importance of the discharge. It was also found that the variation of the relative downstream head compared to the relative upstream head is done in a proportional manner for the different discharge on the PK-Weir A2m.

The PK-Weir with overhangs continued up to the base (A2m) seems to be more exposed to the submergence that the PK-Weir standard type A.

The PK-Weir has the advantage to be better adapted to function in conditions of submerged flow, however, the linear weir is more susceptible to submergence. As well, the importance of the effect of the submergence depends on the type of weir.

REFERENCES

Belaabed, F., Ouamane, A. 2011. Contribution to the study of the Piano Key Weirs submerged by the downstream level, *Labyrinth and piano key weirs-PKW 2011*: 89–95. London: CRC Press.

Dabling, M.R., Tullis, B.P. 2012. Piano Key Weir Submergence in Channel Applications. *International Workshop on Piano Key Weir for In-Stream Storage and Dam Safety (PKWISD-2012)*, New Delhi, India.

Ho Ta Khanh, M. 2012. Utilization of Piano Key Weirs for low barrages. *Hydro 2012*, Bilbao, Spain

Lopes, R., Matos, J., Melo, J.F. 2009. Discharge capacity for free flow and submerged labyrinth weirs. *Proc. 33rd IAHR congress*, Vancouver, Canada.

Ouamane, A., Lempérière, F. 2010. Study of various alternatives of shape of piano key weirs, *HYDRO 2010 – Meeting Demands in a Changing World*, Lisbon, Portugal.

Tullis, B.P. 2006. Predicting submergence effects for labyrinth weirs, *International Symposium on Dams in the Societies of the XXI Century*. Barcelona, Spain.

Tullis, B.P., Young, J.C., Chadler, M. A. 2007. Head-Discharge Relationships for Submerged Labyrinth Weir. *ASCE Journal of Hydraulic Engineering*, 133(3), 248–254.

Villemonte, J.R. 1947. *Submerged weir discharge studies*. Eng. News-Rec.

Labyrinth and Piano Key Weirs II – PKW 2013 – Erpicum et al. (eds)
© 2014 Taylor & Francis Group, London, ISBN 978-1-138-00085-8

Experimental study on vertical velocity and submergence depth near Piano Key weir

N. Sharma & H. Tiwari
Department of Water Resources Development and Management, IIT Roorkee, India

ABSTRACT: Insufficient spillway capacity has been the cause of one-third of all dam failures. An innovative and effective way of increasing the spillway capacity is to use a Labyrinth weir in the modified form of PKW having range of specific flow from 3 to 1000 m^3/s/m. As the flow approaches towards the PK Weir, it was observed that the Z-component of the velocity of water in lower levels increases which enhances the flushing capacity. One of the important observations is that with decrease in discharge from higher flow to medium flow; Z component velocity is getting lowered by about 14 percent. Increase in discharge from lower to medium flow Z component velocity is getting increased by about 40 percent. It indicates that at lower discharge, upward velocity component is appreciably more which may be helpful for significant amount of sediment suspension at lower discharge also. For flood flow, it was found that there has been reduction of around 25–30% of submergence depth with respect to ogee spillway. Induced flow characteristics near PKW and tractive stress exertion by flowing water in the inlet cell are capable to flush out even the coarser sand particles.

1 INTRODUCTION

Spillways play a major role in ensuring the flood safety of dams. Insufficient spillway capacity has been the cause of one-third of all dam failures (Schleiss, 2011). Spillways represent a substantial portion of total project costs and they play a major role in ensuring safety (Modi & Seth, 2011). An effective way of increasing the spillway capacity is to use a Labyrinth weir (Tacail et al., 1990; Tullis et al., 1995). The concept of the Labyrinth weir is to vary the plan shape of the crest to increase the effective crest length (Lempérière & Jun, 2005). A new concept of a Labyrinth weir has been proposed with a new shape like black and white Piano keys when viewed in plan, this new concept was called the Piano Key Weir (Lempérière & Ouamane, 2003). This design solves most of the problem presented by the original Labyrinth weir, and is also more efficient (Anderson, 2011). Some differences between traditional Labyrinth weir and Piano Key Weir are as follows:

- Vertical walls founded on a flat area are replaced by lateral vertical walls and sloping slabs upstream and downstream of the crest. These slabs are partially a cantilever structure, upstream and downstream. Therefore the overall structure is self-balanced.
- The Piano Key Weir can be positioned on the top of the crest of new or existing gravity dams.
- Application can cover a wide range of specific flows, from 3 to 1000 m^3/s/m.
- Piano Key Weir can increase by a factor about 1.50 to 4.00 times than the specific discharge capacity of straight sharp crested weir.
- From a structural point of view, Piano Key Weir is extremely hyper-static structures, which are solid and simple.

The total length of the Piano Key Weir is typically three to seven times the spillway width. The Piano Key Weir is particularly well suited for cases where the length of the structure has to be restricted or for the rehabilitation of existing spillways. A Piano Key Weir can pass large discharge at a relatively low head. Its advantages include relatively low construction and maintenance costs

Figure 1. A typical view of Piano Key Weir (PKW).

and more reliable operation, compared with gated spillways. Construction of the first prototype PK weir, Goulours dam in France, was completed in 2006; construction of the second prototype PK weir, Saint-Marc dam in France, was completed in 2008 (Anderson, 2011). This new Piano-Key Weir technology has already been used by the Himachal Pradesh Power Corporation Ltd. (HPPCL) on Sawra Kuddu Barrage.

2 DEVELOPMENT AND FEATURES OF PIANO-KEY WEIR (PKW)

An initial model investigations and behavior of Piano Key Weir was studied by Lempérière and Ouamane (2003) in terms of a magnification ratio of the Piano Key Weir against sharp-crested linear weir having the same channel width. The results of the test showed that the Piano Key Weirs are simple solutions as safe and easy to operate as traditional free flow spillways and much more efficient. They may improve the flood control by many existing dams. Behavior of Piano Key Weir was studied by Barcouda et al. (2006) in terms of a magnification ratio of the Piano Key Weir flow for a sharp-crested linear weir having the same channel width. The most significant parameters in determining the capacity of a weir are its height relative to the upstream depth, the crest shape and the crest length (Falvey, 2003 & Afshar, 1988). The results of the test show that the Piano Key Weirs are more efficient than the traditional Labyrinth weir and Piano Key Weirs can be an interesting solution for increasing the active storage of reservoir or for improving the safety of dam during extreme flood. It will also help in minimizing the sub-mergence effect. Flushing is vital for the preservation of long-term storage in reservoirs. However, downstream impacts such as turbidity act as a constraint in the planning and operation of sediment flushing. In this study velocity distribution of all the three components of the velocity at different points at up-stream of the weir and over the weir and Velocity vector diagram in up-stream of the weir as the flowing water approaches the P.K. Weir and over the weir as it spills across the crest of P.K. Weir is investigated through laboratory experiment which has a good scope regarding the flushing study in reservoirs. A typical view of PKW is given in Figure 1.

3 LABORATORY ARRANGEMENT FOR VELOCITY COMPONENT STUDY

An overflowing tank of 240 cm × 240 cm was installed at the upstream head-end to ensure the supply of steady discharge into the experimental flume. The flume used was made of mild steel with side walls made of transparent perspex sheet. Flume has an in-built upstream tank of 150 cm × 180 cm dimensions. Diagrammatic scheme is given in Figure 2. The bed of flume is supported on angle

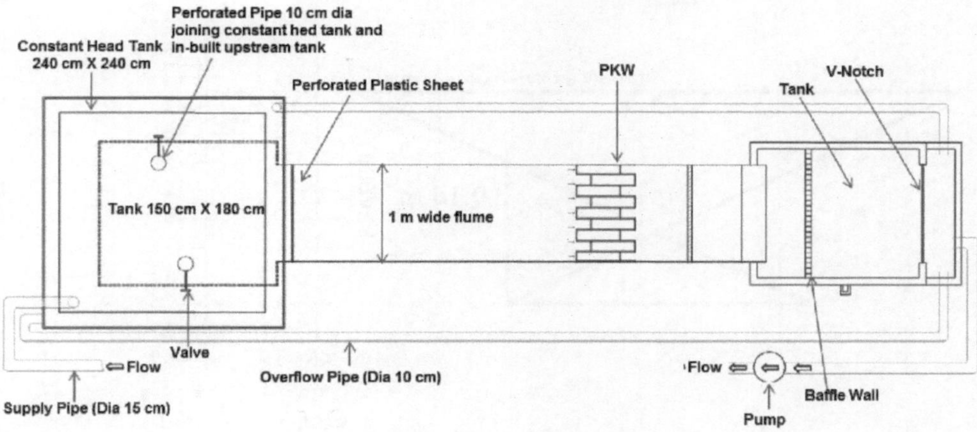

Figure 2. Laboratory Arrangements (PKW).

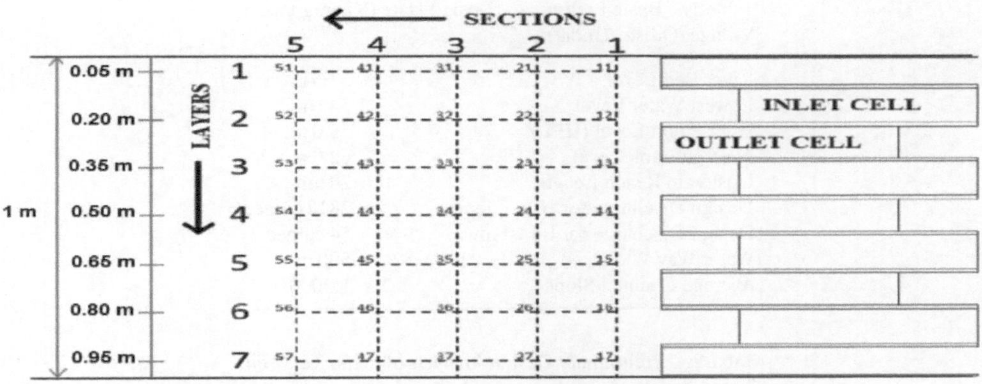

Figure 3. Layout of grid in the space up-stream to the Piano Key weir.

iron sections. The water flowing in the flume, after it passes over the Piano-Key weir, falls into a down-stream tank installed with V-notch discharge measuring devices, which is connected to down-stream of V-notch storage tank. From the down-stream of V-notch storage tank, tank water is lifted with the help of a two 10 HP pumps. 10 cm diameter pipes carry water from the storage tank to the upstream constant head tank. The discharge is regulated with the help of a gate valve placed between the constant head tank and the inbuilt upstream tank. Laboratory Arrangements for vertical velocity measurement was given in Figure 2.

An imaginary 3D mesh-grid was formed in the space available in the up-stream of the Piano-Key weir. This grid was divided into 5 sections along x-direction, 7 layers along the y-direction and 15 levels along the z-direction. The velocity readings were taken on the nodes of this mesh-grid using Acoustic Doppler Velocimeter (ADV). The ADV was fitted with a trolley which can move along the movable hand rail and across the fixed side rails of the flume. The hand rail can also move along the side rails so as to change to the desired position. The position along the z-direction was adjusted with the help of the vernier gauge, through which ADV was connected to the trolley. The position along the y-direction was adjusted by moving the trolley along the movable hand rail. The position along the z-direction was adjusted by a moving handrail itself along the fixed side (also along the direction of flow). The layout of the plan and section of the grid formed is as shown in the Figure 3 and Figure 4. As illustrated by Wang et al. (2011), the velocity measurements obtained with the

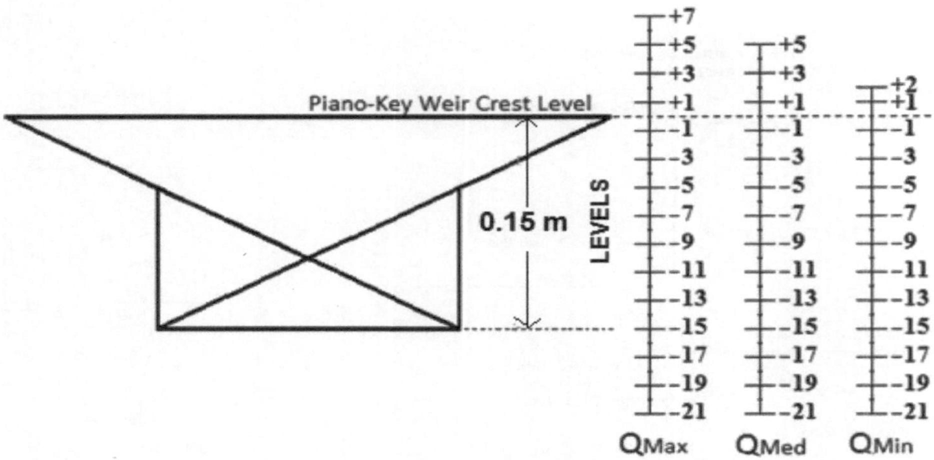

Figure 4. Depths (Levels) of nodes at which velocity readings were taken (@ 69, 50, 31 lps).

Table 1. Basic Preliminary Design Data (Kharagprasad Village, Odisha (India)).

River Bed Level	53.5 m
Lowest Water Level	54.0 m
High Flood Level (HFL)	58.0 m
Average Bank Level	62.0 m
Upstream Reach Length	20 km
Design Discharge for High Flood	2818 cumec
Design Discharge for Low Flow	54 cumec
Water Way Width	500 m
Average Channel Slope	1:2000

Table 2. Preliminary Design Analysis of Piano Key Weir Type Barrage.

Height of Piano Key Weir	4.5 m
Crest Level of Piano Key Weir	58 m
Head Over the Crest on P. K. Weir	1.4 m
For High Flood Discharge	
Maximum Water Level on P. K. Weir	59.4 m
Head Over the Crest on P. K. Weir	5.56 m
For Low Flow	
Minimum Water Level on P. K. Weir	58.05 m

SONTEK ADV, is at 5 cm below the acoustic sensor (of the ADV) with a probe volume of 0.09 cm^3. With a frequency of 25 Hz, a recording time of 40 sec was taken for 1000 measurements required for satisfying time average conditions. A filter condition of Signal to Noise Ratio (SNR) greater than 15 dB and correlation (COR) greater than 70 percent was used as suggested by Volugaris and Trowbridge (1998), and the filtered data only are retained.

4 CASE STUDY FOR BRAHMANI RIVER IN LABORATORY

Table 1 explains field data available and Table 2 give details about design criteria. Model in laboratory was run for high flood as well as low flow condition. This model was also run for

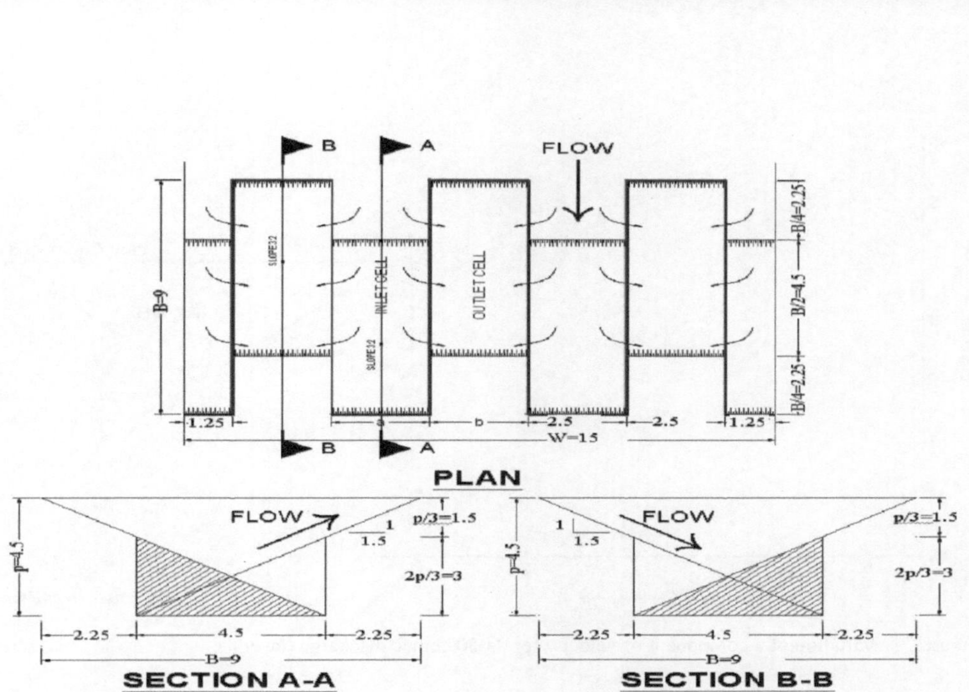

Figure 5. Plan and Sectional View of Piano-Key Weir laboratory arrangement.

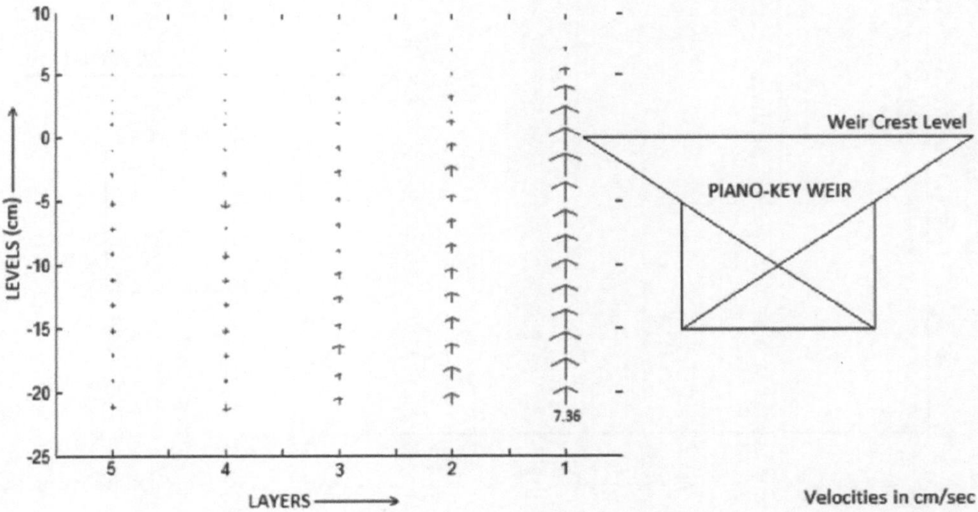

Figure 6. Variation of z component of velocity for 0.069 cumec discharge (layer 4).

a discharge of 20347.22 cumec which was the flood condition when upstream hydraulic structure releases excess water. This PKW at the proposed site is not yet commosioned. Plan and Sectional View of Piano-Key Weir laboratory arrangement details in Figure 5.

5 RESULTS AND DISCUSSIONS

Z component velocity (cm/s) variation which plays important for sediment suspension as well as flushing point of view is shown in Figure 6, Figure 7 and Figure 8 for discharge conditions of 0.069, 0.050 and 0.031 cumec.

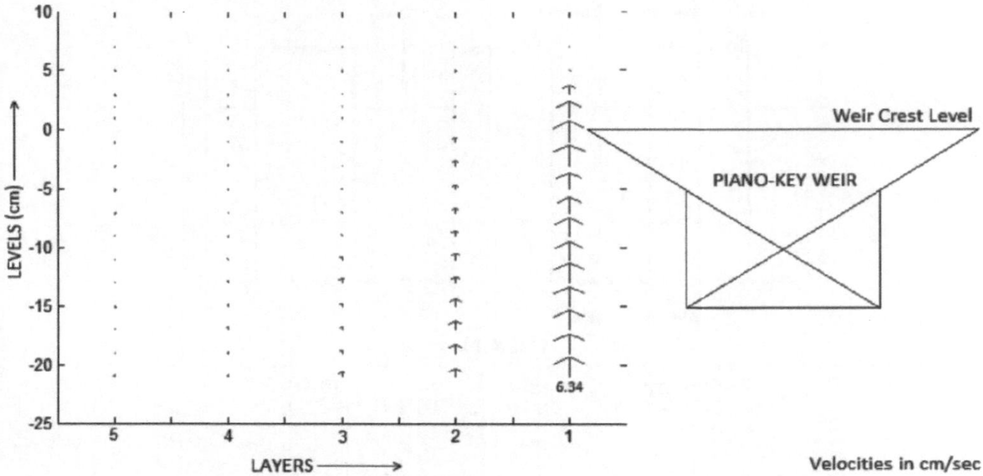

Figure 7. Variation of z component of velocity for 0.050 cumec discharge (layer 4).

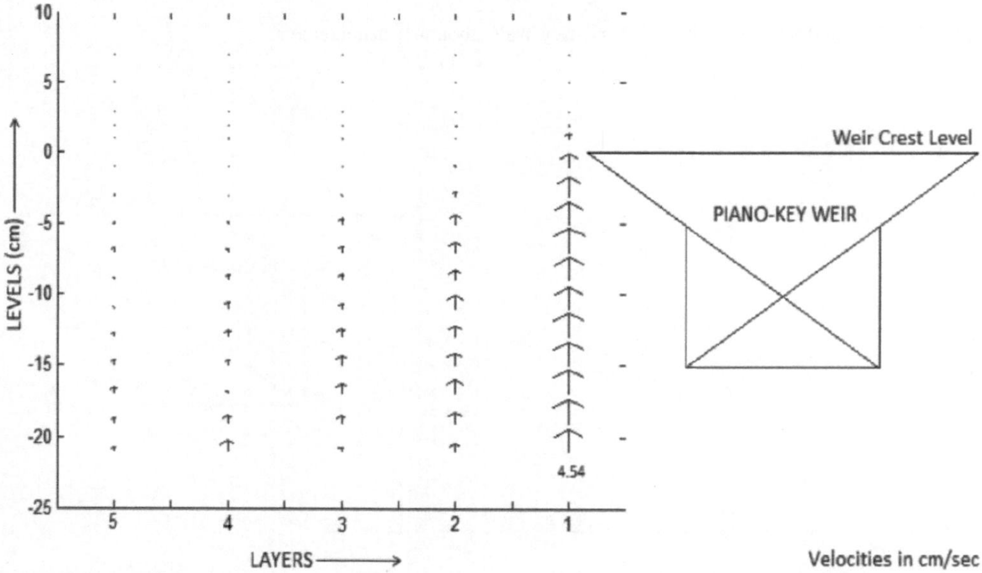

Figure 8. Variation of z component of velocity for 0.031 cumec discharge (layer 4).

As the flow approaches towards the PK Weir, it was observed that the Z-component of the velocity of water in lower levels increases. From this it can be concluded that water gets some uplift as it approaches towards PK weir so as to pass through the inlet cell and get spilled in the downstream side. This observed flow characteristic not only enhances the discharge capacity of the weir but also helps in flushing of the sediment through the inlet cell of the Piano-Key Weir. Due to flow characteristics near PKW and tractive stress exertion by flowing water in the inlet cell is flow competence is enabled to flush out even the coarser sand particles of size up to 2 mm. One of the significant observations is that with a decrease in discharge from to higher flow (0.069 cumec) to medium flow (0.050 cumec); Z component velocity is getting lowered by 14 percent. Increase in

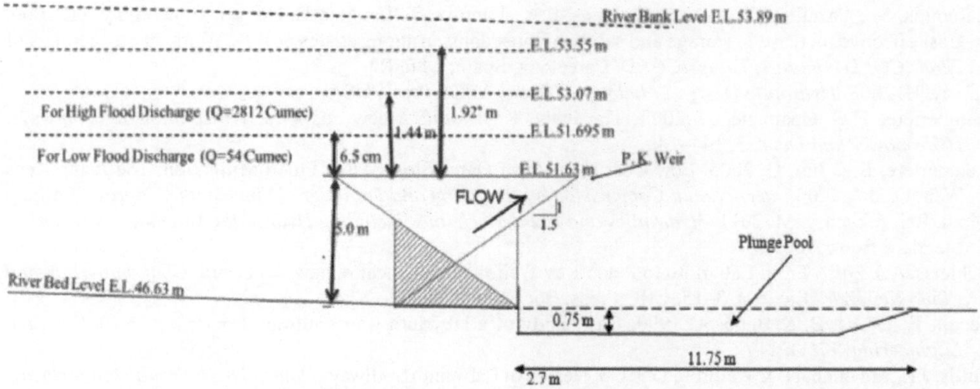

Figure 9. Submergence Depth for PKW compared with Ogee spillway.

discharge from lower flow (0.031 cumec) to medium flow (0.050 cumec); Z component velocity is getting increased by 40 percent. It indicates that at lower discharge upward velocity component is quite more and may be helpful for a significant amount of sediment suspension at lower discharge also.

From Figure 9, study on the geometry of Piano-Key weir, it can be observed that for a given design flood discharge 20,347.22 cumec (high flood condition), the head over the proposed Piano-Key weir (4.2 m) was much lesser than that of a conventional ogee weir (6.173 m) of the same height. Some of the experimental results are given below for the case study of Brahmani River.

- Length of backwater due to incident afflux with P. K. Weir in position (High Flood) = 16 km
- Length of backwater due to incident afflux with P. K. Weir in position (Low Flow) = 9.1 km
- Volume of In-Stream Storage in Low Flow = 10.15 Million Cubic Meter (MCM)
- Water Requirement for Industry/Drinking Water Purpose = 2.2 Cumec
- If minimum flow ceases to exist, the above In-stream storage will cater for 54 days

Because of this, the submergence of the area upstream to the weir can be significantly reduced. So increasing discharge capacity and increment of flushing capability enhances the structural safety as well as reservoir sedimentation. It has also advantages of cost effective solution and it can be installed over existing dams.

6 CONCLUSIONS

At lower discharge ranges, relatively higher vertical velocity component implies presence of sediment in suspension.

The use of PKW can significantly reduce the upstream submergence area in comparison to conventional weir or barrage.

Induced flow characteristics near PKW and tractive stress exertion by flowing water in the inlet cell, the flow competence is enabled to flush out even the coarser sand particles.

REFERENCES

Afshar, A. 1988. The Development of Labyrinth Weir Design, *Water Power and Dam Construction*, 40(5): 36–39.

Anderson, R. 2011. Piano Key Weir Head Discharge Relationships, *Dissertations for Graduate Studies*, School of Master of Science, Utah State University.

Barcouda, M., Cazaillet, O., Cochet, P., Jones, B.A., Lacroix, S., Laugier, F., Odeyer, C. & Vigny, J.P. 2006, Cost effective increase in storage and safety of most dams using fusegates or P. K. Weirs, in proceedings of *22nd ICOLD congress, CIGB/ICOLD*, Barcelona, Spain, Q86, R3.

Falvey, H. 2003. *Hydraulic Design of Labyrinth Weirs*. ASCE Press, USA.

Lempérière, F. & Ouamane, A. 2003. The Piano Keys weir: a new cost-effective solution for spillways. *Hydropower and Dams* 5, 144–149.

Lempérière, F. & Jun, G. 2005. Low Cost Increase of Dams Storage and Flood Mitigation: The Piano Keys Weir. Q.53 R. 2.06 *International Commission on Irrigation and Drainage – Nineteenth Congress,* Beijing.

Modi, P.N. & Seth, S.M. 2011. *Hydraulics and Fluid Mechanics including Hydraulics Machines*. New Delhi. Standard Book House.

Schleiss, A.J. 2011. From Labyrinth to Piano Key Weirs – A historical review. *Proc. Intl. Conf. Labyrinth and Piano Key Weirs* Liège B, 3–15. CRC Press, Boca Raton FL.

Tacail, F., Evans, B. & Babb, A. 1990. Case study of a labyrinth weir spillway. *Canadian Journal of Civil Engineering*. 17(1): 1–7.

Tullis, J.P., Amanian, N. & Waldron, D. 1995. Design of Labyrinth Spillways. *Journal of Hydraulic Engineering,* 121(3), 247–255.

Volugaris, G. & Trowbridge, J.H. 1998. Evaluation of acoustic Doppler Velocimeter (ADV) for turbulence measurements. *Journal of Atmospheric and Oceanic Technology* 15(1):272–289.

Wang, X.Y., Yang, Q.Y., Lu, W.Z. & Wang, X.K. 2011. *Experimental study of near-wall turbulent characteristics in an open channel with gravel bed using an acoustic Doppler velocimeter*, Springer-Verlag.

Labyrinth and Piano Key Weirs II – PKW 2013 – Erpicum et al. (eds)
© 2014 Taylor & Francis Group, London, ISBN 978-1-138-00085-8

Discharge characteristics of Piano Key weirs under submerged flow

G.M. Cicero & J.R. Delisle
EDF – National Hydraulics and Environment Laboratory, Chatou, France

ABSTRACT: The Piano Key Weir (PKW) is a type of labyrinth weir with efficient hydraulic performance especially at low heads. Since most experiments have been carried out under free-flow conditions, EDF has launched a new experimental program including the testing of PKWs under submerged conditions. The 3 classical PKWs types A, B and C have been tested to isolate the overhang effect on both free-flow and submerged flow conditions. After a comparison of the 3 PKWs discharge efficiencies under free-flow conditions, this article focuses on the results under submerged flow conditions. The data-sets analysed with a non-dimensional approach allow obtaining empirical correlations which characterize the PKW sensitivity to submergence. The PKWs were also compared to the submerged behaviours of linear weirs (sharp-crested, broad-crested and Ogee crest). Then, the correlations were applied to compare the discharge efficiencies of the PK and the linear weirs under the same submerged conditions.

1 INTRODUCTION

The Piano Key Weir (PKW) is a type of labyrinth weir with efficient hydraulic performance especially at low heads. PKWs are classified according to the presence or absence of the overhangs (Lemperiere et al. 2011): type A with symmetrical overhangs, types B and C with only up- or downstream overhangs and type D, without overhangs.

The reduced footprint due to the overhangs allows installing PKWs on the top of the crest of existing gravity dams to upgrade their spillway capacities. Consequently, most of the experiments have been carried out to study the hydraulic performance under free-flow conditions. However, the hydraulic behavior of PKWs under submerged flow has also been studied by Ho Ta Khanh et al. (2011), Belaabed & Ouamane (2011), Dabling & Tullis (2012), and Kabiri-Samini & Javaheri (2012).

In 2011, Electricité De France (EDF) started a new experimental program at the National Hydraulics and Environment Laboratory (LNHE) in Chatou to expand the knowledge on some geometric parameters effects such as the overhangs, the crest shape, the dam height and the noses under the outlet keys. This program includes the testing of a non rectangular PKW design, with sidewalls angles, and the testing of all geometries under submerged conditions. This program will also provide a useful database to validate 1D or 3D numerical modeling and empirical correlations.

Since the first submerged PKWs tested in Vietnam (Ho Ta Khanh et al. 2011), Belaabed & Ouamane (2011) and Dabling & Tullis (2012) published data sets applying the Tullis et al. (2007) approach based on normalized up- and downstream head. The same approach was used herein to analyze the behaviors of the PKWs tested at LNHE under submerged conditions.

The purpose of this article is to present the testing of the three PKW types A, B and C and the comparison of the discharge efficiency on both free-flow and submerged conditions.

The standard notation as defined by Pralong et al. (2011) has been used herein.

2 EXPERIMENTAL SET-UP

2.1 *Testing apparatus*

The PKW testing was conducted in a 2 m wide, 1 m deep and 25 m long rectangular channel devoted to the discharge capacity measurements up to 500 l/s.

The weir discharge (Q) was measured by an electromagnetic flow-meter of 1% accuracy and the piezometric heads were measured at ~6 m upstream (h_u) and ~3 m downstream (h_d) of the PKW with an accuracy of +/−0.18 mm. The total upstream (H_u) and downstream (H_d) heads were calculated adding the velocity head corresponding to the average cross-sectional velocity at the level measurement locations.

2.2 *PKWs design*

The objective of this experimental program is to study the overhang effects on the discharge efficiency of the PKWs type A, B and C, both in submerged and free-flow conditions. The PKW geometry was the design recommended by Lemperiere et al. (2011) for the type B, i.e. with the same inlet and outlet key widths ($W_i = W_0$). As a result, the type C (without upstream overhang) was the type B (without downstream overhang) just reversed in the channel. The PKWs were fabricated in PVC and the main geometric parameters (Table 1) are strictly the same except for the lengths of the upstream and downstream overhangs (B_o, B_i) and consequently for the bottom slopes of the outlet and the inlet keys (S_o, S_i).

The PKW width (W) was the total channel width to maximize the number of PKW units (N_u) and to avoid side effects. The PKWs have a flat top crest and the side wall thickness ($T_s = 2$ cm) was chosen wide enough to enable various crest shapes testing on the type A PKW. Since Vermeulen et al. (2011) have recommended a 35 cm wall thickness for prototypes, the geometric scale of the models was around 1/20.

To allow a wider range of submergence conditions, the PKWs were installed close to the bottom, on top of a platform in concrete, so that the crest level was at 0.355 m ($P_d/P = 0.6$). The PKWs were tested without noses under the upstream overhangs.

2.3 *Testing procedure*

For the tests under free flow conditions, the flow rate was increased step by step to measure the discharge and the piezometric upstream level when the flow conditions were stabilized. For submerged testing, the tailwater level was increased step by step while maintaining a constant flow rate to measure the discharge, the upstream and the downstream piezometric levels at stabilized flow conditions.

For both free-flow and submerged testing, the discharge and the water levels were recorded for a 3 min period at a 0.5 Hz frequency. As a result, the discharge and level measurements used for analysis were the average and standard deviation of 90 temporal values.

Table 1. Main geometric parameters of the PKWs tested at LNHE (dimensions in meters).

PKW	N_u	W_i	W_o	P	B_i	B_o	B	S_i	S_o	T_s	n
Type A	6.5	0.133	0.133	0.222	0.154	0.154	0.573	0.556	0.556	0.02	4.6
Type B	6.5	0.133	0.133	0.222	0	0.307	0.573	0.902	0.402	0.02	4.6
Type C	6.5	0.133	0.133	0.222	0.307	0	0.573	0.402	0.902	0.02	4.6

3 EXPERIMENTAL RESULTS

3.1 *Free-flow results*

The discharge efficiency of the 3 PKW types were compared using the non-dimensional discharge coefficient C_w related to the classical weir discharge equation:

$$C_W = \frac{Q}{W\sqrt{2g}H^{1.5}} \tag{1}$$

The non-dimensional rating curves $C_w(H/P)$, of the 3 types are given in Figure 1a, with error bars indicating the measurement uncertainties. To avoid scale effects due to viscosity and surface tension that may occur at very small heads, it is recommended to consider the measured values for H > 3 cm (Leite Ribeiro et al. 2012).

The experimental data were very well correlated ($R^2 > 0.999$) by polynomial interpolations given by Equation 2 with the coefficients (a_i) of Table 2. Within their validity range of H/P, these equations allowed to predict the measured discharge with a maximum error of 1.3% for type A, 0.9% for type B and 1.0% for type C. Note that these equations are only valid for the PKW geometries of Table 1 (and for $P_d/P = 0.6$).

$$C_W = a_0 + a_1(H/P) + a_2(H/P)^2 + a_3(H/P)^3 + a_4(H/P)^4 \tag{2}$$

These equations allow comparing the discharge efficiency of the 3 PKW types at the same upstream head (Fig. 1b). The type B can be 10% to 5% more efficient than the type A, whereas the type C can be 15% less efficient than the type A. These results are consistent with experiments of Ouamane & Lemperiere (2006) who observed that the type B was 9% to 12% more efficient than the type A.

Anderson & Tullis (2011) have compared the discharge efficiency of a type D (with no overhang) to a type A. The geometry was also the design recommended by Lemperiere (2011) but the ratios $W_i/W_o = 1.25$ and $P/T_s = 15.5$ were different on the PKWs tested at LNHE. The results (Fig. 1b) of the Utah Water Research Laboratory (UWRL) show that that the type D is less efficient than the

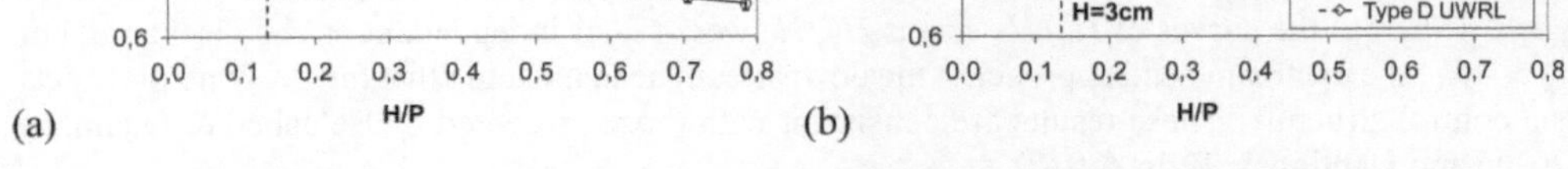

Figure 1. Experimental rating curves $C_w(H/P)$ (a) and gain in discharge relative to the type A (b).

Table 2. Coefficients of Equation 2 to predict the rating curves $C_W(H/P)$ under free flow conditions.

PKW	a_0	a_1	a_2	a_3	a_4	validity	$\text{Error}_{\max}$
Type A	1.63	0.59	−11.56	21.72	−12.46	$0.1 < H/P < 0.71$	1.3%
Type B	1.54	3.27	−22.15	37.32	−20.37	$0.1 < H/P < 0.72$	0.9%
Type C	1.91	−3.85	4.17	−1.56	0	$0.1 < H/P < 0.79$	1.0%

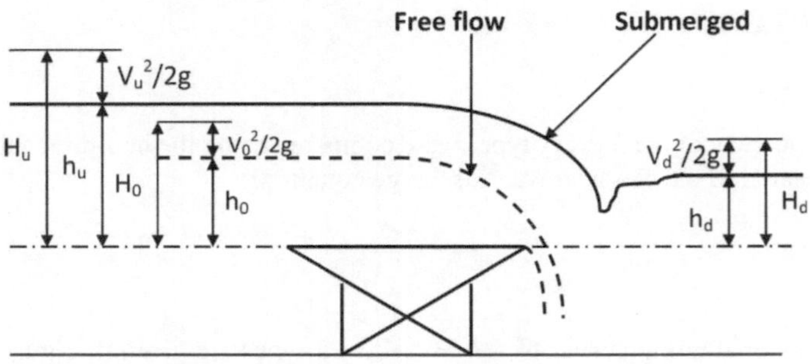

Figure 2. Hydraulic parameters under submerged and free-flow conditions.

type A and more efficient than the type C for most of the upstream head range. Thus, the free-flow performance can be classified: Type B > Type A > Type D > Type C.

3.2 *Submerged flow results*

Figure 2 presents the main hydraulic parameters under submerged and free-flow conditions. The initial approach to study weir submergence was based on the flow reduction factor Q_s/Q_f versus the submergence factor $S = H_d/H_u$ (Villemonte, 1947). Tullis et al. (2007) proposed an alternative method to study the submergence of labyrinth and linear weirs based on the use of the total upstream head (H_u) and the total downstream head (H_d) normalized by (H_0) the total upstream head for the same discharge under free-flow conditions. This method already used by Belaabed & Ouamane (2011) and Dabling & Tullis (2012) for PKW submergence studies was applied herein to analyse the data of the 3 types of PKW tested at LNHE.

For each PKW type, the submerged testing was carried out at 5 constant discharges from 100 l/s to 470 l/s, with the following procedure. For each discharge, starting from free-flow conditions with a tailwater level below the crest elevation, the downstream level was raised by steps of 2 cm to isolate the modular submergence limit, when the free-flow conditions still applies (i.e., $H_u = H_0$) whereas the downstream level exceeds the crest elevation ($H_d > 0$). Above this modular limit, the downstream level was raised by steps of 5 cm.

3.2.1 *Sensitivity to submergence*

For the 3 PKW types, the submerged flow occurs when the downstream level exceeds the crest level and the modular submergence limit increases with the flow rate. For each PKW, we plotted the upstream head ratio (H_u/H_0) versus the downstream head ratio (H_d/H_0) measured at constant discharge Q_S. The upstream head for free flow H_0 is computed according to Q_S by Equation 1 and Equation 2 with the associated coefficients of Table 2. For type A (Fig. 3a), type B (Fig. 4a) and type C (Fig. 5a) the curves of H_u/H_0 versus H_d/H_0 were rather independent on the discharge. For $H_d > 4H_0$, the upstream head approaches the downstream head meaning that the PKW no more acts as a control structure. These results are consistent with those presented by Belaabed & Ouamane (2011) and Dabling & Tullis (2012).

The data were analysed to obtain a relationship between Q_s, H_u and H_d which allows to solve for the third parameter when two of them are known. Plotting the reverse normalized upstream head H_0/H_u (bounded by 1) versus the submergence factor S, the data sets were correlated by exponential interpolations given by the general Equation 3 associated to Table 3. According to this equation, the modular submergence limit S_m can be defined by $H_0 < 0.99H_u$ for $S > S_m$.

$$\frac{H_0}{H_u} = 1 - 0.01e^{\alpha(S - S_m)} \quad (3)$$

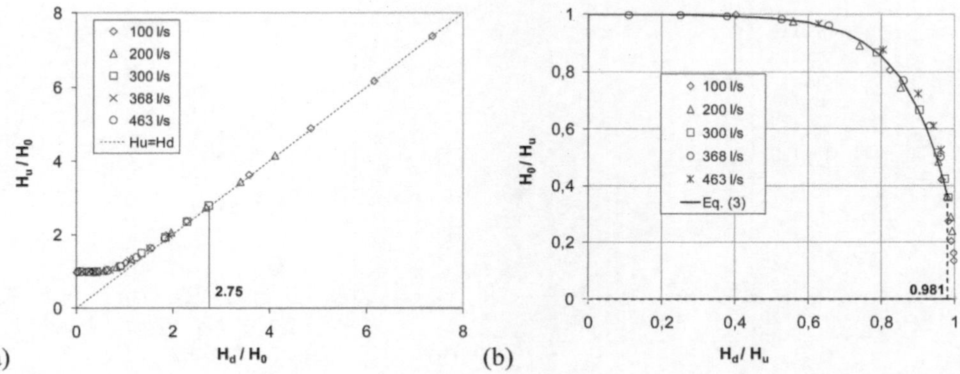

Figure 3. PKW Type A: dimensionless submerged head relationships measured at constant discharges: H_u/H_0 versus H_d/H_0(a), and H_0/H_u versus H_d/H_u (b).

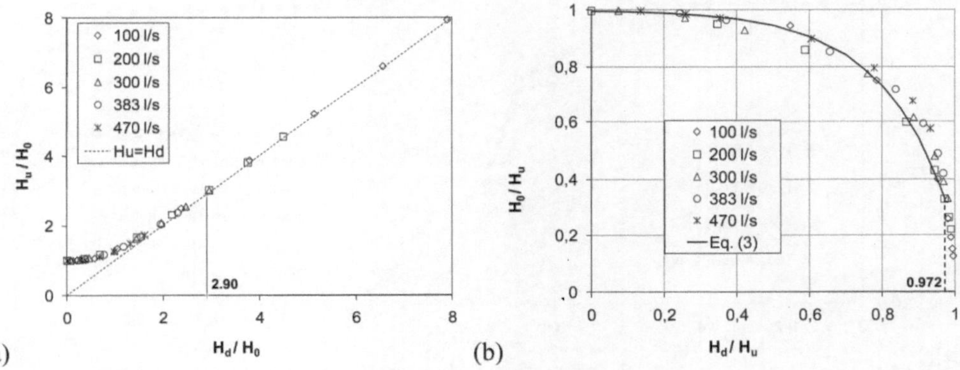

Figure 4. PKW Type B: dimensionless submerged head relationships measured at constant discharges: H_u/H_0 versus H_d/H_0(a), and H_0/H_u versus H_d/H_u (b).

Table 3. Coefficients of Equation 3 to predict the PKW sensitivity under submerged conditions.

PKW	α	S_m	validity	R^2	$\text{Error}_{\text{average}}$	$\text{Error}_{\text{max}}$
Type A	8.225	0.475	$0 < S < 0.981$	0.990	2.4%	14.6%
Type B	5.322	0.181	$0 < S < 0.972$	0.969	3.1%	21.0%
Type C	10.854	0.585	$0 < S < 0.959$	0.940	3.1%	15.4%

However, these equations cannot be applied when the submergence factor S tends to 1, since the discharge becomes undefined if $H_d/H_0 > 2.75$ for the type A (Fig. 3b), if $H_d/H_0 > 2.90$ for the type B (Fig. 4b) and if $H_d/H_0 > 2.40$ for the type C (Fig. 5b). The corresponding limitations of S associated to Equation 3 are respectively $S < 0.981$ for the type A, $S < 0.972$ for the type B and $S < 0.959$ for the type C.

The validity of Equation 3 was checked by comparing the computed and the measured values of H_0/H_u. Within the limitations of S, the ratio H_0/H_u can be predicted with an average error of 2% or 3% (Table 3). Then the discharge $Q_s(H_0)$ were predicted by Equation 2 for the measured values of H_u and H_d with an average error of 3% for types A and B, and of 5% for type C. Note that the error on H_u/H_0 could reach 15% to 21% because the correlation coefficients associated

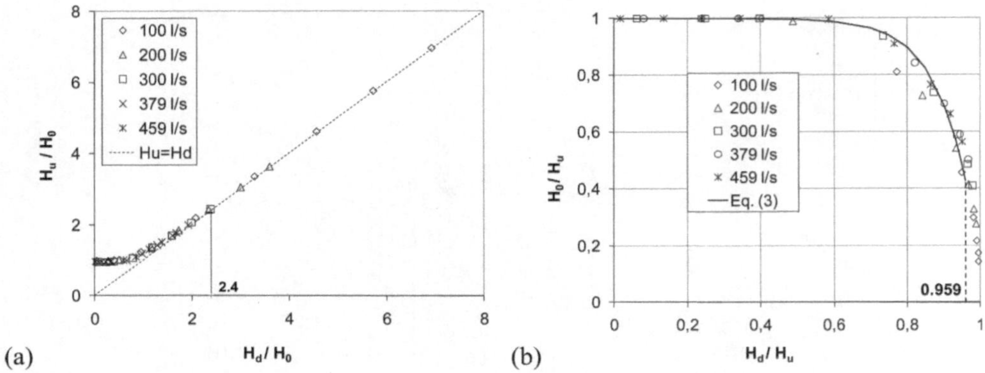

Figure 5. PKW Type C: dimensionless submerged head relationships measured at constant discharges: H_u/H_0 versus H_d/H_0(a), and H_0/H_u versus H_d/H_u (b).

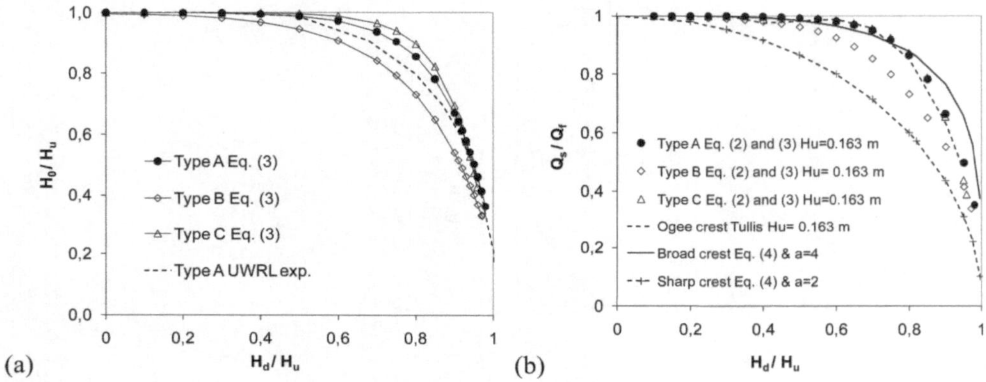

Figure 6. Comparison of the sensitivity to submergence of PKWs (a) and of PK and linear weirs (b).

to Equation 3 were rather low ($0.94 < R^2 < 0.99$) compared to the correlation coefficients ($R^2 \sim 1$) associated to Equation 2 for free-flow conditions.

The "sensitivity to submergence" of a PK or linear weir can be characterized by the upstream head increase H_u/H_0 for a constant discharge, or the discharge reduction Q_s/Q_f for the same upstream head, versus the submergence ratio H_d/H_u. Thus, the sensitivity of the 3 PKW types were compared (Fig. 6a) using Equation 3 with the associated coefficients of Table 3. The type C was found less sensitive than the type A, which was less sensitive than the type B.

The data of Dabling & Tullis (2012) were used to plot on Figure 6b the sensitivity of the type A PKW tested at UWRL. Both PKWs type A tested at UWRL and LNHE have a similar modular submergence limit ($S_m \sim 0.5$) but the LNHE model is less sensitive than the UWRL model. This could be due to the differences of geometries mentioned before, i.e. the ratios $W_i/W_o = 1.25$ (UWRL) and $P/T_S = 15.5$ (UWRL).

Hager (1986) published relationships of the discharge reduction Q_s/Q_f versus S for various linear weirs. The general Equation 4 can be used for both a sharp crested weir and a long broad crested weir with $a = 2$ and $a = 4$ respectively.

$$\frac{Q_s}{Q_f} = (1 - S^a)^{1/a} \tag{4}$$

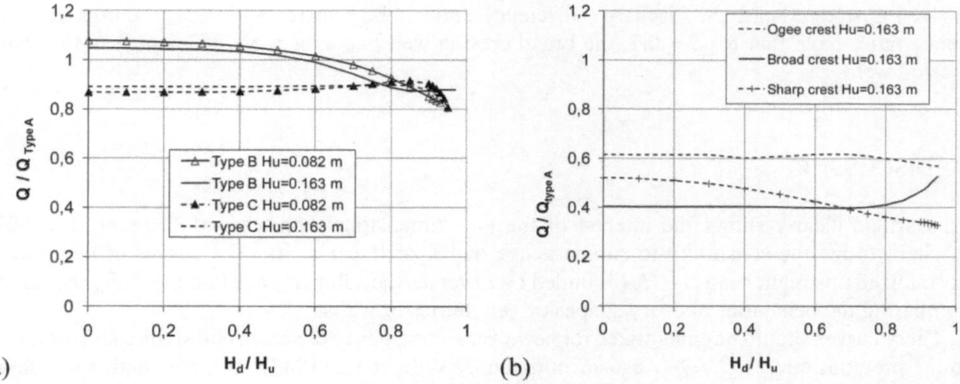

(a) (b)

Figure 7. Submerged discharge efficiencies of PKWs type B and type C (b) and of linear weirs (a), compared to the PKW type A under the same upstream and downstream heads.

For submerged Ogee crest, Tullis (2011) published charts of Q_s/Q_f versus S, depending on the free-flow discharge (Q_f) and the vertical distances (P_{uo}, P_{do}) from the upstream and downstream aprons to the weir crest. The Hager's Equation 4 for both the sharp ($a=2$) and the broad crested ($a=4$) weirs, and the Ogee crest data corresponding to the design discharge, $H_u/P_{uo}=0.46$ and $H_u/P_{do}=0.46$ were plotted on Figure 6b. The PKWs values reported on this figure, were computed by Equations 2 and 3, for $H_u=0.163$ m, to compare the PKWs and the Ogee crest at the same upstream head and same crest elevation (i.e. $P_{uo}=P_{do}=P+P_d=0.355$ m). As a result, the sharp crested weir is the most sensitive to submergence, and the broad crested weir is the least sensitive (less than the PKWs). For $H_u=0.163$ m, the Ogee crest weir and the PKWs of type A and C have the same sensitivity to submergence.

3.2.2 *Hydraulic efficiency under submerged flow conditions*

If H_d and H_u are known, the submerged Equation 3 can be solved for H_0 and the submerged discharge Q_s is a function of H_0 via Equation 2. If H_d and Q_s are known, Equation 2 can be solved for H_0 prior to computing H_u by Equation 3.

The first method was applied to calculate the discharges Q_s over the type A, B and C under the same submerged conditions H_d and H_u. Note that for each PKW type, the free-flow Equation 2 and the submerged Equation 3 were used on their validity domain. The results (Fig. 7a) clearly show that for a given upstream head, the submerged discharge efficiency depends on the submergence ratio $S=H_d/H_u$ and consequently on the downstream head. For the chosen values of H_u, type C is less efficient than type A, type B is more efficient than type A for $S>0.5$, and type B is more efficient than type C for $S<0.75$. Thus, the PKW performance under submergence cannot be classified as easily as for free-flow conditions since the discharge efficiency depends on both the "sensitivity to submergence" and the free-flow discharge efficiency.

The Hager's Equation 4 and the Ogee crest "sensitivity curve" presented in Figure 6b were used to compare the discharge efficiency of the linear weirs to the type A tested at LNHE. The weir discharges were computed for the same width as the PKWs ($W=1.99$ m), and for an Ogee crest of the same crest elevation ($P_{uo}=P_{do}=0.355$ m). The submerged (Q_s) and the free-flow (Q_f) discharges were computed for the same upstream head ($H_u=0.46P_{uo}=0.163$ m). For the linear weirs, Q_f was computed by Equation 1 with $C_W=0.327$ (minimum value for long broad crested weir), $C_W=0.42$ (average value for sharp crested weir) and $C_W=0.495$ (value at design head for Ogee crest weir).

The results (Fig. 7b) show that the PKW type A is 40% more efficient than the Ogee crest, 50% to 70% more efficient than the sharp crested weir, and 50% to 60% more efficient than the broad crested weir. Thus, the type B and C will be more efficient than linear weirs. As for PKWs, for a given upstream head, the discharge efficiency under submergence will depend on the submergence ratio. Note that for $S > 0.7$, the broad crested weir becomes more efficient than the sharp crested weir.

4 DISCUSSION

This article clearly shows the interest of the non-dimensional approach of Tullis et al. (2007) to characterize the sensitivity to submergence of PK or linear weirs. The curves of the reverse normalized upstream head H_0/H_u (bounded by 1) versus the submergence factor $S = H_d/H_u$, allow comparing the behaviour of every shapes or geometries of weirs.

These curves should be generalized for new weir submergence experimental studies and preferred to the previous curves Q_s/Q_f versus S initiated by Villemonte (1947). Actually, both approaches need the free-flow head-discharge characteristics to estimate the submerged discharge Q_s, and both approaches enable to solve for Q_s if H_d and H_u are known. But, the Tullis approach also enables to solve for H_u (resp. H_d) if H_d (resp. H_u) and Q_s are known whereas it is not possible with the Villemonte approach.

5 CONCLUSIONS

The testing of the 3 PKWs types A, B and C in the 2 m wide channel at the LNHE allows isolating the overhang effects on the discharge efficiency under free flow and submerged conditions, for the geometries recommended by Lemperiere et al. (2011), with $W_i/W_o = 1$ and $P/T_s \sim 10$.

Under free-flow conditions, the discharge of the types A, B and C can be predicted by Equation 2 and Table 2 with a maximum error of 1.5%. The type B is 10% to 5% more efficient than the type A, whereas the type C is 15% less efficient. The type D (without overhangs) tested at UWRL (with $W_i/W_o = 1.25$ and $P/T_s \sim 15$) is also less efficient than the type A but more efficient than the type C.

The submerged flow data were analysed applying the non-dimensional approach of Tullis (2007). The "sensitivity to submergence" of the types A, B and C were characterized by Equation 3 and Table 3. The modular submergence limit seemed to increase with the length of the downstream overhang since it was observed $S_m \sim 0.2$ for the type B, $S_m \sim 0.5$ for the type A, and $S_m \sim 0.6$ for the type C. The type C was shown to be less sensitive to submergence than the type A, which was less sensitive than the type B.

The discharge efficiencies of the 3 types of PKWs were compared under the same submerged conditions (H_u and H_d). Type C was less efficient than type A for $0 < H_d < 0.9H_u$, whereas type B was more efficient than type A for $H_d < 0.5H_u$, and type B was more efficient than type C for $H_d < 0.8H_u$.

In addition, the 3 types of PKWs were compared to linear weirs of same width and crest elevation as the PKWs, for the same submerged conditions $H_u \sim 0.7$P and $0 < H_d < 0.9H_u$:

- The PKWs were more sensitive to submergence than the long broad crested weir, but less sensitive than the sharp crested weir. The Ogee crest weir had the same sensitivity as the types A and C.
- The linear weirs were much less efficient than the PKWs. The loss in discharge relative to the type A was around 40% for the Ogee crest weir, from 50% to 70% for the sharp crested weir, and from 50% to 60% for the broad crested weir.

REFERENCES

Anderson, R.M. & Tullis, B.P. 2011. Influence of Piano Key Weir geometry on discharge. *Proc. International Workshop on Labyrinths and Piano Key Weirs PKW 2011, CRC Press,* 75–80.

Belaabed, F. & Ouamane, A. 2011. Contibution to the study of the Piano Key Weir submerged by the downstream level. *Proc. International Workshop on Labyrinths and Piano Key Weirs PKW 2011, CRC Press,* 89–96.

Dabling, M.R. & Tullis, B.P. 2012. Piano Key Weir submergence in channel applications. *Proc. International Workshop on Piano Key Weir for In-Stream Storage and Safety, New Delhi, India.* 13–23.

Hager, W.H. 1986. Discharge Measurements Structures. *Communication1,R. Sinniger, ed. Chaire de Constructions hydrauliques,* Dep. de Génie Civil. EPFL, Lausanne (Switzerland).

Ho Ta Khanh, M., Chi Hien, T. & Thanh Hai, N. 2011. Main results of the P.K. weir model tests in Vietnam (2004 to 2010). *Proc. International Workshop on Labyrinths and Piano Key Weirs PKW 2011, CRC Press,* 191–198.

Kabiri-Samani, A. & Javaheri, A. 2012. Discharge coefficient for free and submerged flow over Piano Key Weirs. *Journal of Hydraulic Research,* 50(1): 114–120.

Leite Ribeiro, M. Pfister, M., Schleiss, A.J. & Boillat, J.L. 2012. Hydraulic design of A-type Piano Key Weirs. *Journal of Hydraulic Research,* 50(4): 400–408.

Lemperiere, F., Vigny, J.P. & Ouamane, A. 2011. General comments on Piano Kew Weirs: The past and present. *Proc. International Workshop on Labyrinths and Piano Key Weirs PKW 2011, CRC Press,* 17–25.

Ouamane, A. & Lemperiere, F. 2006. Design of a new economic shape of weir. *Proceedings of the International Symposium on Dams in the Societies of the 21st Century, Barcelona.*

Ouamane, A. 2011. Nine years of study of the Piano Key Weir in the university laboratory of Biskra "lessons and reflections". *Proc. International Workshop on Labyrinths and Piano Key Weirs PKW 2011, CRC Press,* 51–58.

Pralong, J., Vermeulen, J., Blancher, B., Laugier, F., Erpicum, S., Machiels, O., Pirotton, M., Boillat, J.L., Leite Ribeiro, M. & Schleiss, A.J. 2011. A naming convention for the Piano Key Weirs geometrical parameters. *Proc. International Workshop on Labyrinths and Piano Key Weirs PKW 2011, CRC Press,* 271–278.

Tullis, B.P., Young, J.C. & Chandler, M.A. 2007. Head-Discharge Relationships for Submerged Labyrinth Weirs. *Journal of Hydraulic Engineering,* 133(3): 248–254.

Tullis, B.P. 2011. Behavior of Submerged Ogee Crest Weir Discharge Coefficients. *Journal of Irrigation and Drainage Engineering,* 133(3): 677–681.

Vermeulen, J., Laugier, F., Faramond, L. & Gilles, C. 2011. Lessons learnt from design and construction of EDF first Piano Key Weirs. *Proc. International Workshop on Labyrinths and Piano Key Weirs PKW 2011, CRC Press,* 215–224.

Villemonte, J.R. 1947. Submerged-Weir Discharge Studies. *Engineering News-Record,* 139(26): 54–56.

Labyrinth and Piano Key Weirs II – PKW 2013 – Erpicum et al. (eds)
© 2014 Taylor & Francis Group, London, ISBN 978-1-138-00085-8

Comparative analysis of the energy dissipation on a stepped spillway downstream of a Piano Key Weir

A. Silvestri, P. Archambeau, M. Pirotton, B. Dewals & S. Erpicum
Research group HECE, ArGEnCo Department, University of Liege, Liege, Belgium

ABSTRACT: A systematic experimental study has been carried out to compare the residual energy at the toe of a stepped spillway of varied length considering a classical ogee-crested weir (OCW) or two different piano key weirs (PKW) at the top of the structure. Four spillway lengths have been considered on a wide range of discharge. An indirect method has been used to compute the residual flow energy at the spillway toe from water depth measurements in a horizontal channel downstream. The results indicate that uniform flow conditions, i.e. flow energy at the spillway toe independent of the spillway length, are reached on significantly shorter spillways downstream of a PKW than downstream of an OCW. However, the uniform flow energy at constant discharge on the spillway is found to vary depending on the weir type. In particular, it is smaller downstream of the OCW than downstream of the PKWs.

1 INTRODUCTION

Stepped spillways are used in dam construction to dissipate flow energy along a spillway, and thus to reduce the length or eliminate the need for a stilling basin at the dam toe. By providing natural aeration, stepped spillways also decrease the risk of cavitation for small specific discharges. They are usually built downstream of standard ogee-crested weirs (OCW).

Such spillways are generally designed to create skimming flow regime, i.e. for water to flow as a coherent stream over the pseudo bottom formed by the outer step edges, without air pockets under the jets and with development of intense recirculating zones in the triangular cavities formed by the step faces and the pseudo bottom (Boes & Hager, 2003). The vortices, maintained by the transmission of shear stress from the coherent stream over the pseudo bottom, contribute significantly to the energy dissipation for this type of regime.

In skimming flow regime, as depicted by Chanson (1994), the free-surface of the flow is smooth in the early steps and no air entrainment occurs. Close to the bottom, turbulence is generated. When the outer edge of the turbulent boundary layer reaches the free surface, air entrainment commences. Downstream of this point of inception of air entrainment, the flow becomes rapidly aerated and the mixture of air and water extends gradually through the fluid to the pseudo bottom. Farther downstream, the flow becomes uniform. From this moment, air concentration, average flow velocity and water depth remain nearly constant along the spillway. The flow energy is thus also constant, as equilibrium between the dissipation rate and the gravity effect has been reached.

A criterion has been proposed by Boes et Hager (2003) to predict the minimum vertical distance $H_{dam,u}$ from spillway crest needed to reach uniform flow depending on the critical depth h_c and the slope ϕ of the spillway:

$$\frac{H_{dam,u}}{h_c} = 24(\sin\phi)^{2/3} \tag{1}$$

In uniform flow conditions, Boes and Hager (2003) propose to evaluate the energy $E_{f,u}$ at the spillway end using the following equation

$$\frac{E_{f,u}}{E_{\max}} = \frac{F}{\frac{H_{dam}}{h_c} + \frac{3}{2}} \tag{2}$$

where E_{max} is maximum reservoir energy head, H_{dam} is the dam height and F is a parameter related to the bottom roughness. This equation, derived from experimental results gained on a stepped spillway equipped with a jet box upstream (no weir), should be independent of the type of weir upstream of the stepped spillway. Indeed, uniform flow characteristics are only related to the discharge and the spillway geometry and materials.

Labyrinth and Piano Key weirs (PKW) are known to create basically three dimensional flows and to entrain air at the base of the structure, with formation of different interacting jets, shockwaves and spray (Lopes et al., 2011; Erpicum et al., 2011; Ho Ta Khanh et al., 2011). These observations suggest that the use of a PKW to control the flow upstream of a stepped spillway may shorten the length needed to develop a fully aerated flow and reach uniform flow condition compared to classical linear weirs.

Following preliminary tests carried out by Erpicum et al. (2011), an experimental study has been realized at the Laboratory of Engineering Hydraulics, University of Liège, to compare, in an idealized environment, the energy dissipation on a stepped spillway downstream of a PKW and downstream of a standard OCW. This paper presents the general results of this study.

2 EXPERIMENTAL FACILITY

The experimental facility previously used by Erpicum et al. (2011) had a relatively long spillway. As a consequence, previously tested H_{dam}/h_c ratios ranged from 16.1 to 58.8 and uniform flow conditions were reached at the spillway toe for almost the whole range of tested discharges (Erpicum et al., 2011). For the present study, this facility has been modified to reduce the spillway length. Four lengths have been considered. They were equal to 0.64 m (position I), 1.07 m (position II), 1.29 m (position III) and 1.71 m (position IV), respectively.

The modified facility considered in the present study (Fig. 1) is made of a 0.494 m wide stepped spillway with a slope of 52° and regular steps of 2.4 cm in length and 3 cm in height. The spillway is linked to an upstream reservoir and a 4.20 m long downstream horizontal channel. All the walls are made of steel, PVC or Acrylic in order to minimize sidewalls friction effects. The upstream reservoir is fed by a regulated pump connected to a pressurized pipes network. The downstream extremity of the horizontal channel is equipped with an adjustable vertical sill.

Discharge was measured in the upstream pipe using an electromagnetic flow meter (accuracy of 1% at full span). The water depths in the upstream reservoir and along the horizontal channel have been measured using several ultrasonic probes (accuracy of +/−2%). Because of small free surface oscillations, water depth measurements have been averaged for a period of 90 s to define a single water depth value in each point (Silvestri et al., 2013).

Three different weirs have been tested upstream of the stepped spillway: a standard OCW (USBR profile with a design head of 216 mm) and two geometries of PKW, noted PKW_1 and PKW_2 (Erpicum et al., 2011). The dimensions of the PKW_2 are generally 1.6 times smaller than PKW_1, except the spillway width and the wall thickness which are constant. PKW_1 represents 1.5 inlets and 1.5 outlets while PKW_2 represents 2.5 inlets and 2.5 outlets. They are both "modified type-A PKWs", i.e. with downstream and upstream overhangs of different lengths.

Tested discharges ranged from 10 l/s to 80 l/s (discharge per unit length q from 0.02 $m^3/s/m$ to 0.16 $m^3/s/m$).

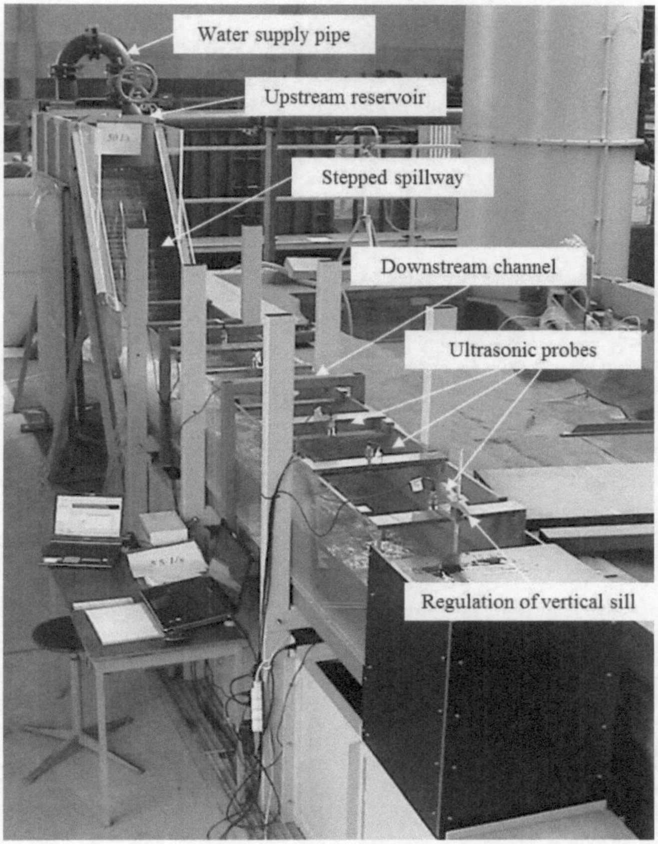

Figure 1. Experimental facility.

3 METHODOLOGY

The experiments depicted in this paper consisted in measuring and comparing, for constant discharges, the residual energy E_t at the spillway toe depending on the type of upstream weir and spillway length.

Assuming a constant flow velocity on the cross section at the spillway toe, the flow energy can be evaluated directly from the water depth h_t and discharge per unit length q as

$$E_t = h_t + \frac{(q/h_t)^2}{2g} \tag{3}$$

Because of air entrainment and high flow velocity at the spillway toe, the direct measurement of the water depth at this location is difficult. An indirect method has therefore been used to evaluate the water depth h_t at the spillway toe, as depicted by Matos & Quintela (2004), Shvainshtein (1999) and Erpicum et al. (2011). The method consists in creating a hydraulic jump in the downstream horizontal channel by means of the downstream sill adjustment, to measure the water depth h_2 downstream of the jump (second conjugate depth), where flow depth varies less and aeration rate is smaller, and to calculate the corresponding upstream water depth h_1 using the conjugate depths equation (Silvestri et al., 2013). To avoid the need for the evaluation of friction losses in the horizontal channel, the hydraulic jump has been located exactly at the spillway toe, at the same position for all tests (Fig. 2). Consequently, the water depth h_t at the spillway toe is the computed value h_1.

Figure 2. Location of the hydraulic jump at the spillway toe.

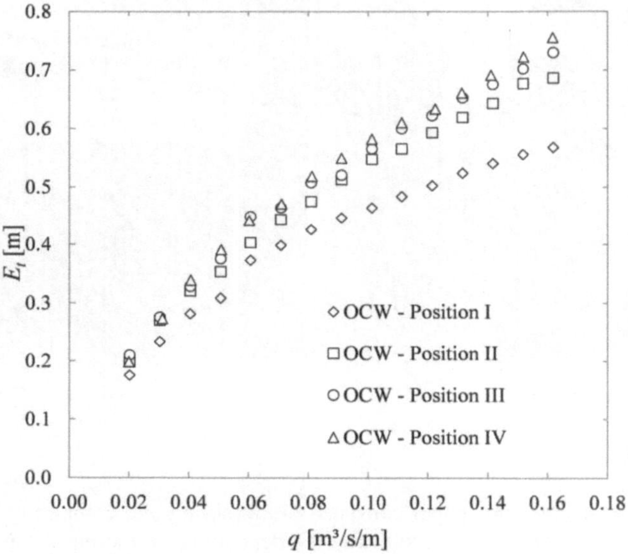

Figure 3. Residual energy E_t at the spillway toe as a function of the discharge per unit length q – OCW.

With respect to the measurements accuracy, the repetition of the same tests shows a variation of +/−2% in the value of the flow energy computed at the spillway toe from measured h_2 values.

4 RESULTS

Figure 3 to Figure 5 show, for each weir, the variation of the residual energy E_t at the spillway toe as a function of the specific discharge q, considering the four lengths of the spillway.

Typically, the residual flow energy at the spillway toe increases with the discharge and the length of the spillway. For the smallest spillway length (position I), the alignment of almost the whole of the experimental points for PKW_1 (biggest dimensions – Fig. 4) and of the points corresponding to higher discharges for PKW_2 (smallest dimensions – Fig. 5) is not similar to the one observed for other results. These points correspond to tests where the flows coming out of the PKW inlets were impacting the spillway close to the spillway toe (Fig. 6). As a consequence, spray entered the

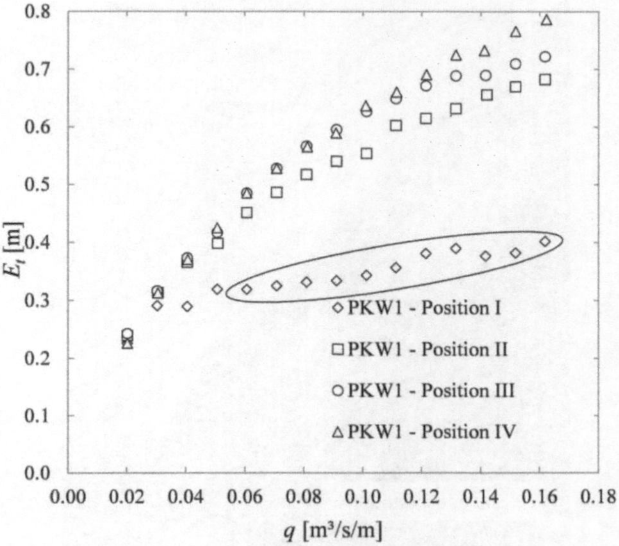

Figure 4. Residual energy E_t at the spillway toe as a function of the discharge per unit length q – PKW_1.

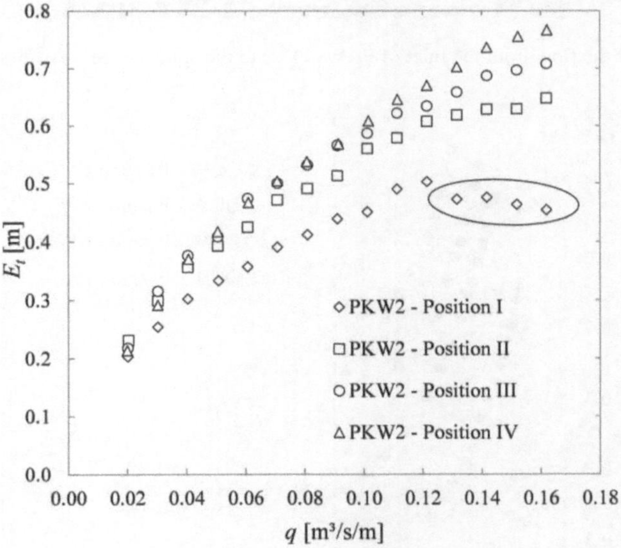

Figure 5. Residual energy E_t at the spillway toe as a function of the discharge per unit length q – PKW2.

hydraulic jump and affected the positioning of the latter, so that the experimental results are probably wrong. These specific points have thus not been considered in the following steps of the analysis.

Similar values of E_t are obtained for constant discharges and varied spillway length (small discharges and important lengths) depending on the type of weir (Fig. 7 and 9). For these discharges and spillway lengths, uniform flow conditions have been reached.

From these results, the critical value provided by equation 1 for the relative dam height necessary to reach uniform flow conditions downstream of an OCW seems adequate ($H_{dam}/h_c = 20.5$ – Fig. 7). Critical dam height to reach uniform flow condition downstream of PKWs is undoubtedly lower (around 17.5 – Fig. 8 and Fig. 9).

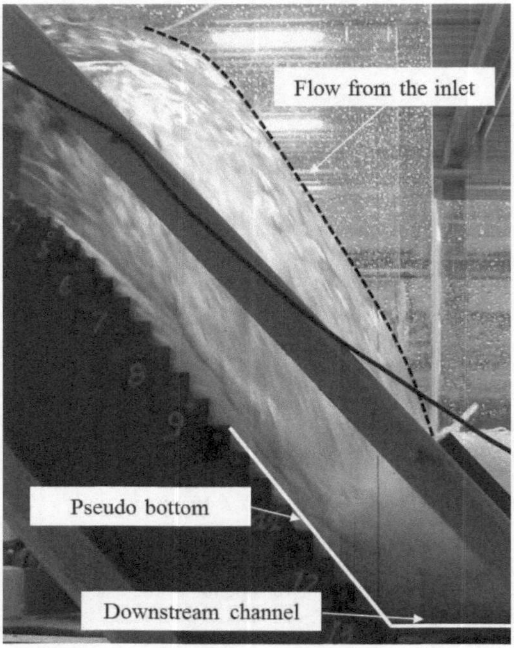

Figure 6. Impact of the flow from the inlet of PKW_2 close to the spillway toe – Position I and discharge of 0.13 $m^3/s/m$.

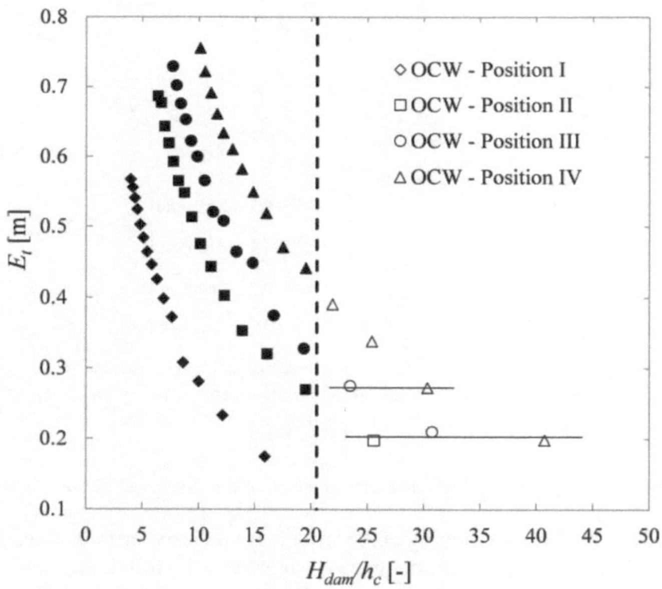

Figure 7. Residual energy E_t at the spillway toe as a function of the relative dam height H_{dam}/h_c – OCW.

5 DISCUSSION

From the ogee crested weir results in uniform flow conditions, it is possible to compute the local head loss due to the slope transition between the spillway and the downstream horizontal channel.

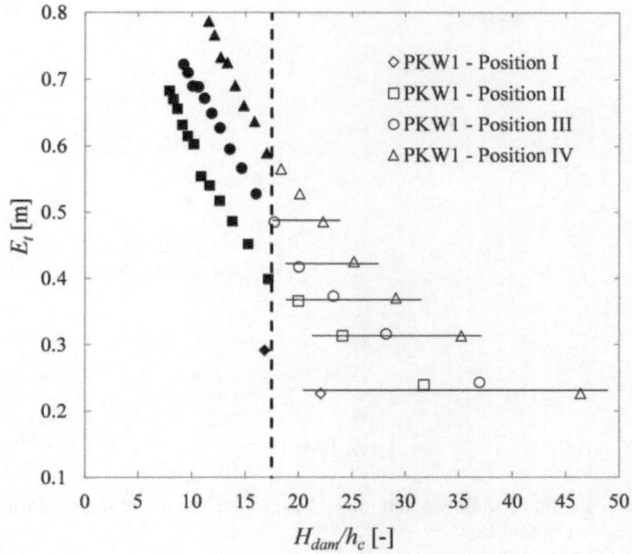

Figure 8. Residual energy E_t at the spillway toe as a function of the relative dam height H_{dam}/h_c – PKW_1.

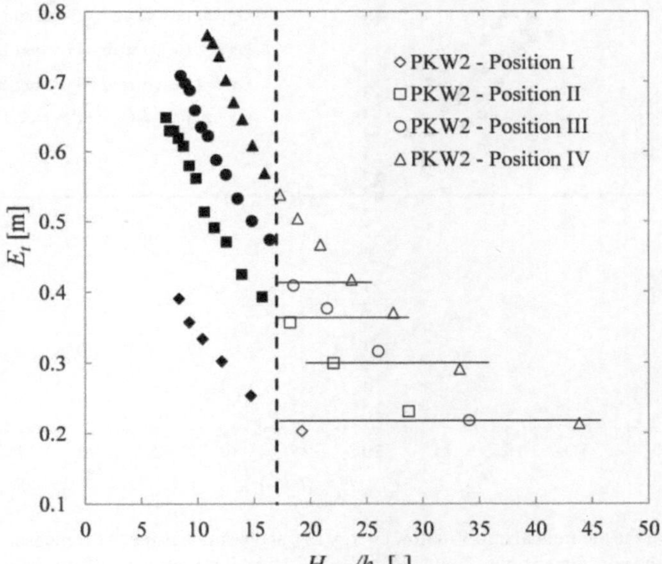

Figure 9. Residual energy E_t at the spillway toe as a function of the relative dam height H_{dam}/h_c – PKW_2.

Indeed, in this specific flow conditions, the local head loss is the difference between the residual energy E_p at the spillway toe and the theoretical uniform flow energy $E_{f,u}$ at the spillway end predicted using equation (2). Assuming a classical form for the local head loss equation computed as a function of the downstream flow velocity, a constant value of 0.787 is found for the local head loss coefficient k, with h_V the kinetic term of the downstream flow (Fig. 10).

$$\Delta E_{toe} = k\frac{(q/h_t)^2}{2g} = k\,h_V \tag{4}$$

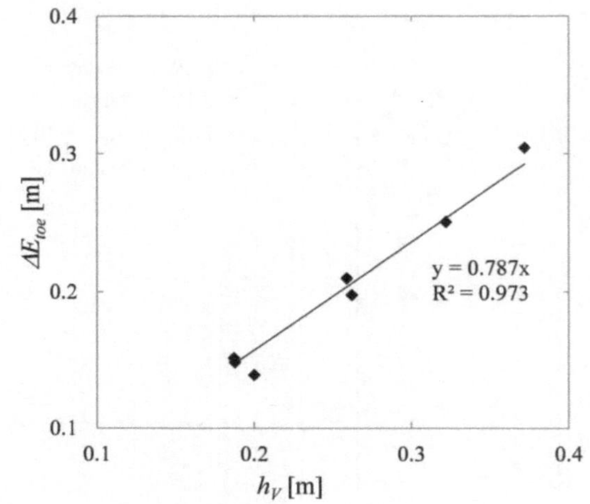

Figure 10. OCW results in uniform flow conditions – Local head loss at the slope transition as a function of the kinetic term of the downstream flow.

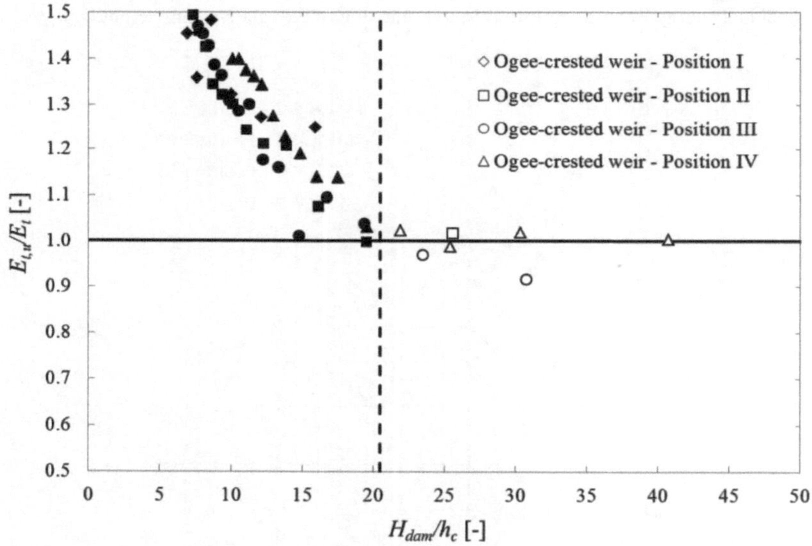

Figure 11. Ratio of the theoretical uniform flow energy $E_{t,u}$ at the spillway toe to the measured residual energy E_t as a function of the non-dimensional dam height H_{dam}/h_c – White symbols: uniform flow conditions; black symbol: non-uniform flow conditions defined from Figure 7 – OCW.

Considering the same local head loss equation and constant k value whatever the flow conditions on the spillway, the theoretical uniform flow energy $E_{t,u}$ at the spillway toe can be computed for all the geometries of the upstream weir and spillway lengths as the sum of the theoretical uniform flow energy $E_{f,u}$ at the spillway end provided by equation 2 and the local head loss at the spillway toe (Eq. 4). The comparison of these theoretical energy values to the measured values is shown on Figure 11 to Figure 13.

With the OCW, the energy ratio $E_{t,u}/E_t$ tends logically to one for longer spillways. For both PKWs however, the energy ratio $E_{t,u}/E_t$ becomes lower than 1 for important spillway lengths. Uniform

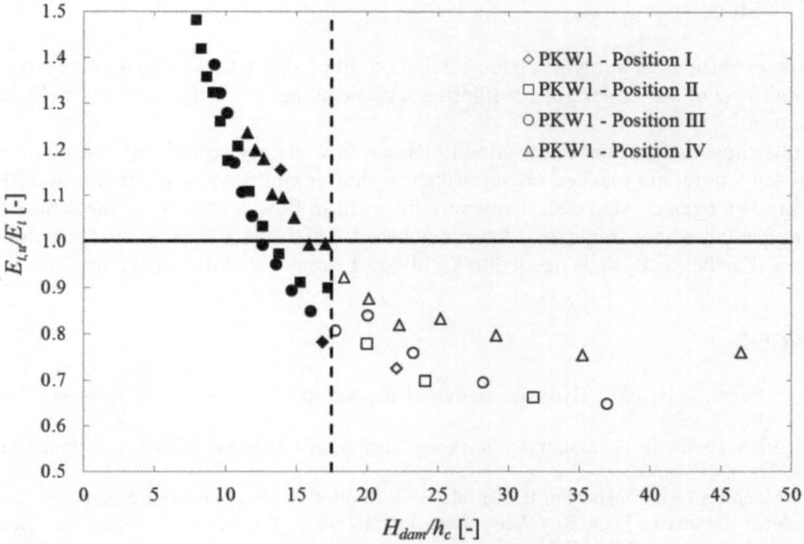

Figure 12. Ratio of the theoretical uniform flow energy $E_{t,u}$ at the spillway toe to the measured residual energy E_t as a function of the non-dimensional dam height H_{dam}/h_c – White symbols: uniform flow conditions; black symbol: non-uniform flow conditions defined from Figure 8 – PKW_1.

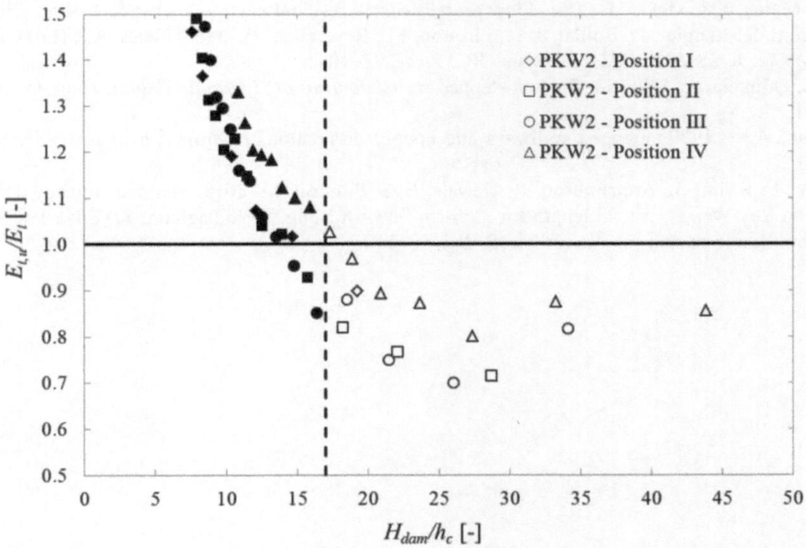

Figure 13. Ratio of the theoretical uniform flow energy $E_{t,u}$ at the spillway toe to the measured residual energy E_t as a function of the non-dimensional dam height H_{dam}/h_c – White symbols: uniform flow conditions; black symbol: non-uniform flow conditions defined from Figure 9 – PKW_2.

flow characteristics and/or local head loss are thus not the same as with the OCW. In particular, at constant discharge, the uniform flow energy on the stepped spillway is higher downstream of the PKWs than downstream of the OCW. This may be explained by non-uniform discharge repartition over the spillway width downstream of PKWs or an increased level of turbulence. Further analyses are needed to highlight the reasons of this difference.

6 CONCLUSION

A systematic experimental study has been carried out to compare the residual energy at the toe of a stepped spillway of varied length considering a classical ogee-crested weir or PKWs at the top of the structure.

The results show that uniform flow conditions, i.e. flow energy at the spillway toe independent of the spillway length, are reached on significantly shorter spillways downstream of a PKW than downstream of an ogee-crested weir. However, the uniform flow energy is not the same depending on the weir type. It has been found to be lower downstream of the ogee-crested weir than downstream of the PKWs. Further analysis is needed to highlight the reasons of this difference.

REFERENCES

Boes, R.M. & Hager, W.H. 2003. Hydraulic design of stepped spillways. *Journal of Hydraulic Engineering ASCE.*, 129(9), 671–679.

Chanson, H. 1994. Hydraulics of skimming flows over stepped channels and spillways. *Journal of Hydraulic Research*, 32(3), 445–460.

Erpicum S., Machiels O., Archambeau P., Dewals B. & Pirotton M. 2011. Energy dissipation on a stepped spillway down-stream of a Piano Key Weir – Experimental study. In Erpicum, S., Laugier, F., Boillat, J.-L., Pirotton, M., Reverchon, B. and Schleiss, A.J. (Eds), *Labyrinth and piano key weirs-PKW 2011*, London: CRC Press, 105–111.

Ho Ta Khanh, M., Truong, C.H. & Nguyen, T.H. 2011. Main results of the PK Weir model tests in Vietnam (2004 to 2010). In Erpicum, S., Laugier, F., Boillat, J.-L., Pirotton, M., Reverchon, B. and Schleiss, A.J. (Eds), *Labyrinth and piano key weirs-PKW 2011*, London: CRC Press, 191–198.

Lopes, R., Matos, J. & Melo, J.F. 2011. Flow properties and residual energy downstream of labyrinth weirs. In Erpicum, S., Laugier, F., Boillat, J.-L., Pirotton, M., Reverchon, B. and Schleiss, A.J. (Eds), *Labyrinth and piano key weirs-PKW 2011*, London: CRC Press, 97–104.

Matos, J. & Quintela, A. 1995. Jet flow on stepped weirs. *Journal of Hydraulic Engineering, ASCE*, 121(5), 443–444.

Shvainshtein, A.M. 1999. Stepped spillways and energy dissipation. *Hydrotechnical construction*, 33(5), 275–282.

Silvestri, A., Erpicum, S., Archambeau, P., Dewals, B. & Pirotton, M. 2013. Stepped spillway downstream of a Piano Key Weir – Critical length for uniform flow. In Bung, D. & Pagliara, S. (Eds), IWLHS 2013, Karlsruhe: Bundesanstalt für Wasserbau (BAW), 99–110.

C. Hydraulic and structural design

Labyrinth and Piano Key Weirs II – PKW 2013 – Erpicum et al. (eds)
© 2014 Taylor & Francis Group, London, ISBN 978-1-138-00085-8

Comparison of Piano Key Weirs with labyrinth and gated spillways: Hydraulics, cost, constructability and operations

G.S. Paxson
Schnabel Engineering, West Chester, Pennsylvania, USA

B.P. Tullis
Utah Water Research Laboratory, Utah State University, Logan, Utah, USA

D.J. Hertel
Engineering Solutions, LLC, Bozeman, Montana, USA

ABSTRACT: Similar to labyrinth weirs, piano key weirs (PKW) are folded in plan to increase discharge capacity for a given spillway channel width. Because of their configuration, PKW are better suited than labyrinth weirs for applications where the weir footprint (length and/or width) dimensions are restricted (e.g., crest of gravity dams). PKW can facilitate a significant amount of weir length relative to their footprint size. For channel applications without significant footprint restrictions, labyrinth and PKW may both represent viable spillway options. Gated spillways can typically pass a higher unit discharge than passive flow control structures (e.g., labyrinth and PKW) due to the resulting low crest elevation that occurs when the gates are opened. Gated spillways are effective at passing flood events but rely on manual or automated operation. This can be particularly challenging for manually operated gates at remote dam sites. Labyrinth, PK, and gated weirs all have their place in spillway design. This paper evaluates some of the advantages and disadvantages of each in channel applications for dam rehabilitation with respect to economic, constructability, operational, and hydraulic considerations. The cost estimates included in the evaluation consider the complexity of each structure and the application to different constraints. Comparisons are limited to a single PKW geometry.

1 INTRODUCTION

The piano key weir (PKW) concept was developed by Ouamane of the University of Briska (Algeria) and Lempérière of Hydrocoop in France (Lempérière and Ouamane, 2003). Similar to labyrinth weirs, PKW are folded in plan to increase discharge; however, the PKW has a smaller foundation footprint than a labyrinth weir, due to their cantilevered apex geometry, making them a more practical alternative for applications where the footprint is limited (e.g., crest of gravity dams). More than 100 PKW model studies have been performed (Lempérière, 2009) and the first prototype weirs were constructed in France at Goulors and Saint-Marc dams in 2006 and 2008, respectively (Laugier, 2007, Laugier et al., 2009).

The typical PKW has a rectangular crest layout (in plan) with inclined inlet and outlet cycle floors that are cantilevered beyond the spillway foundation footprint. Figure 1 shows a typical PKW geometry.

In the USA, labyrinth weirs are relatively common; however, the authors are not aware of any PKW installations. The PKW may prove to be a comparable alternative for both new dams and the rehabilitation of existing dams with inadequate spillway capacity to meet design flood requirements.

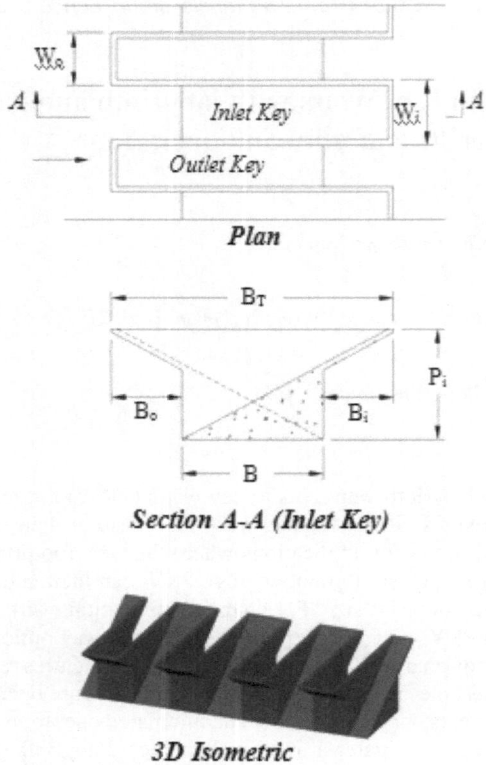

Figure 1. PKW geometry.

2 HYDRAULICS

2.1 *Overview of PKW hydraulics*

Similar to labyrinth weirs, PKW are hydraulically efficient at low upstream heads relative to the weir height and this efficiency decreases with increasing upstream head. Anderson and Tullis (2013) performed model studies for numerous PKW geometries and developed discharge coefficients (C_d) for estimating discharge (Q) using Eq. (1). An alternative head-discharge relationship [Eq. (2)] has been suggested by Parlong et al. (2011).

$$Q = 2/3C_d\sqrt{2g}LH_u^{3/2} \quad (1)$$

$$Q = C_{dW}W\sqrt{2g}H_u^{3/2} \quad (2)$$

In the Eqns. (1) and (2), g = gravitational constant; L = total length of weir crest; W is the cannel width, C_d and C_{dW} are discharge coefficients that apply to Eqs (1) and (2) respectively, and H_u = total upstream head measured relative to the weir crest, sometimes referred to as H_t.

For this paper, the PKW geometry presented in Anderson and Tullis (2013) with an inlet to outlet key width ratio of 1.25 (PK1.25) was selected for evaluation and comparison with other spillway types. Figure 2 shows the relationship between C_d and the dimensionless upstream head ratio (H_t/P, where P represents the weir height).

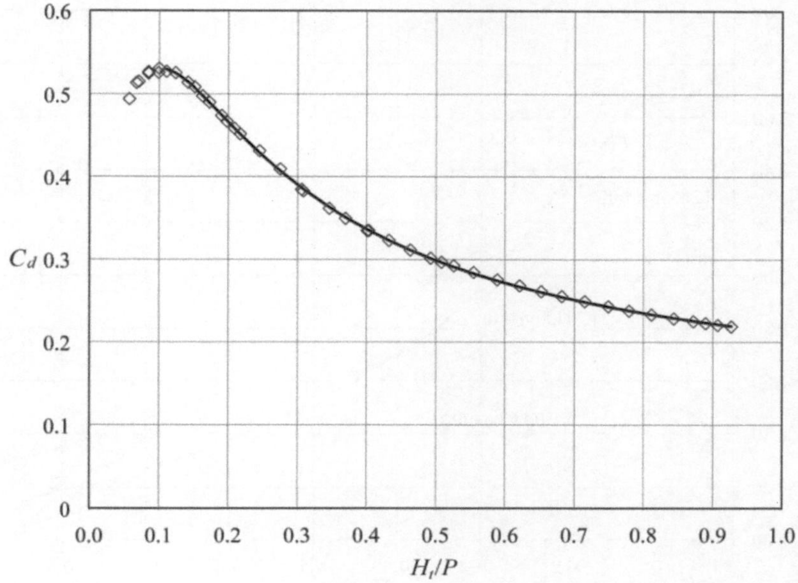

Figure 2. Discharge coefficient for PK1.25: inlet-to-outlet key width ratio of 1.25 (from Anderson and Tullis, 2013).

2.2 *Comparison with labyrinth weir hydraulics*

The hydraulic performance of a selected PKW geometry can be compared to labyrinth weir performance as presented in Crookston and Tullis (2013). For this comparison, the cycle efficiency concept introduced by Crookston and Tullis (2013) can be applied. Cycle efficiency is representative of the discharge per cycle (labyrinth) or unit (PKW) and could also be considered unit discharge (discharge per unit width of spillway). Figure 3 shows this comparison for labyrinth weirs with a quarter round crest shape and sidewall angles between 10 and 20 degrees (d). For this example, the PK1.25 has an L/W ratio of 5.1 compared to L/W ratios for the labyrinth weirs ranging from 4.8 for the 10 degree labyrinth to 2.5 for the 20 degree labyrinth.

Figure 3 demonstrates the efficiency of the PKW at low H_t/P ratios and shows that for H_t/P ratios greater than about 0.4, the efficiency of the PK1.25 weir is comparable to a labyrinth weir with a sidewall angle of 15 degrees. In other words, for a given total spillway width, a PK1.25 weir would have similar discharge capacity to a labyrinth weir with a 15 degree sidewall angle for a given weir height and total upstream head.

2.3 *Comparison with gated spillway hydraulics*

Spillway gates can be categorized as underflow or overflow; underflow gates lift up and water flows beneath the gate and over the fixed crest of the spillway. For overflow gates, the gate crest lowers to the fully open position (flow over fixed crest).

Because of their ability to provide reservoir storage volume above the spillway fixed crest elevation, gated spillways can provide more discharge capacity than fixed crest weirs due to the additional head over the spillway when the gates are opened.

PKW can be compared to a gated spillway having a weir length equal to the total width of the PKW by assuming that the gates are completely opened for the design head. To accomplish this, the fixed crest elevation and gate height were set to match the discharge capacity of a given PKW geometry. The gate height (P_{gate}) is computed as a function of the PKW height (P_i) and plotted

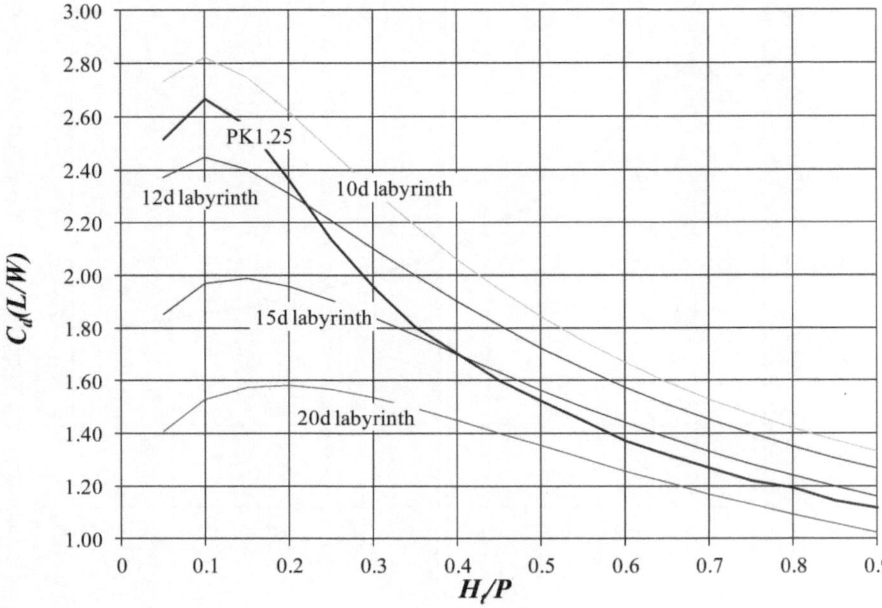

Figure 3. Cycle Efficiency versus H_t/P for quarter round labyrinth weirs and PKW with inlet to outlet key ratio of 1.25.

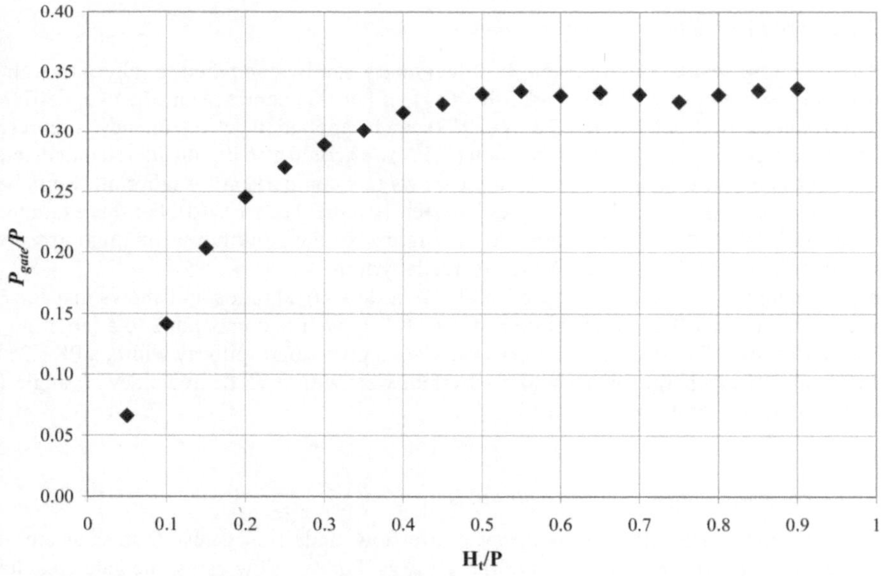

Figure 4. Gate height as a function of PKW height versus H_u/P_i for a gated spillway with fixed crest weir coefficient equal to 0.71 and a PKW with an inlet to outlet ratio of 1.25.

versus H_t/P_i, assuming the gates are fully open (i.e., the head over the fixed crest of the gated spillway is equal to $H_u + P_{gate}$). Figure 4 shows this relationship for a gated spillway with a fixed crest discharge coefficient (C_d) equal to 0.71 and the PK1.25 weir.

Figure 4 illustrates that for the given spillway width, gated spillway discharge coefficient ($C_d = 0.71$) and PKW geometry (PK1.25), a gate height equal to about one-third of the PKW height would be required to pass the same discharge for design water surface elevations resulting in H_t/P_i equal to 0.5 and higher.

3 CONSTRUCTION COST

Estimated construction costs for spillways with comparable hydraulic performance can be compared to assist in the selection of a spillway type. The comparative costs presented herein are based on a simplified model to demonstrate the potential cost savings associated with PKW construction where applicable. Because spillway foundations can vary significantly, those elements were not compared in this model. Foundation elements could include foundation improvements, anchors, demolition, and other measures.

3.1 *Evaluation of PKW elements for quantity and cost estimating*

For the selected example PKW, methods were developed to estimate quantities based on the model with an inlet to outlet key ratio of 1.25 used for Anderson and Tullis (2013). PKW are typically constructed of concrete, either cast-in-place or precast, depending upon the size of the structure and other site conditions. The cost model presented herein utilizes cast-in-place concrete. The weir components were further classified as either "mass" or "reinforced" concrete elements and representative unit prices were applied to each group. Generally, the unit cost of mass concrete is less than that of reinforced concrete since there is less forming and reinforcing required per unit volume of concrete.

3.2 *Comparison of PKW with labyrinth weir*

Using the the methods presented in Crookston and Tullis (2013), a 2-cycle labyrinth weir with a 15° sidewall angle (α), was selected for comparison with a 4-key PKW with similar head-discharge characteristics. For both weirs, the weir height (P) was 3 m, the total spillway width was 14.1 m for the labyrinth and 14.2 m for the PKW. The computed discharge for each spillway for a design head (H_t) of 2.4 m was computed to be 186 m^3/s. The total weir length for the labyrinth and PKW are 44.8 m and 72.3 m, respectively. The longitudinal PKW crest (B_t) and footprint (B) dimensions are 7.4 m and 3.8 m, respectively, compared to the labyrinth weir longitudinal footprint dimension of 10.8 m.

The concrete volumes for the PK and labyrinth weirs were estimated. In addition to the weir wall and sloping floor concrete volume estimates, estimated concrete volumes for the training walls and slabs were also included. The training walls were 5.4 m high to accommodate the maximum $H_t = 2.4$ m. The slab for the labyrinth spillway included the entire weir footprint, while for the PKW, the slab extended from the downstream end of the base to the downstream end of the weir.

Unit prices (USA$ per m^3) were developed for the concrete elements of the PK and labyrinth weir based upon forming, material, and other applicable construction costs. Concrete types, quantities and cost estimates for the PK and labyrinth weirs are presented in Tables 1 and 2.

For this example, the PKW with similar head-discharge characteristics to the labyrinth weir requires 35% less concrete volume, primarily due to the significantly reduced upstream to downstream footprint of both the foundation and weir. The cost estimates indicate that cost of the PKW is about 45% less than the labyrinth weir. These comparisons are for a design H_t/P of 0.8, which the authors understand to be higher than "typical" PKW applications. However, as presented in Paxson et al. (2013), it is relatively common practice in the USA to design labyrinth spillways for H_t/P ratios of 0.7 and greater.

It should be noted that for channel applications where the length of the spillway channel from the inlet to the outlet is fixed and the entire channel requires a concrete lining (slab) and training walls, the quantities and costs for the PK and labyrinth weirs would be more comparable. Table 3

Table 1. PKW quantity and cost estimate (concrete only).

Element	Type of Concrete	Volume m^3	Unit Cost USA\$/m^3	Total Cost USA\$
Base	Mass	56.5	\$250	\$14,125
Weir Walls	Reinforced	11.5	\$800	\$9,200
Overhangs	Reinforced	6.4	\$800	\$5,120
Training Walls	Reinforced	35.0	\$700	\$24,500
Slab	Reinforced	15.8	\$300	\$4,740
Total		125		\$57,685

Table 2. Labyrinth weir quantity and cost estimate (concrete only).

Element	Type of Concrete	Volume m^3	Unit Cost USA\$/m^3	Total Cost USA\$
Weir Walls	Reinforced	50.4	\$800	\$40,320
Training Walls	Reinforced	50.8	\$700	\$35,560
Slab	Reinforced	86.9	\$300	\$26,070
Total		188		\$101,950

Table 3. Labyrinth and PKW concrete volumes and costs for Fixed-channel dimensions.

Spillway	Concrete Volume m^3	Total Cost USA\$
PKW	212	\$90,075
Labyrinth Weir	188	\$101,950

shows the estimated concrete volumes and costs for each spillway assuming the concrete volumes for the slab and training wall for the PKW are replaced with those required for the labyrinth weir.

This comparison demonstrates that the channel lining and training walls make up a significant portion of the concrete volume, resulting in more concrete for the PKW. However, the estimated cost of the PKW is still about 10% less than the labyrinth, primarily due to the lower unit cost of the mass concrete base.

3.3 *Comparison of PKW with gated spillway*

A gated spillway could be sized to match the design discharge (186 m^3/s) of the PKW presented in the previous section, assuming similar total widths for the PKW and gated spillway. To match the design peak stage, the head over the fixed crest is equal to the gate height (P_{gate}) plus H_t. For a 14.2 m wide gated spillway with a fixed crest discharge coefficient of 0.71, a gate height of 1 m would result in a computed discharge of 187 m^3/s for the gates completely opened (total head over fixed crest equals 3.4 m), which is comparable to the PKW presented in Section 3.2. For a channel application, a slab would be required for mounting the gate; training walls would also be required. For estimating, the slab was assumed to have a longitudinal footprint dimension of 3 m and the required training wall height is 3.4 m ($P_{gate} + H_u$). A single leaf steel hinged crest gate (1 m

Table 4. Gated spillway concrete volumes/costs and gate costs.

Element	Type of Concrete	Volume m^3	Unit Cost USA\$/m^3	Total Cost USA\$
Training Walls	Reinforced	8.2	$700	$5,740
Slab	Reinforced	21.3	$300	$6,390
Subtotal		29.5		$12,130
Gate Furnish & Install				$250,000
Total				$262,130

high by 14.2 m long) was used in cost estimating (based upon manufacturer provided estimates). The quantity and cost estimate for the gated spillway is presented in Table 4.

This estimate indicates that this application of a gated spillway would be significantly more costly than a PKW, primarily due to the cost to furnish and install the gates. Note that if a larger gate height could be accommodated, this spillway alternative may be more cost competitive with the PKW. If a gate height equal to the PKW height (3 m) were selected, the total spillway width could be reduced to about 7 m, reducing the concrete and gate costs.

4 OPERATIONS

Where feasible, passive spillways such as PK and labyrinth weirs are preferred over gated spillways due to the operation and maintenance requirements for gated spillways. For manually operated gates, the ability to pass large discharges is dependent upon the operator. For automated gates, operation may rely on electrical or mechanical systems. There are also concerns with gate failure and/or operator (manual or automatic) error, resulting in large releases that could cause downstream damages.

In some cases, it may be beneficial to use a PKW in parallel with gated spillway structures. During a flood, the initial inflow hydrograph will pass over the PKW (passive control) and if the flood exceeds the capacity of the PKW, the gated spillway can be operated. Passive and gated spillway structures in parallel would be particularly beneficial for remote dam locations without a full-time dam tender. The discharge capacity of the PKW would lengthen the required response time of operators to reach the dam for gate operation.

5 CONCLUSIONS

PKW present a unique solution to provide significant hydraulic capacity for new dams and for the rehabilitation of existing dams. As presented herein, PKW are considered to be cost competitive with labyrinth and gated spillways and may be a more economical solution in many situations. In addition, PKW may be a preferable alternative to gated spillways due to their ability to pass large flows without the operation and maintenance issues and costs related to gates. Since each project presents its own set of opportunities and constraints, the applicability for a particular spillway type should be evaluated on a site by site basis.

REFERENCES

Anderson, R.M. and B.P. Tullis. Piano Key Weir hydraulics and labyrinth weir comparison, *Journal of Irrigation and Drainage Engineering*, ASCE, 2013.139:246–253.

Crookston, B.M. and B.P. Tullis. Hydraulic design and analysis of labyrinth weirs. I: Discharge relationships, *Journal of Irrigation and Drainage Engineering*, ASCE, 2013.139:363–370.

Laugier, F. Design and construction of the first Piano Key Weir spillway at Goulours Dam, *Hydropower and Dams*, 2007 (issue 5).

Laugier, F., A. Lochu, C. Gille, M. Leite Ribeiro, and J.-L. Boillat. Design and construction a labyrinth PKW spillway at St-Marc dam, France. *Hydropower & Dams*. Issue 5, 2009. pp. 100–107.

Lempérière, F. and A. Ouamane. The Piano Keys weir: A new cost-effective solution for spillways, *Hydropower and Dams*, 2003 (5), 144–159.

Lempérière, F. 2009. New labyrinth weirs triple the spillways discharge (http://www.hydrocoop.org/publications/Labyrinth_%20weirs_09[1].2009.pdf).

Paxson, G.S., J.S. Monroe, B.M. Crookston, and D. Campbell. Balancing site considerations with hydraulic efficiency for labyrinth spillways, ICOLD 2013 International Symposium, Seattle, Washington.

Pralong, J., J. Vermeulen, B. Blancher, F. Laugier, S. Erpicum, O. Machiels, M. Pirotton, J.-L. Boillat, M. Liete Ribeiro, and A. Schleiss. A naming convention for the Piano Key Weirs geometrical parameteres. Workshop on Labryinth and Piano Key Weirs-2011, University of Liege, Belgium.

Labyrinth and Piano Key Weirs II – PKW 2013 – Erpicum et al. (eds)

An analytical approach for Piano Key weir hydraulic design

O. Machiels
ARCADIS Belgium sa, Liège, Belgium

B. Dewals, P. Archambeau, M. Pirotton & S. Erpicum
Research group HECE, ArGEnCo Department, University of Liege, Liège, Belgium

ABSTRACT: A hydraulic design equation for Piano Key Weirs is set up using the results of a large experimental parametric study. Comparison of the analytical formulation with experimental data from various physical modeling studies, considering idealized as well as real projects Piano Key Weirs configurations, shows a 15% accuracy in unit discharge prediction on a large range of geometric parameters values. Comparison of the proposed formulation to two other recently proposed equations highlights the importance to consider such analytical approaches inside the range of parameter values for which they have been validated.

1 INTRODUCTION

Hydraulic design stage is still challenging when looking at Piano Key Weirs (PKW) development. Three approaches may be considered to define a suitable geometry of the structure: physical modeling, numerical modeling and analytical modeling.

The physical modeling approach is perhaps the most widely used (Ho Ta Khanh et al. 2011, Vermeulen et al. 2011). It is at the basis of all the studies dedicated to PKW hydraulics. It consists in building a scale model of the project and enables to reproduce, in a controlled environment making a qualitative as well as quantitative analysis possible, the whole complexity of the flows occurring in the studied systems. Scale models are however time and cost consuming. This usually limits the possibilities to optimize a design as the number of tested geometries may be reduced.

Numerical modeling is more and more often used to design a first "optimal" geometry before final optimization and validation using a physical model. It is also used for parametric studies. 3D (Pralong et al. 2011a) and 1D (Erpicum et al. 2010) numerical approaches have already been applied to PKW hydraulic design. Numerical modeling is usually less expensive than physical modeling. 3D modeling may be as accurate as physical modeling, but this approach is still time consuming. 1D modeling is faster but less accurate.

Analytical modeling is certainly the fastest and the simplest method to design a structure, as far as validated formulations are available. Such design equations have been recently proposed for PKW (Kabiri-Samani & Javaheri, 2012, Leite Ribeiro et al., 2012). In this paper, an equation derived from Machiels (2012) study is detailed and compared to various experimental data as well as prototype data. It is also compared to both other already published equations.

2 DESIGN EQUATION

The design equation presented in this paper has been developed on the basis of the results of extensive parametric tests carried out considering 27 physical models of PKW with varied geometries and 5 models of inclined sharp crested weirs (Machiels, 2012).

These tests highlighted the main influence of local approach flow conditions to the crest on the global discharge capacity. Accordingly, the discharge per unit length q (specific discharge) on a

PKW is computed as the sum of independent contributions from the upstream, side and downstream crests:

$$q = q_u \frac{W_o}{W_u} + q_d \frac{W_i}{W_u} + q_s \frac{2B}{W_u} \tag{1}$$

where q_u, q_d and q_s are, respectively, the specific discharge on the upstream, the downstream and the side crest, W_i and W_o are the inlet and outlet key widths, W_u is the unit width and B is the upstream/downstream length of the crest defined according to Pralong et al. (2011b).

The specific discharges on the various sections of the crest are evaluated from the SIA formulation for sharp-crested weirs (Swiss Society of Engineers and Architects, 1926), modified to take into account local conditions identified during the experimental campaign.

For the upstream crest, the specific discharge is calculated as:

$$q_u = 0.374\left(1 + \frac{1}{1000H + 1.6}\right)\left(1 + 0.5\left(\frac{H}{H + P_T}\right)^2\right)\sqrt{2gH^3} \tag{2}$$

where H is the upstream head and P_T is the crest height computed as the sum of the inlet key height P_i and the and the dam height P_d.

For the downstream crest, the specific discharge is calculated as:

$$q_d = 0.445\left(1 + \frac{1}{1000H + 1.6}\right)\left(1 + 0.5\left(\frac{H}{H + P_i}\right)^2\right)\sqrt{2gH^3} \tag{3}$$

where P_i is the inlet key height.

For the side crest, the specific discharge is calculated as:

$$q_s = 0.41\left(1 + \frac{1}{833H + 1.6}\right)\left(1 + 0.5\left(\frac{0.833H}{0.833H + P_e}\right)^2\right)\left(\frac{P_e{}^{\alpha} + \beta}{(0.833H + P_e)^{\alpha} + \beta}\right)K_{W_i}K_{W_o}\sqrt{2gH^3} \tag{4}$$

where P_e is the mean side crest height calculated by:

$$P_e = P_T\frac{B_o}{B} + P_i\frac{1 - \frac{B_o}{B}}{2} \tag{5}$$

with B_o the upstream overhang length and α and β coefficients whose expression was fitted on experimental results to take into account the effect of the water depth variation along the inlet:

$$\alpha = \frac{0.7}{S_i{}^2} - \frac{3.58}{S_i} + 7.55 \text{ and } \beta = 0.029e^{\frac{-1.446}{S_i}} \tag{6}$$

where S_i is the inlet key slope.

K_{Wi} is a coefficient taking into account the effect of the inlet key width W_i on the side crest efficiency. It is calculated as:

$$K_{W_i} = 1 - \frac{\gamma}{\gamma + W_i{}^2} \tag{7}$$

where γ is a parameter whose expression was empirically defined from the experimental results as:

$$\gamma = 0.0037\left(1 - \frac{W_i}{W_o}\right) \tag{8}$$

K_{Wo} is a coefficient taking into account the reduction of the effective side crest length because of the local submergence induced by the outlet key. The discharge capacity of the outlet key is mainly controlled by its slope S_o and width W_o. K_{Wo} is thus computed as:

$$K_{W_o} = 1 \qquad for\ \frac{H}{W_o} \leq \delta_1$$

$$\begin{aligned} K_{W_o} = & \frac{2}{(\delta_2-\delta_1)^3}\left(\frac{H}{W_o}\right)^3 - \frac{3(\delta_2+\delta_1)}{(\delta_2-\delta_1)^3}\left(\frac{H}{W_o}\right)^2 \\ & + \frac{6\delta_2\delta_1}{(\delta_2-\delta_1)^3}\frac{H}{W_o} + \frac{\delta_2{}^2(\delta_2-3\delta_1)}{(\delta_2-\delta_1)^3} \end{aligned} \qquad for\ \delta_1 < \frac{H}{W_o} \leq \delta_2 \tag{9}$$

$$K_{W_o} = 0 \qquad for\ \frac{H}{W_o} > \delta_2$$

where δ_1 and δ_2 are two thresholds whose analytical expression was empirically defined from the experimental results:

$$\delta_1 = -0.788 S_o{}^{-1.88} + 5 \tag{10}$$

$$\delta_2 = 0.236 S_o{}^{-1.94} + 5 \tag{11}$$

3 VALIDATION

3.1 *Data from idealized PKW configurations*

The unit discharges predicted by the analytical formulation are compared in figure 1 to the experimental results obtained on the 27 physical models of PKW used to set up the design equation. They are compared in figure 2 to the experimental results gained on 9 other physical models of PKW considered by Machiels (2012) during his PhD studies.

In addition, the analytical formulation is compared in figure 3 with experimental data from other studies and hydraulic laboratories. 46 additional geometries are considered (32 from Ecole Polytechnique de Lausanne, Switzerland (Le Doucen et al. 2009; Leite Ribeiro et al. 2011), 1 from University of Biskra, Algeria (Ouamane, 2006), 4 from Electricité de France, France (Boursiquot, 2003), 4 from University of Liege, Belgium (Karelle, 2008) and 5 from Ho Chi Minh University, Vietnam (Truong Chi Hien & Huynh Hung, 2005, Truong Chi Hien & Ho Ta Khanh, 2007)).

Table 1. Non dimensional geometric parameters of idealized PKW models considered for the validation of the analytical formulation.

Source	Number of models	Number of tests	L/W	P_i/W_u	B_o/B_i	W_i/W_o	P_i/P_o
Present study (Machiels, 2012)	27	724	5	0.33 … 2	0 … ∞	0.46 … 2.18	1
Others from Machiels (2012)	9	223	4.15 … 5	0.33 … 1.56	1	0.64 … 1.57	1
Other studies	46	397	2.98 … 7	0.29 … 1.07	0 … ∞	0.5 … 2.6	0.72 … 1.38

Source	B/P_i	B_i/B & B_o/B	P_r/P_i	P_d/P_i	T/W_u	Crest shape	H/P_i
Present study (Machiels, 2012)	1 … 6	0 … 0.67	0	0.33 … 2	0.05	flat topped	0.08 … 3.2
Others from Machiels (2012)	1 … 4.67	0.26 … 0.33	0 … 0.43	0.32 … 4.37	0.0375	flat topped	0.09 … 2.47
Other studies	1.52 … 4.61	0 … 0.67	0	0 … 5.21	0.01 … 0.06	variable	0.25 … 2.79

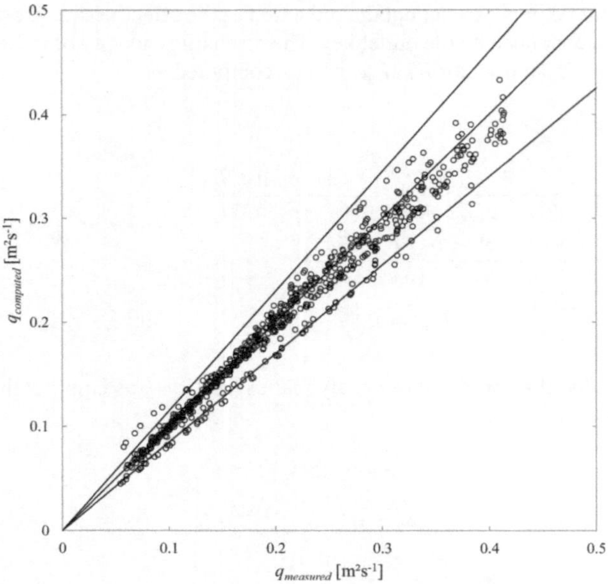

Figure 1. Comparison between specific discharges computed with the analytical formulation proposed in this paper and the experimental results from the authors – Plain line = perfect agreement; dotted lines = ±15% variation.

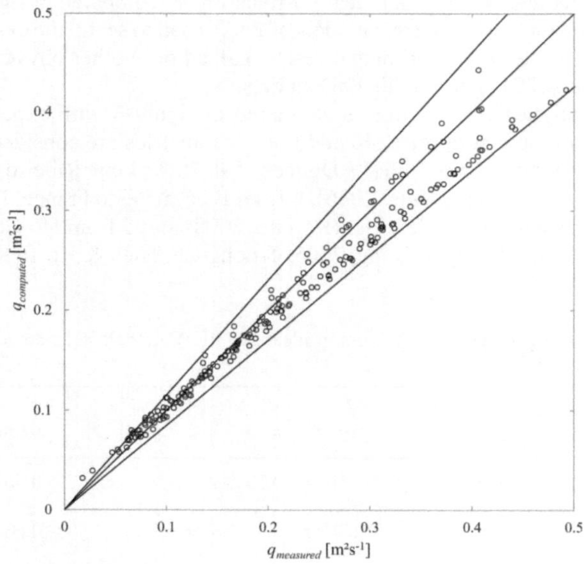

Figure 2. Comparison between specific discharges computed with the analytical formulation proposed in this paper and some experimental results from Machiels (2012) – Plain line = perfect agreement; dotted lines = ±15% variation.

The range of the main geometric non dimensional ratios of these physical models is given in table 1, where L is the developed crest length, W is the weir width, B_i is the downstream overhang length, P_o is the outlet key height, P_r is the parapet wall height and T is the sidewall thickness. To avoid scale effects, the dataset has been limited to upstream heads higher than 0.05 m, as suggested by Leite Ribeiro et al (2012). 1344 tests are thus considered for the comparison.

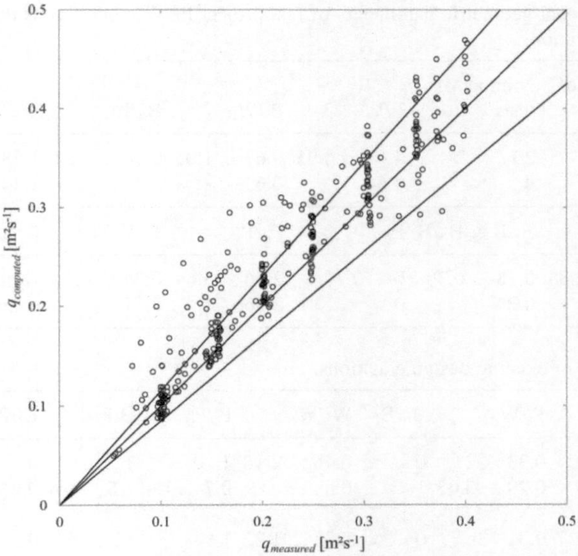

Figure 3. Comparison between specific discharges computed with the analytical formulation proposed in this paper and experimental results from other laboratories – Plain line = perfect agreement; dotted lines = ±15% variation.

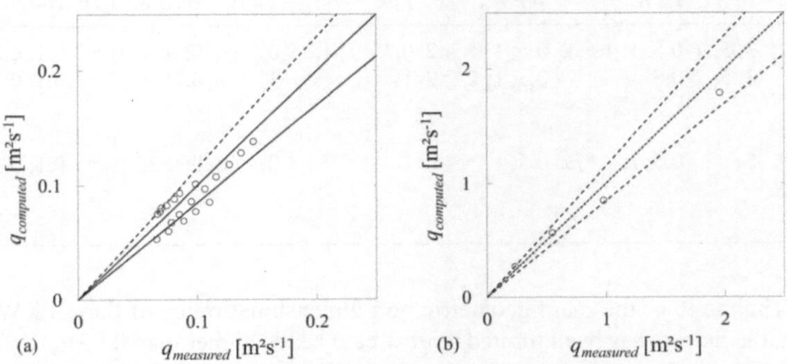

Figure 4. Comparison between specific discharges computed with the analytical formulation proposed in this paper and (a) experimental results from real projects physical models and (b) prototype data – Plain line = perfect agreement; dotted lines = ±15% variation.

The accuracy of the proposed formulation is generally better than 15% for geometric parameters ratios in the range of validity of the formula (Fig. 1 and Fig. 2). Considering an extended range of parameters values (Fig. 3), and especially lower and higher L/W ratio, the accuracy of the formula is significantly deteriorated for a limited number of tests and is still around 15% for most of them.

3.2 *Data from real PKW projects*

The proposed formulation has been compared to the results gained during the physical modeling studies of 4 real projects (Fig. 4a): the Raviège dam project (1:35 scale model – 2 PKW configurations – Noel et al. 2012), the Ouldjet Mellegue project (1:50 scale model – Machiels et al. 2011) and the Escouloubre PKW (1:7 scale model – Machiels, 2012). For the lattest, prototype measurements are also available (Machiels, 2012). They are compared to the analytical results

Table 2. Non dimensional geometric parameters of real project PKW models considered for the validation of the analytical formulation.

Source	Number of models	Number of tests	L/W	P_i/W_u	B_o/B_i	W_i/W_o	P_i/P_o
Scale models	4	20	4.64 … 6.95	0.63 … 1.05	0.90 … 1.23	1.28 … 1.50	1
Prototype	1	4	4.64	0.63	1	1.44	1

Source	B/P_i	B_i/B & B_o/B	P_r/P_i	P_d/P_i	T/W_u	Crest shape	H/P_i
Scale models	2.38 … 2.88	0.18 … 0.30	0 … 0.21	2.56 … 5.64	0.04 … 0.11	variable	0.06 … 0.72
Prototype	2.88	0.24	0	2.56	0.11	chanfered	0.06 … 0.27

Table 3. Validity field of existing design equations.

Source	L/W	P_i/W_u	B_o/B_i	W_i/W_o	P_i/P_o	B/P_i	B_i/P_i & B_o/P_o	H/P_i
Present study	5	0.33 … 2	0 … ∞	0.46 … 2.18	1	1 … 6	0 … 2.67	0.08 … 3.2
Leite Ribeiro et al. (2012)	3 … 7	0.29 … 0.65	1	0.5 … 2	0.7 … 1.4	1.5 … 4.6	0.25 … 4	0.1 … 2.8
Kabiri-Samini & Javaheri (2012)	2.5 … 7	0.3 … 3	0 … ∞	0.33 … 1.22	1	1 … 2.5	0 … 0.65	0.1 … 0.6

Table 4. Values of the non-dimensional geometric ratios considered to compare existing design equations.

Parameter set	L/W	P_i/W_u	W_i/W_o	P_i/P_o	B_o/B_i	B_i/B & B_o/B	H/P_i
All	3; 5; 7	0.5; 1; 1.5; 2	0.5; 1; 1.5; 2	0.75; 1; 1.25	0… ∞	0; 0.25; 0.67	0.1; 0.5; 1; 1.5; 2; 3
Common with Leite Ribeiro et al. (2012)	3; 5; 7	0.5	0.5; 1; 1.5; 2	1	1	0.25; 0.67	0.1; 0.5; 1; 1.5; 2
Common with Kabiri-Samini & Javaheri (2012)	5	0.5; 1; 1.5; 2	0.5; 1	1	0… ∞	0; 0.25	0.1; 0.5

in figure 4b. The range of the main geometric non dimensional ratios of these PKW is given in table 2. Again, the dataset has been limited to upstream heads higher than 0.05 m.

Despite the specificities of real projects configurations (reservoir flow conditions, abutment and end keys geometry…), the accuracy of the proposed formulation is again around 15% for the considered projects.

4 COMPARISON TO OTHER PUBLISHED FORMULATIONS

Leite Ribeiro et al. (2012) and Kabiri-Samani & Javaheri (2012) both recently proposed a hydraulic design equation for PKW in non-submerged flow conditions. The validity field of these equations is presented in Table 3. Moreover, the study by Leite Ribeiro et al. (2012) considered half-circular crests and the one by Kabiri-Samani & Javaheri (2012) sharp-crested weirs.

In order to compare the 3 formulations, more than 7 700 test cases have been automatically generated according to geometric non-dimensional ratio values given in table 4. The comparison of the unit discharges predicted by the formulation proposed in this paper and Leite Ribeiro et al. (2012) and Kabiri-Samani & Javaheri (2012) equation is shown, respectively, in figure 5a and figure 5b, considering the whole test only the ones respecting the validity fields of both equations.

For geometric parameter outside the validity field of one equation, the difference in computed values provided by two different equations may be very important (more than 100%). For parameters

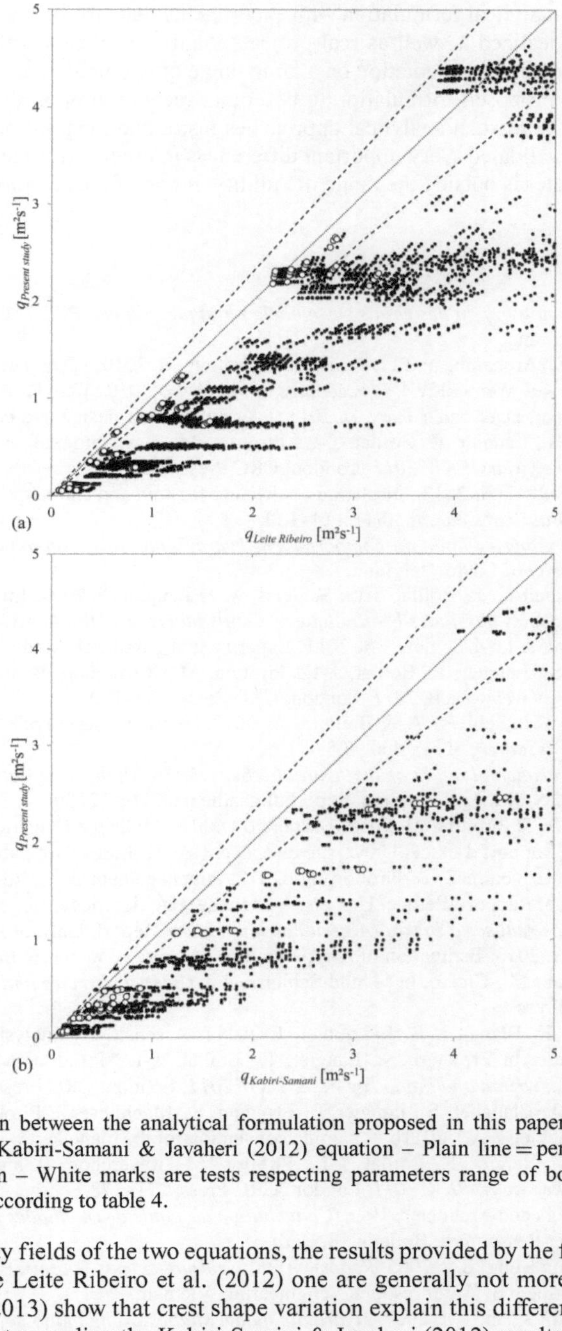

Figure 5. Comparison between the analytical formulation proposed in this paper and (a) Leite Ribeiro et al. (2012), (b) the Kabiri-Samani & Javaheri (2012) equation – Plain line = perfect agreement; dotted lines = ±15% variation – White marks are tests respecting parameters range of both formulations; black marks are other tests according to table 4.

respecting the validity fields of the two equations, the results provided by the formulation proposed in this paper and the Leite Ribeiro et al. (2012) one are generally not more than 15% different. Pfister & Schleiss (2013) show that crest shape variation explain this difference. The difference is much more important regarding the Kabiri-Samini & Javaheri (2012) results.

5 CONCLUSION

A comprehensive hydraulic design equation for Piano Key Weirs has been set up using the results of a large experimental parametric study.

Comparison of the analytical formulation with experimental data from varied physical modeling studies, considering idealized as well as real projects Piano Key Weirs configurations, shows a 15% accuracy in unit discharge prediction on a large range of geometric parameters values.

Comparison of the proposed formulation to two other recently proposed equations highlights the importance to consider such analytical approaches inside the range of parameters values for which they have been validated. Very important differences in predicted values may indeed appear when only one parameter is outside the range of validity of one of the equations.

REFERENCES

Boursiquot, D. 2003. *Compte-rendu des essais sur modèle physique. Report, P75/03/078*, EDF R&D, LNHE Department, Chatou, France.

Erpicum, S., Machiels, O., Archambeau, P., Dewals, B. & Pirotton, M. 2010. 1D numerical approach to model the flow over a Piano Key Weir (PKW). *Proceedings of SimHydro 2010, Nice*. Paris: SHF.

Ho Ta Khanh, M., Sy Quat, D. & Xuan Thuy, D. 2011. P.K. weirs under design and construction in Vietnam (2010). In Erpicum, S., Laugier, F., Boillat, J.-L., Pirotton, M., Reverchon, B. and Schleiss, A.J. (Eds), *Labyrinth and piano key weirs-PKW 2011*, London: CRC Press, 225–232.

Kabiri-Samani, A. & Javaheri, A. 2012. Discharge coefficient for free and submerged flow over Piano Key weirs. *Journal of Hydraulic Research*, 50(1), 114–120.

Karelle, Y. 2008. *Etude expérimentale du comportement hydraulique des déversoirs en touches de piano*. Master thesis, University of Liège, Belgium.

Le Doucen, O., Leite Ribeiro, M., Boillat, J-L., Schleiss, A. & Laugier, F. 2009. Etude paramétrique de la capacité des PKW. *Modèles physiques hydrauliques – Outils indispensables du XXIe siècle*. Paris: SHF

Leite Ribeiro, M., Boillat, J.-L. & Schleiss, A. 2011. Experimental parametric study for hydraulic design of PKWs. In Erpicum, S., Laugier, F., Boillat, J.-L., Pirotton, M., Reverchon, B. and Schleiss, A.J. (Eds), *Labyrinth and piano key weirs-PKW 2011*, London: CRC Press, 183–190.

Leite Ribeiro, M., Pfister, M., Schleiss, A. & Boillat, J.-L. 2012. Hydraulic design of A-type Piano Key Weirs. *Journal of Hydraulic Research*, 50(4), 400–408.

Machiels, O. 2012. *Experimental study of the hydraulic behavior of Piano Key Weirs*. PhD thesis, HECE research unit, University of Liège, Belgium. http://hdl.handle.net/2268/128006.

Machiels, O., Erpicum, S. & Pirotton, M. 2011. Barrage d'Ouldjet Mellegue (Algérie) – Etude hydraulique sur modèle physique. Rapport d'étude. Université de Liège, HACH: Liège. Not published.

Noel, D., Machiels, O., Erpicum, S. & Pirotton, M. 2012. Aménagement de la Raviège – Modèle réduit hydraulique du barrage. Rapport d'étude. Université de Liège, HACH: Liège. Not published.

Ouamane, A. 2006. *Essais réalisés à Biskra (Algérie) en mars 2006*. Report, Univ. of Biskra, Algeria.

Pfister, M. & Schleiss, A. 2013. Estimation of A-type Piano Key weir rating curve. In Erpicum, S., Laugier, F., Pfister, M., Pirotton, M., Cicéro, G.M. and Schleiss, A.J. (Eds), *Labyrinth and piano key weirs-PKW 2013*, London: CRC Press

Pralong, J., Montarros, F., Blancher, B. & Laugier, F. 2011a. A sensitivity analysis of Piano Key Weirs geometrical parameters. In Erpicum, S., Laugier, F., Boillat, J.-L., Pirotton, M., Reverchon, B. and Schleiss, A.J. (Eds), *Labyrinth and piano key weirs-PKW 2011*, London: CRC Press, 133–139.

Pralong, J., Vermeulen, J., Blancher, B., Laugier, F., Erpicum, S., Machiels, O., Pirotton, M., Boillat, J.-L., Leite Ribeiro, M. & Schleiss, A.J. 2011b. A naming convention for the piano key weirs geometrical parameters. In Erpicum, S., Laugier, F., Boillat, J.-L., Pirotton, M., Reverchon, B. and Schleiss, A.J. (Eds), *Labyrinth and piano key weirs-PKW 2011*, London: CRC Press, 271–278.

Swiss Society of Engineers and Architects, 1926, *Contribution à l'étude des méthodes de jaugeage*, Report of Swiss Bureau of Water Ressources, Bulletin 18, Bern.

Truong Chi Hien & Huynh Hung, B.E. 2005. *Report on P.K. weir model tests*. Report, Ho Chi Minh University of Technology, Department of Water resources engineering, Vietnam.

Truong Chi Hien & Ho Ta Khanh, M. 2007. *Hydraulic model studies of the piano keys weir type B*. Report, Ho Chi Minh University of Technology, Department of Water resources engineering, Vietnam.

Vermeulen, J., Laugier, F., Faramond, L. & Gille, C. 2011. Lessons learnt from design and construction of EDF first Piano Key Weirs. In Erpicum, S., Laugier, F., Boillat, J.-L., Pirotton, M., Reverchon, B. and Schleiss, A.J. (Eds), *Labyrinth and piano key weirs-PKW 2011*, London: CRC Press, 215–224.

Labyrinth and Piano Key Weirs II – PKW 2013 – Erpicum et al. (eds)
© 2014 Taylor & Francis Group, London, ISBN 978-1-138-00085-8

Estimation of A-type Piano Key weir rating curve

M. Pfister & A.J. Schleiss
Laboratory of Hydraulic Constructions (LCH), Ecole Polytechnique Fédérale de Lausanne (EPFL), Lausanne, Switzerland

ABSTRACT: The Piano Key weir (PKW) is a hydraulically attractive alternative to linear overflow weirs, increasing the unit discharge for similar heads and spillway widths. This advantage allows operating dam reservoirs on an increased level and provides thereby an enhanced retention volume. It is a result of the non-linear nature of PKWs being folded back and forth to make repeating cycles or keys. As PKWs are relatively novel structures, only few design equations are available, so that normally physical model tests of prototypes have to be conducted assuring and optimizing their hydraulic characteristics. Nevertheless, first comprehensive and systematic model test series were conducted in several laboratories. Based on such a test series, a general design equation for A-type PKWs is proposed and discussed. Considering furthermore data of other laboratory studies, the latter is validated. It turned out that main and secondary parameters exist regarding the relative effect on the rating curve. The main parameters having a significant effect on the capacity are the relative developed crest length and the relative head. The secondary parameters of small but not negligible effect comprise the ratio of the inlet and outlet key widths, the ratio of the inlet and outlet key heights, the relative overhang lengths, and the relative height of the parapet walls.

1 INTRODUCTION

The concept of the Piano Key weir (PKW) is a further development of the Labyrinth weir (Tullis 1995, Falvey 2003). Detailed historical reviews on the evolution from the Labyrinth weir to the PKW are given by Schleiss (2011) and by Lempérière et al. (2011). Two main advantages of PKWs as compared with Labyrinth weirs are: (1) Reduced structural footprint allowing for the installation on top of gravity dams (Lempérière and Ouamane 2003); and (2) Increased discharge capacity, as the inclined bottom of the keys instead of the horizontal-vertical arrangement improves the hydraulic efficiency (Laugier et al. 2009, Anderson and Tullis 2011, 2012).

Due to increased flood discharges and severe specifications of the dam safety, a large number of existing dams requires spillway rehabilitation to improve their hydraulic capacity. The PKWs are then often an interesting option (Laugier 2007, Leite Ribeiro et al. 2009, Laugier et al. 2009). Tests performed on scale models as well as numerical simulations contributed to the recent increase of knowledge about their hydraulic characteristics and constructive aspects. Nevertheless, the design of most PKW prototypes is still supported by physical modelling (Laugier 2007, Dugué et al. 2011, Erpicum et al. 2011a, Cicero et al. 2011).

Lempérière and Ouamane (2003) were among the first to present systematic PKW tests of type A and B, proposing rough design criteria. Ouamane and Lempérière (2006) extended their 2003 study for different dimensionless parameters. Results highlight the relevance of the ratio between the developed crest length L and the transverse width W. Lempérière et al. (2011) summarize different types of PKWs which have been studied by Hydrocoop since 1998, classified according to the presence or absence of overhangs. In type A, up- and downstream overhangs are provided. Types B and C include only up- or downstream overhangs. Although type D has inclined bottoms, it does not contain overhangs.

Although the flow over a PKW is highly three-dimensional, Erpicum et al. (2011b) present a simplified one-dimensional numerical modeling for preliminary designs with an accuracy of some 10%. The model is based on cross-section-averaged equations of mass and momentum conservation with only the upstream discharge as boundary condition. By comparing PKWs of type A and D, Anderson and Tullis (2011) confirm that the presence of the overhangs has a positive effect on the PKW discharge capacity. Upstream overhangs increase the inlet flow area and wetted perimeter, which results in lower energy losses.

Kabiri-Samani and Javaheri (2012) were among the first to publish general equations to derive the rating curve of PKWs. They conducted extensive model tests in a 12 m long and 0.4 m wide channel, operated with specific discharges between 25 and 175 l/sm. The test program included totally 600 tests resulting in some 3000 data points. The tested PKWs included types A, B, and C, all of sharp-crested type. They investigated the PKW performance under free flow and submerged conditions, i.e. with an effect of the tailwater. The Poleni equation was taken as reference, and the rating curve expressed in terms of the discharge coefficient. Leite Ribeiro et al. (2012a) performed systematic model tests in a channel of 3 m length and 0.5 m width. In total, 380 tests were conducted, considering 49 different PKW geometries of type A, with cylindrical weir crests. The specific model discharge varied between 26 and 440 l/sm. Uniquely free overflow without submergence from the tailwater was tested. The rating curve was defined relative to a linear sharp-crested weir. The present paper focuses on these tests. Machiels et al. (2012) finally conducted systematic model tests in a 7.2 m long and 1.2 m wide flume. For each parameter set investigated, a 0.3 m wide PKW model including 2.5 cycles was inserted into the channel and charged with specific discharges between 13 and 400 l/sm. The PKWs were of broad-crested type. Again, the results are only valid for free overflow conditions. Starting with an analytical approach, Machiels et al. (2012) finally gives the specific PKW discharges as the sum of the specific discharges on the downstream, the upstream, and the side wall.

The common notation (Fig. 1) as defined by Pralong et al. (2011) is applied herein, with B = streamwise length, L = developed crest length, P = vertical height, R = parapet wall height, T_S = wall thickness, and W = width. Furthermore, subscript i refers to the inlet key, i.e. the key that is filled with water for a reservoir surface at the PKW crest elevation, and subscript o to the

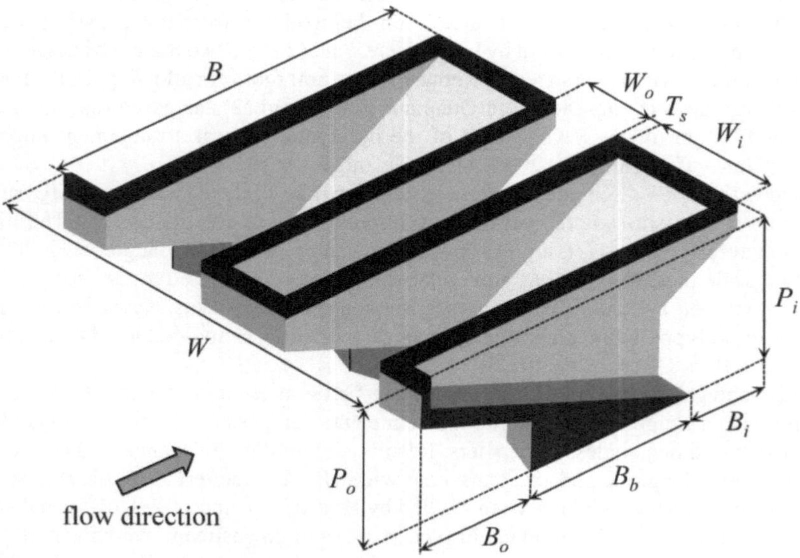

Figure 1. Common PKW notation according Pralong et al. (2011).

outlet key, i.e. the 'dry' key for the latter reservoir level. The developed crest length L is defined as $L = N(W_i + W_o + 2T_S + 2B)$ or alternatively as $L = W + (2NB)$, with N = number of PKW cycles.

Mainly two approaches are common to describe the PKW discharge Q_P: by (1) deriving a specific discharge coefficient using the Poleni equation (Kabiri-Samani and Javaheri 2012, Anderson and Tullis 2011, Ouamane et al. 2006), or (2) comparing the PKW discharge Q_P with that of a linear sharp-crested weir Q_S (Leite Ribeiro et al. 2012a). The second option is chosen herein as it seems better physically based (Falvey 2003, Schleiss 2011). The discharge Q_P as measured in the model is thereby compared with the theoretical value Q_S of a linear sharp-crested weir of width W, both under the same H. Then, the discharge enhancement ratio r follows as

$$r = \frac{Q_P}{Q_S} \tag{1}$$

The discharge Q_S over a linear sharp-crested weir serving as reference is

$$Q_S = C_S W \sqrt{2g} H^{1.5} \tag{2}$$

with $C_S = 0.42$ as discharge coefficient for a sharp-crested weir (e.g. Hager and Schleiss 2009).

2 PHYSICAL MODEL TESTS

In the frame of the study conducted by Leite Ribeiro et al. (2012a), systematic physical model tests were carried out at the Laboratory of Hydraulic Constructions (LCH) of EPFL. A straight rectangular channel with a length of 40 m, a width of 2 m and a height of 1 m was used. The effective reach including the PKWs was reduced to $W = 0.5$ m and of 3 m length, by inserting vertical parallel plates. The model was of sectional type including 1.5 PKW units (Fig. 2). To exclude an effect of the latter set-up on the rating curve, preliminary tests were conducted with more cycles, resulting in analogue rating curves (Leite Ribeiro et al. 2012b). Different arrangements with similar up- and downstream overhangs were tested, all according to type A. The thickness of the walls was constant

Figure 2. PKW model flow for configurations including 1.5 cycles, view from downstream on left and in flow direction on right.

with $T_S = 0.02$ m, and the overflow crest shape was half-circular (cylindrical weir). Semi-circular noses were mounted below the outlet keys on the reservoir side. All tests were conducted for free overfall conditions, so that no backwater effect influenced the rating curve.

Totally, 49 different PKW configurations were analyzed, resulting in 380 data points. The parameter variation included (in absolute dimensions): $1.50\,\text{m} \leq L \leq 3.50\,\text{m}$, $0.33\,\text{m} \leq B \leq 1.00\,\text{m}$, $0.10\,\text{m} \leq W_i \leq 0.20\,\text{m}$, $0.10\,\text{m} \leq W_o \leq 0.20\,\text{m}$, $0.10\,\text{m} \leq P_i \leq 0.28\,\text{m}$, $0.10\,\text{m} \leq P_o \leq 0.28\,\text{m}$, $0.00\,\text{m} \leq P_d \leq 0.62\,\text{m}$, $0.07\,\text{m} \leq B_i \leq 0.40\,\text{m}$, $0.07\,\text{m} \leq B_o \leq 0.40\,\text{m}$, $0.00\,\text{m} \leq R \leq 0.06\,\text{m}$, and $0.02\,\text{m} \leq H \leq 0.27\,\text{m}$. The parameters $T_s = 0.02$ m and $W = 0.50$ m were kept constant. Here, H = total approach flow energy head above the PKW crest, and P_d = channel height below the PKW foot, i.e. below P_i to investigate the effect of the approach flow velocity. The bottom slopes of the inlet key $(P_i - R)/(B_i + B_b)$ and of the outlet key $(P_o - R)/(B_o + B_b)$ were between 0.34 and 0.84. The model discharge was varied between $0.013\,\text{m}^3/\text{s} \leq Q_P \leq 0.220\,\text{m}^3/\text{s}$. Expressed in relative terms, the parameter variation included $3.0 \leq L/W \leq 7.0$, $0.1 \leq H/P_i \leq 2.8$, $1.5 \leq B/P_i \leq 4.6$, $0.7 \leq P_i/P_o \leq 1.4$, $0.2 \leq B_i/B = B_o/B \leq 0.4$, and $0.5 \leq W_i/W_o \leq 2.0$. The values Q_P were measured with a magnetic-inductive flow meter, and H using a point gauge in a lateral and stagnant region. Values $H < 0.05$ m and $H > 0.5(P_i + P_d)$ were not considered for the data analysis, as they might be influenced by scale effects or by the approach flow velocity.

3 RESULTS

3.1 *Main parameters*

The developed crest length L per linear weir width W is significant to describe the discharge capacity of a PKW. Herein, these parameters are expressed as $(L - W)/W$ to account for the relative capacity increase as compared to a linear sharp-crested weir. Accordingly, if $L = W$, then $(L - W)/W = 0$, so that no capacity increase as compared to the latter is observed. Furthermore, the parameter P_i/H is known to be relevant for Labyrinth weirs (Tullis 1995), and also significantly affects the PKW efficiency. The data analysis conducted by Leite Ribeiro et al. (2012a) finally let to the principal normalization

$$\delta = \left(\frac{(L-W)P_i}{WH}\right)^{0.9} \tag{3}$$

Figure 3a shows the values r derived from measurements versus δ, including data of Leite Ribeiro et al. (2012a) and some data of Machiels et al. (2012). Note that the data of Machiels et al. (2012) with a bottom slope $(P_i - R)/(B_i + B_b) > 0.7$ were not collapsing with the proposed normalization, so that these were ignored and the application of Eq. (3) accordingly limited. Nevertheless, it can be seen that the data according to the limitations of the physical modeling essentially collapse, supporting the relevance of δ. The data trend is described with a linear function as (Leite Ribeiro et al. 2012a)

$$r = 1 + 0.24\delta \tag{4}$$

The coefficient of determination between the measured values and the prediction according to Eq. (4) is $R^2 = 0.964$ for the data of Leite Ribeiro et al. (2012a) and $R^2 = 0.975$ for those of Machiels et al. (2012). The constant "1" in Eq. (4) assures equality with the sharp-crested weir ($r = 1$ for $\delta = 0$), and the term 0.24δ stands for the efficiency increase generated by the PKW if $\delta > 0$.

3.2 *Secondary parameters*

It is known from several PKW model studies that some geometrical ratios, which are not included in the pragmatic approach according to Eq. (4), slightly increase or decrease the discharge efficiency

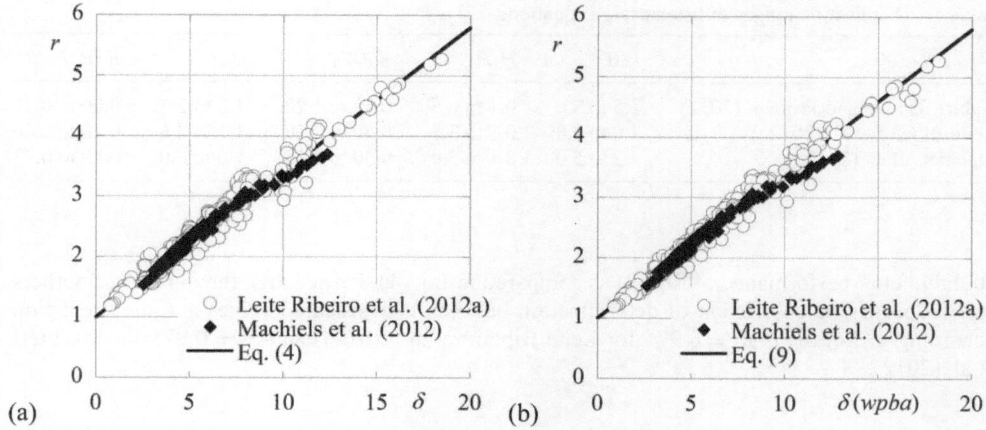

Figure 3. Values r derived from measurements versus (a) δ, compared with prediction following Eq. (4), and (b) $\delta(wpba)$, compared with prediction following Eq. (9).

for similar heads. These ratios were not addressed in Eq. (4) and are specified herein, by introducing four correction factors. The latter are equivalent to individual factors between 0.92 and 1.20, for the geometries tested by Leite Ribeiro et al. (2012a). The individual correction factors are defined as given below, namely for the ratio between inlet and outlet key width as

$$w = \left(\frac{W_i}{W_o}\right)^{0.05} \tag{5}$$

for the ratio between outlet and inlet key height as

$$p = \left(\frac{P_o}{P_i}\right)^{0.25} \tag{6}$$

for the ratio between key overhangs to the base length as

$$b = \left(0.30 + \frac{B_o + B_i}{B}\right)^{-0.50} \tag{7}$$

and for the relative parapet wall height as

$$a = 1 + \left(\frac{R_o}{P_o}\right)^{2.00} \tag{8}$$

To include the secondary parameters, Eq. (4) is completed with all correction factors, resulting in the comprehensive expression

$$r = 1 + 0.24\delta(wpba) \tag{9}$$

For Eq. (9), the same range of validity applies as for Eq. (4) which is mentioned in chapter 2. Figure 3b shows the measured values r, normalized with $\delta(wpba)$. The data analysis indicated a

Table 1. Application ranges of listed design equations.

	L/W	H/P	W_i/W_o	B/P	B_i/B, B_o/B
Kabiri-Samani and Javaheri (2012)	2.5 to 7.0	0.1 to 0.6	0.33 to 1.22	1.0 to 2.5	0.00 to 0.26
Leite Ribeiro et al. (2012a)	3.0 to 7.0	0.1 to 2.8	0.50 to 2.00	1.5 to 4.6	0.20 to 0.40
Machiels et al. (2012)	4.2 to 5.0	0.1 to 5.0	0.50 to 2.00	1.0 to 6.0	0.29 to 0.33

slightly better performance of Eq. (9) as compared to Eq. (4). Particularly, the number of outliers was reduced. The coefficient of determination between the measured values and the prediction according to Eq. (9) is $R^2 = 0.976$ for Leite Ribeiro et al. (2012a) and $R^2 = 0.975$ for Machiels et al. (2012).

4 COMPARISON (CASE STUDY)

A exemplarily comparison of the so far available design equations related to rating curves of PKWs (Kabiri-Samani and Javaheri 2012, Leite Ribeiro et al. 2012a, and Machiels et al. 2012) is given by Pfister and Schleiss (2013) and reproduced herein. An explicit comparison between the prediction according to Leite Ribeiro et al. (2012a) and the data of Machiels et al. (2012) is included in Fig. 3, as these individual values were available. As for the equation of Kabiri-Samani and Javaheri (2012), no measurement data are accessible, so that an implicit comparison based on a virtual site is necessary.

Such a comparison in the frame of a virtual prototype has to be defined in accordance with the limits of the individual studies as given in Table 1, as far as possible. The chosen prototype remains somehow arbitrary in terms of its characteristics, limiting the informative value of the comparison.

The following virtual site is used as base of the comparison, including

1. A Roller Compacted Concrete (RCC) gravity dam with a $W = 100$ m wide chute spillway on its downstream face, a dam height of $P_d = 30$ m below the PKW foundation, free overfall conditions at the weir without tailwater submergence, and ignoring the effect of the distal weir ends.
2. A design discharge of $Q_D = 2500$ m^3/s, according to a specific discharge of 25 m^2/s (as common on stepped spillways, Pfister et al. 2006).
3. A symmetrical A-type PKW, mounted at the dam crest as unregulated control structure, with $B = 8.00$ m as total streamwise length, $P = P_i = P_o = 5.00$ m as vertical height, $T_S = 0.35$ m as wall thickness, $R = 0$ m (without parapet walls), $W_i = 1.80$ m as inlet key width, $W_o = 1.50$ m as outlet key width, and $B_i = B_o = 2.00$ m as overhang lengths (Fig. 1). The crest shape remains undefined.

The following characteristics follow from the above chosen PKW geometry: cycle width $W_u = W_i + W_o + 2T_S = 4.00$ m, number of cycles $N = W/W_u = 25$, developed crest length $L = W + (2NB) = 500$ m, $L/W = 5.00$, $B/P = 1.60$, $W_i/W_o = 1.20$, $B_i/B = B_o/B = 0.25$, and $S_i = S_o = 0.83$. The latter does not respect the limitations of the prediction proposed by Leite Ribeiro et al. (2012a), requiring $S_i = S_o < 0.70$. The related data analysis showed that their equation then slightly overestimates the discharge capacity. To derive the discharge capacity, furthermore only water heads between 0 to 6 m were considered.

Given that the three compared equations base on model studies using different crest shapes, their effect has to be taken into account. Kabiri-Samani and Javaheri (2012) used a sharp-crested weir, Leite Ribeiro et al. (2012a) a cylindrical weir crest, and Machiels et al. (2012) a broad-crested weir. The discharges were thus first computed according to the proposed equation of the

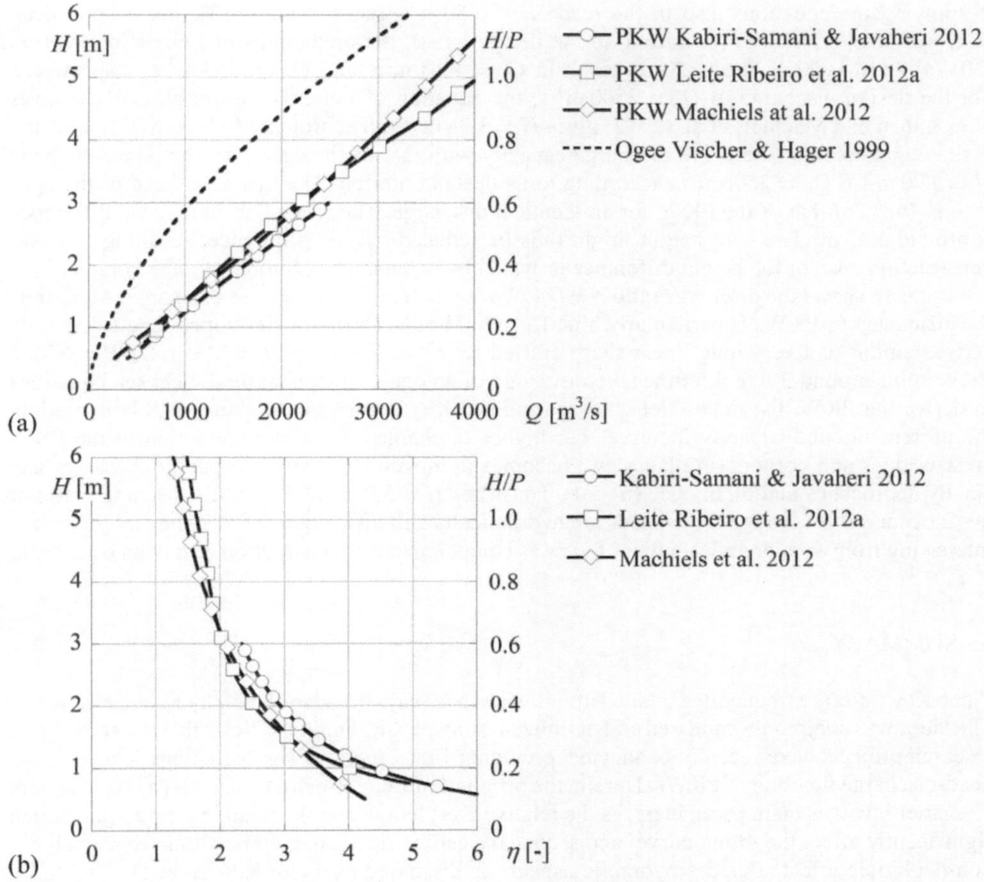

Figure 4. Comparisons of (a) rating curves for PKW and ogee crest weirs, (b) relative discharge capacity of PKW with ogee as reference (ogee capacity as $\eta = 1$).

aforementioned studies, and then normalized to the broad-crested weir. The discharge coefficients C_d were derived with

- Hager and Schwalt (1994) for the broad-crested weir, where C_d tends from $C_d = 0.33$ for $H < 0.2$ m to $C_d = 0.42$ for $H > 1$ m.
- Hager and Schleiss (2009) for the sharp-crested weir, where $C_d = 0.42$ for all H.
- Castro-Orgaz (2012) for cylindrical weir crests if $H < 0.5$ m ($C_d = 0.54$ at $H = 0.5$ m), then linearly approaching $C_d = 0.42$ at $H = 4.4$ m as observed by Ramamurthy and Vo (1993), and finally $C_d = 0.42$ for all $H > 4.4$ m.

The resulting rating curves for the three herein compared design equations are shown in Fig. 4a, also respecting their application limits in terms of H/P. As can be seen, the rating curves of the three PKW studies are similar. In general, the empiric equation of Kabiri-Samani and Javaheri (2012) predicts the highest discharge capacity. Additionally, the rating curve of a linear standard crest profile (ogee) is shown in Fig. 4a, derived from Vischer and Hager (1999) considering a design head of $H_D = 5.00$ m for Q_D.

All PKW curves start at typically $H/P = 0.1$ according to $H = 0.50$ m because of the lower model limitation to avoid scale effects. The error following from scale effects is probably small if applying the equations also in the reach $H < 0.50$ m (Pfister et al. 2013). For a large head of $H = 5.00$ m ($H/P = 1$) (according to the design head), the predictions of Leite Ribeiro et al. (2012a) and Machiels et al. (2012) result in $Q_P = 4041$ m^3/s and $Q_P = 3594$ m^3/s, respectively. For the design discharge of $Q_D = 2500$ m^3/s, the equation of Leite Ribeiro et al. (2012a) gives $H = 3.36$ m and Machiels et al. (2012) gives $H = 3.38$ m (both at around $H/P = 0.67$). Interesting is the comparison of the PKW discharge capacity with that of the ogee. For the latter, the head $H = 5.00$ m for $Q_D = 2500$ m^3/s according to its design criterion. The hydraulic head of the ogee is thus 148% of that of the PKW, for an identical discharge. The related absolute head difference is around 1.63 m. The dam height might thus be reduced by this difference, resulting in lower construction cost, or the height difference is available to generate additional storage volume.

Figure 4b shows the discharge ratio $\eta = Q_P/Q_O$ versus H, with Q_O = ogee discharge. As visible, the efficiency of PKWs is particularly high for small heads. Then, the developed length L is fully active, similar to a very long linear sharp crested weir. For a head of $H = 1.00$ m ($H/P = 0.2$), a PKW spills around 3.5 to 4.5 times the discharge of an ogee, dependent on the chosen equations to derive the PKW discharge. Note that the ratio $L/W = 5.00$ herein, a value slightly exceeding the aforementioned capacity increase. For higher discharges, the flow interaction in the PKW crest wedges and corners (in plan view) becomes dominant, so that the capacity decreases and finally approaches that of an ogee ($\eta \rightarrow 1$). For heads $H > 3.0$ m ($H/P > 0.6$) a discharge ratio of $\eta \leq 2$ occurs. Even though these heads are hydraulically still efficient at PKWs, they might be less interesting from an economical point of view as compared to the construction cost of an ogee crest.

5 SUMMARY

Piano Key weirs are frequently included in spillways these days because of their hydraulic efficiency. The latter was commonly validated and optimized using physical models. Nevertheless, some basic research projects were recently conducted, providing first general design equations related to the head-discharge function of PKWs. Herein, the proposal of Leite Ribeiro et al. (2012a) is presented. The latter involves main parameters, as the relative crest length and the relative weir height, which significantly affect the rating curve; and secondary geometrical parameters with have a small but non-negligible effect. Besides hydraulic aspects as discussed by Leite Ribeiro et al. (2012a), a constructional and economic consideration is given by Machiels et al. (2012).

PKWs are particularly efficient in conveying large specific discharges under small heads if compared to ogee crests. An exemplarily comparison of these two unregulated spillway inlets is given and discussed. Furthermore, the herein compared equations to predict the PKW rating curves give similar, but not identical results. The difference might be explained with the different crest types of the physical models, which were used to derive the design equations.

ACKNOWLEDGMENTS

The authors thank Mr Olivier Le Doucen (e-dric Engineering, Lausanne) for having conducted the model tests in the frame of his Master Thesis at EPFL. The excellent support of Dr Jean-Louis Boillat (formerly at LCH, EPFL) and Dr Marcelo Leite Ribeiro (now at Stucky, Lausanne; formerly at LCH, EPFL) is also kindly acknowledged.

REFERENCES

Anderson, R.M., Tullis, B. 2011. Influence of Piano Key Weir geometry on discharge. *Proc. Intl. Conf. Labyrinth and Piano Key Weirs* Liège B, 75–80. CRC Press, Boca Raton, FL.

Anderson, R.M., Tullis, B. 2012. Comparison of Piano Key and rectangular Labyrinth Weir hydraulics. *J. Hydraulic Eng.*, 138(4), 358–361.

Castro-Orgaz, O. 2012. Discussion Overflow characteristics of circular-crested weirs. *J. Hydraulic Res.* 50(2): 241–243.
Cicero, G.M., Menon, J.M., Luck, M., Pinchard, T. 2011. Experimental study of side and scale effects on hydraulic performances of a Piano Key Weir. *Proc. Intl. Conf. Labyrinth and Piano Key Weirs* Liège B, 167–172. CRC Press, Boca Raton, FL.
Dugué, V., Hachem, F., Boillat, J.-L., Nagel, V., Roca, J.-P., Laugier, F. 2011. PKWeir and flap gate spillway for the Gage II Dam. *Proc. Intl. Conf. Labyrinth and Piano Key Weirs* Liège B, 35–42. CRC Press, Boca Raton, FL.
Erpicum, S., Nagel, V., Laugier, F. 2011a. Piano Key Weir design of Raviege dam. *Proc. Intl. Conf. Labyrinth and Piano Key Weirs* Liège B, 43–50, CRC Press, Boca Raton, FL.
Erpicum, S., Machiels, O., Archambeau, P. Dewals, B., Pirotton, M. 2011b. 1D numerical modeling of the flow over a Piano Key Weir. *Proc. Intl. Conf. Labyrinth and Piano Key Weirs* Liège B, 151–158. CRC Press, Boca Raton, FL.
Falvey, H.T. 2003. *Hydraulic design of labyrinth weirs*. ASCE Press, Reston VA.
Hager, W.H., Schleiss, A.J. 2009. *Constructions hydrauliques, Ecoulements stationnaires*. Traité de Génie Civil, Vol. 15, Presses Polytechniques et Universitaires Romandes, Lausanne, Switzerland.
Hager, W.H., Schwalt, M. 1994. Broad-crested weir. *J. Irrigation and Drainage Engng* 120(1): 13–26.
Kabiri-Samani, A., Javaheri, A. 2012. Discharge coefficient for free and submerged flow over Piano Key weirs. *J. Hydraulic Res.*, 50(1), 114–120.
Laugier, F. 2007. Design and construction of the first Piano Key Weir (PKW) spillway at the Goulours dam. *Intl. J. Hydropower and Dams*, 13(5), 94–101.
Laugier, F., Lochu, A., Gille, C., Leite Ribeiro, M., Boillat, J.-L. 2009. Design and construction of a labyrinth PKW spillway at St-Marc Dam, France. *Intl. J. Hydropower and Dams*, 15(5), 100–107.
Leite Ribeiro, M., Bieri, M., Boillat, J-L., Schleiss, A., Delorme, F., Laugier, F. 2009. Hydraulic capacity improvement of existing spillways – Design of piano key weirs. 23rd ICOLD *Congress*. Q90, R43, 25–29. Brasilia, Brazil.
Leite Ribeiro, M., Pfister, M., Schleiss, A.J., Boillat, J.-L. 2012a. Hydraulic design of A-type Piano Key weirs. *J. Hydraulic Res.* 50(4): 400–408.
Leite Ribeiro, M., Pfister, M., Boillat, J.-L., Schleiss, A.J., Laugier, F. 2012b. Piano Key weirs as efficient spillway structure. 24th ICOLD *Congress*, Kyoto (J), Q.94 – R.13.
Lempérière, F., Ouamane, A. 2003. The Piano Keys weir: a new cost-effective solution for spillways. *Intl. J. Hydropower and Dams*, 10(5), 144–149.
Lempérière, F., Vigny, J.P., Ouamane, A. 2011. General comments on Labyrinth and Piano Key weirs: The past and present. *Proc. Intl. Conf. Labyrinth and Piano Key Weirs* Liège B, 17–24. CRC Press, Boca Raton, FL.
Machiels, O., Pirotton, M., Archambeau, P., Dewals, B., Erpicum, S. 2013. Experimental parametric study and design of Piano Key weirs. *J. Hydraulic Res.* in press.
Ouamane, A., Lempérière, F. 2006. Design of a new economic shape of weir. Proc. Intl. Symp. *Dams in the Societies of the 21st Century*, 463–470. Barcelona, Spain.
Pfister, M., Battisacco, E., De Cesare, G., Schleiss, A.J. 2013. Scale effects related to the rating curve of cylindrically crested Piano Key weirs. *Proc. Intl. Conf. Labyrinth and Piano Key Weirs* Chatou F, CRC Press, Boca Raton.
Pfister, M., Schleiss, A.J. 2013. Comparison of hydraulic design equations for A-type Piano Key weirs. Proc. Intl. Conf. *Water Storage and Hydropower Development for Africa* (Africa 2013), Addis Ababa (Ethiopia), CD 13.05.
Pfister, M., Hager, W.H., Minor, H.-E. 2006. Bottom aeration of stepped spillways. *J. Hydraulic Eng.*, 132(8), 850–853.
Pralong, J., Vermeulen, J., Blancher, B., Laugier, F., Erpicum, S., Machiels, O., Pirotton, M., Boillat, J.L., Leite Ribeiro, M., Schleiss, A.J. 2011. A naming convention for the Piano Key weirs geometrical parameters. *Proc. Intl. Conf. Labyrinth and Piano Key Weirs* Liège B, 271–278. CRC Press, Boca Raton, FL.
Ramamurthy, A.R., Vo, N.D. 1993. Characteristics of circular-crested weir. *J. Hydraulic Eng.* 119(9), 1055–1062.
Schleiss, A.J. 2011. From labyrinth to piano key weirs: A historical review. *Proc. Intl. Conf. Labyrinth and Piano Key Weirs* Liège B, 3–15. CRC Press, Boca Raton FL.
Tullis, J.P. 1995. Design of Labyrinths spillways. *J. Hydraulic Eng.* 121(3), 247–255.
Vischer, D., Hager, W.H. 1999. *Dam Hydraulics*. Wiley, Chichester UK.

D. Planned projects

Labyrinth and Piano Key Weirs II – PKW 2013 – Erpicum et al. (eds)
 ISBN 978-1-138-00085-8

Design of a Piano-Key Weir for Giritale Dam spillway in Sri Lanka

H.M. Jayatillake
Ministry of Irrigation & Water Resources, Colombo, Sri Lanka

K.T.N. Perera
Irrigation Department, Colombo, Sri Lanka

ABSTRACT: Giritale Reservoir in Sri Lanka (2 300 hm^3), irrigates 3 077 ha in two seasons of the year and its capacity is about 1/2 the seasonal irrigation demand. Working as an on-line reservoir, it is mainly fed by diversions from Mahaweli River System. Since recent, water shortages have been experienced towards end of cropping seasons severely affecting the livelihoods of people settled in the scheme. If the reservoir capacity can be increased, additional diversions from Mahaweli are possible to mitigate the problem. Reservoir is located in a developed area and inundation is a serious limitation when increasing the capacity. With the adoption of PKW, capacity could be increased by 15% without increasing dam height and with minimum inundation. An optimized stepped slope configuration for energy dissipation was incorporated in downstream to minimize tail works. Economy of increasing storage using PKWs against construction of new reservoirs is also presented.

1 INTRODUCTION

Sri Lanka has been experiencing effects of climate change for some time and more severely from beginning of the latter part of last century (Jayatillake et al. 2005). Changes in monsoonal rain and inter-monsoonal rain regimes with increased spells of high intensity rainfalls and increased occurrence of dry spells have been observed. Floods in early 2011 and again in early 2012 followed by severe drought lasted from April to September and then floods again in December in the same year bear example. Giritale Reservoir Scheme located in North-Central Province of Sri Lanka is among many which were seriously affected by these phenomena.

Giritale Reservoir with a capacity of 23 hm^3 (19 000 Acft)[1] amounting to around or less than 1/2 of its seasonal irrigation demand is fed by diversions from Mahaweli river system, in addition to yield from its own catchment. Working as an on-line reservoir, it irrigates 3 077 ha (7 600 Acs) of land for two seasons of the year, namely, the Yala Season lasting from April to August/September and the Maha Season lasting from October to March. Since its restoration in early 1950s, the average land holding size per farm family has dropped from 2 ha (5 Acs) to less than 1 ha (1–2 Acs) due to increase of population. Some families get their turn of cropping only in every third season viz. once in one and half years. Since the main livelihood of people living there is from agriculture any water shortage affects their livelihoods severely as happened during drought in 2012.

The drought 2012 lasted during almost the entire Yala Season and the bulk of the crop was failed due to water shortage. Thus the possibility of increasing reservoir capacity was investigated with a view to provide adequate water security with increased diversions of water, when available, from Mahaweli River system.

[1] Previous designs are in FPS units.

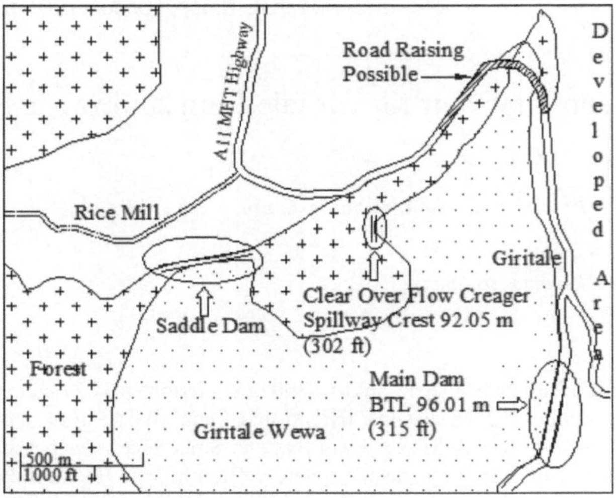

Figure 1. Main features of existing Giritale reservoir (Superimposed on Google Map).

2 THE ORIGINAL GIRITALE RESERVOIR SCHEME

Main features of the original Giritale reservoir consists of a 518 m (1 700 ft) long main dam with a maximum height of 14 m (46 ft) and a Bund Top Level (BTL) of 96.01 m (315.00 ft) ASL[2]; a 670 m (2 200 ft) long saddle dam with a BTL of 94.49 m (310.00 ft) and a 38 m (125 ft) long Creager type clear overflow concrete spillway with a crest level at 92.05 m (302.00 ft) (Fig. 1). Both dams are earth embankments and the spillway does not have gates. The current reservoir capacity is 23.4 hm^3 (19 000 Acft) at Full Supply Level (FSL) and the design High Flood Level (HFL) is 92.96 m (305.00 ft).

The reservoir which had been originally built in 623 AD by ancient kings of Sri Lanka has been restored from time to time and by 1951 it has had a capacity of 9.9 hm^3 (8 000 Acft) irrigating 316 ha (780 Acs) of land. The natural spillway with a crest level at 87.17 m (286.00 ft) then has been at the Saddle Dam.

In early 1950s, the Irrigation Department of Sri Lanka has augmented the reservoir to a capacity of 21.5 hm^3 (17 400 Acft) increasing the irrigated area to 2 632 ha (6 500 Acs) incorporating a new spillway at the current location with its crest level at 91.44 m (300.00 ft) in addition to the 107 m (350 ft) long natural spillway by the Saddle Dam. With this improvement no raising of the Main Dam has been done but the Saddle Dam has been raised to 94.49 m (310.00 ft) from an existing BTL of 91.44 (300.00 ft) ASL.

Again in 1963, the capacity of the reservoir has been further raised to 23.4 hm^3 (19 000 Acft) with the spillway crest level being raised by 0.6 m (2 ft) to 92.05 m (302.00 ft) from the existing level of 91.44 m (300.00 ft) ASL. The dams have not been raised at this stage but the natural spillway that existed by the Saddle Dam has been done away with replacing it with two over flow sections of 15 m (50 ft) each on either side of the concrete spillway. In 2012, under the ongoing Dam Safety & Water Resources Planning Project, the Saddle Dam too was raised to the same height as the Main Dam with the proposal for increasing capacity of the reservoir by adopting a PK Weir.

The national A11 Maradankadawala-Habarana-Thirukkondaiodimadu (MHT) Highway runs around the North-Eastern side of the reservoir periphery and the area above the highway except at the bend is developed. A Garment Factory, some tourist hotels, an army facility and some housing

[2]ASL = Above Sea Level.

Figure 2a. Existing Spillway at Giritale.

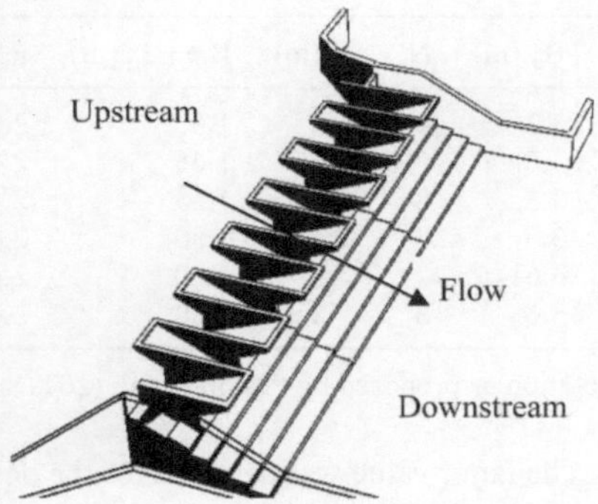

Figure 2b. Existing Spillway at Giritale.

are located in the area and any raising of the reservoir has to be limited so that the new HFL will not cause inundation of this area. Thus raising of the reservoir capacity had to be limited so that the new HFL will not exceed around 93.4 m (306.50 ft) ASL which will involve raising of only a short stretch of the highway at the valley.

On the other hand, in the North, the downstream of the Saddle Dam is heavily developed with large rice milling complex and housing etc. and incorporation of any additional spillway as existed prior to 1960s at this location is now not possible. The potential for increasing the length of existing spillway on either side was also limited due to topography and foundations.

Under the circumstances there was no option other than raising the existing spillway and adoption of a device that gives a higher discharge capacity without increasing the HFL beyond the allowable limit. After study, improvement of existing spillway with the adoption of a Piano-Key Weir with a site specific design found to be providing an effective and viable solution.

3 DESIGN OF THE PIANO-KEY WEIR

3.1 *The existing spillway*

A photograph of the existing spillway is shown in Figure 2a. Raising of the spillway in 1960s has been done without increasing the base width which is approximately 2.7 m (9 ft). Maximum height of concrete section was around 3.7 m (12 ft). The 38.1 m (125 ft) long spillway consists of three blocks of each 12.7 m (41.7 ft) long, separated by contraction joints. The spill tail canal immediately downstream of the spillway consists of soft rock with some hard rock out crops.

The old concrete was of Grade 15 and that on top appeared to be weaker with the poor joint between the two stages being clearly seen while evidence of water seeping through was also seen. Thus it was necessary that the concrete cap on top, done during 1960s be completely demolished and adequate cover be provided for the existing concrete on both upstream and downstream in the new design. Figure 2b gives an isometric view of the proposed PKW at Giritale spillway.

3.2 *PK Weir configuration and parameters*

The PKW parameters for the new design were selected on the following bases:

- Basic shape: Taking into account the site topography, over hangs on both sides were possible and hence a type A shape (Lemperiere & Ouamane, 2003) with symmetrical upstream and downstream overhangs was selected.
- The base length (B_b): With a cover of 600 mm on upstream and with the downstream foundation conditions, a base length (B_b) around 3.3 to 3.6 m was found to be possible after excavation.

Table 1. PKW configurations studied.

	B_b (m)	N_u	B (m)	$B_i = B_o$ (m)	$P_i = P_o$ (m)	P_m (m)	W_i/W_o	W_i (m)	W_o (m)	n = L/W
1	3.3	6	6.6	1.65	2.4	1.6	1.25	3.0	2.4	3.1
2	3.3	7	6.6	1.65	2.4	1.6	1.25	2.5	2.0	3.5
3	3.3	8	6.6	1.65	2.4	1.6	1.25	2.0	1.6	4.0
4	3.6	6	7.2	1.80	2.4	1.6	1.25	3.0	2.4	3.3
5	3.6	7	7.2	1.80	2.4	1.6	1.25	2.5	2.0	3.7
6	3.6	8	7.2	1.80	2.4	1.6	1.25	2.0	1.6	4.3

Notation as proposed by Pralong et al. (2011).

The latter value was selected for the design considering the higher discharge enhancement ratio it gives (see Section 3.3).

- The overhang: An overhang equals to 1/2 the size of base length as proposed Lemperiere (2009) was adopted.
- Height of inlet entrance (P_i): Selection of P_i was based on the existing approach level and a height of backfill that would be required to create active pressure on the upstream face when analyzing over turning. Accordingly, P_i was selected as 2.4 m.
- Inlet-outlet key ratio: As it has been observed that there is practically no increase in discharge enhancement ratio with the increase of W_i/W_o beyond 1.25 (Leite Ribeiro et al. 2011) a ratio of 1.25 was used.
- Sidewall thickness: The sidewall thickness (T_s) was selected as 300 mm to enable placement of reinforcement.
- Number of PKW units (N_u): As the two abutments will have to be raised their cross sections were widened by 600 mm into the weir leaving the two outer blocks shorter by that amount. Number of PKW units for trial design was selected so that L/W ratio is above three and configurations with N_u equal to six, seven and eight were studied. With seven units, alignment of one face of a wall, along the contraction joints was possible while giving a satisfactory L/W ratio. A discharge enhancement ratio above four is also possible over the operating range of heads (See Fig. 4) and the construction will be somewhat easier than that with eight units. Hence the configuration with seven units was preferred.
- Gated key: An overshot flap gate with a sill at former crest level along with provision for stop logs was provided at the outlet key on the Right Bank (RB) end of the weir in order to enable to keep the reservoir water level between the present and new full supply levels for observation purposes during initial filling.
- PKW Crest: Anticipated heads above crest over the operating range is relatively smaller and hence the shape of the weir crest will have some effect on discharge and different shapes are being studied. In order to reduce the roughness coefficient grinding and painting with epoxy is recommended. With the parapet wall the final elevation of the crest will be 93.12 m (305.50 ft) ASL.

The configurations studied are given in Table 1 and the parameters of PK Weir adopted are given in Figures 3a, b and c.

3.3 *Discharge enhancement ratio*

Configurations in Table 2 were studied for discharge enhancement ratio using Universal formula (Leite Ribeiro et al. 2011).

Discharge over standard weir,

$$Q_w = C_d W \sqrt{2g} H^{\frac{3}{2}} \quad (1)$$

where C_d = coefficient of discharge, W = width of weir and H = hydraulic head above crest.

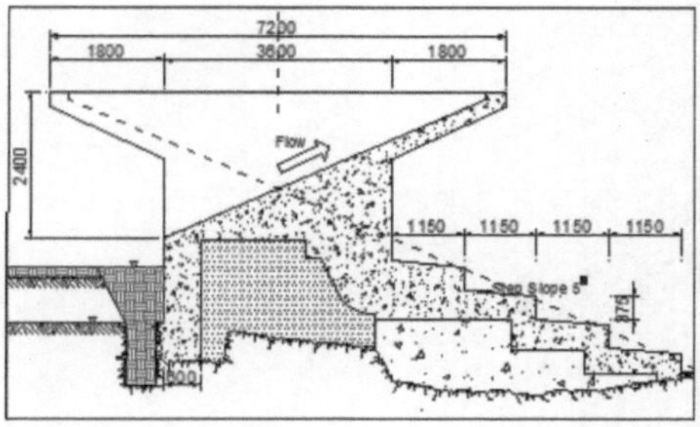

Figure 3a. Cross section of inlet key.

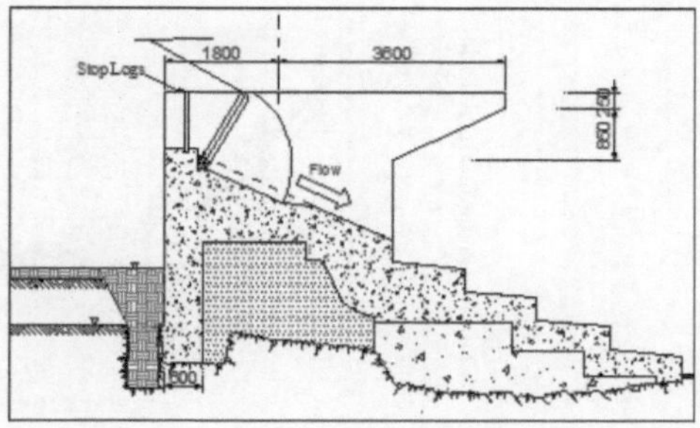

Figure 3b. Cross section of gated outlet key.

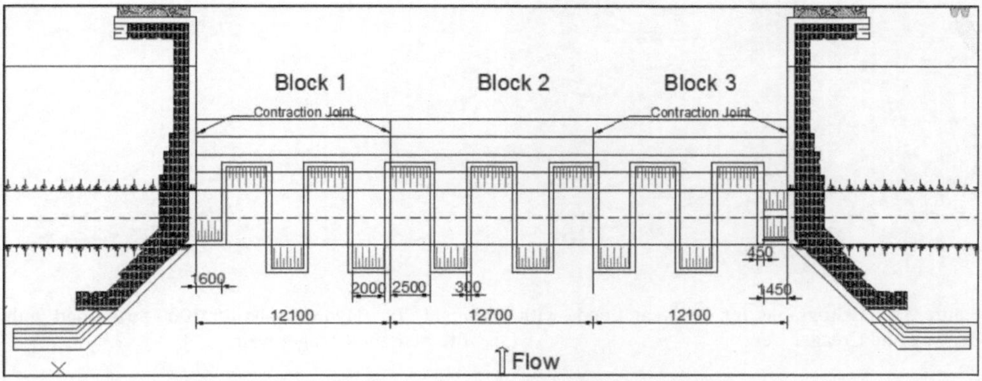

Figure 3c. Plan of PK Weir (All dimensions in Figures 3a, b and c are in millimeters).

Then the discharge enhancement ratio (r) between PKW and standard weir is given by,

$$r = \frac{Q_{PKW}}{Q_W} = \frac{C_{dL}\, L\, \sqrt{2g}\, H^{\frac{3}{2}}}{C_d W\, \sqrt{2g}\, H^{\frac{3}{2}}} \tag{2}$$

where Q_{PKW} = discharge over PKW, L = total developed length along the overflowing PKW crest axis of PKW and C_{dL} = discharge coefficient related to PKW developed length.

Universal formula for discharge enhancement ratio (r) for L/W > 3 (Leite Ribeiro et al. 2011),

$$r = e^{\left(\begin{array}{l} -0.25945\left(\frac{P}{W_i}\right)^{1.4}\left(\frac{H}{P}\right)^{0.15} + 1.0056\left(\frac{L}{W_i}\right)^{0.1}\left(\frac{P}{W_i}\right)^{0.5}\left(\frac{H}{P}\right)^{0.7} \\ +\,0.067404\left(\frac{L}{W_i}\right)^{0.3}\left(\frac{P}{W_i}\right)^{0.1}\left(\frac{W_i}{W_o}\right)^{0.25}\left(\frac{H}{P}\right)^{0.2} \\ +\,13.9156\left(\frac{L}{W}\right)^{0.35}\left(\frac{H}{P}\right)^{0.15} - 14.0239\left(\frac{L}{W}\right)^{0.35}\left(\frac{H}{P}\right)^{0.2} \\ +\,0.094 \end{array}\right)} - 1 \tag{3}$$

Discharge enhancement ratios (r) against head above crest (H) for the different configurations studied are given Figure 4.

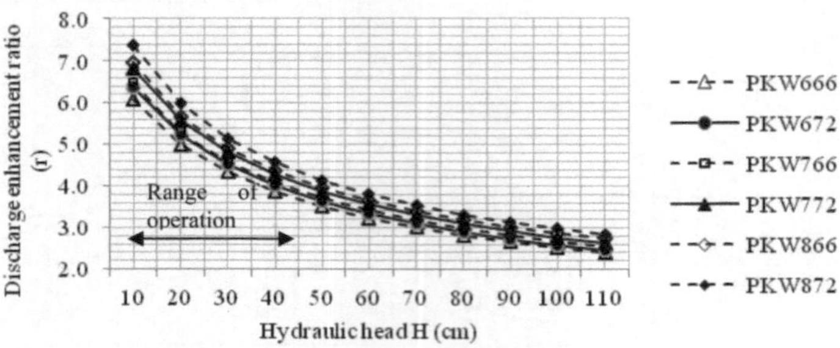

Figure 4. Discharge enhancement ratio (r) vs. hydraulic head (H).

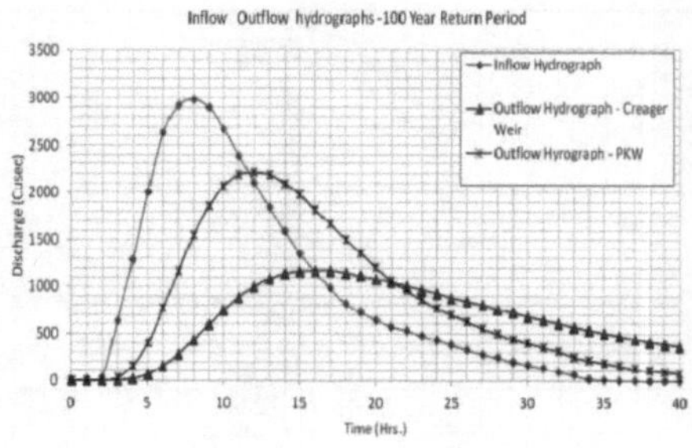

Figure 5. Hydrographs for 100 year floods with PK weir and Creager weir.

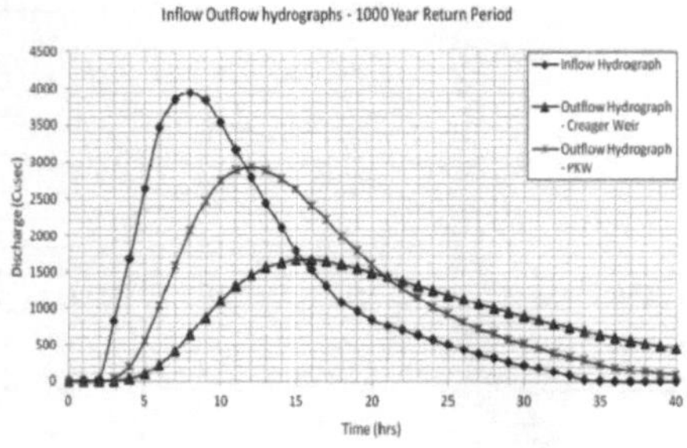

Figure 6. Hydrographs for 1000 year flood with PK weir and Creager weir.

4 FLOOD ROUTING

Spillway crest levels of 92.96 m (305.00 ft) and 93.12 m (305.50 ft) ASL were considered for the design and flood routing was done to estimate the anticipated afflux with 100 year and 1000 year return period floods which were arrived at using Snyder's technique.

It was found that the anticipated afflux for 1000 year flood is within 50 cm (1.5 ft) while that for 100 year flood is around 30 cm (1 ft) for spillway crest at 93.12 m (305.5 ft) ASL. Since this is within the limit for inundation, spillway crest level of 93.14 m (305.5 ft) ASL was proposed for the final design. Results of flood routing with Creager weir and PK Weir for 100 year and 1000 year floods are shown in Figures 5 and 6 respectively.

5 PHYSICAL MODEL TESTS AND DESIGN OF STEPPED ENERGY DISSIPATER

Physical model tests were done along with the Hydraulics Research Laboratory of Irrigation Department of Sri Lanka (Hydraulic Research Laboratory, Irrigation Department, Sri Lanka, 2012). With those results, the discharge enhancement ratio calculated using equation (3) was found to be within 10% test results. In particular model tests were useful in designing the stepped energy dissipater. The height of steps was selected as 375 mm to suit construction and the total depth available. In the initial design, steps were incorporated protruding above the slope line and during model tests it was seen that the flow is missing the steps at certain discharges (Fig. 7a). Hence after several tests steps were lowered below the slope line, their tread widths increased and a tread slope of 5° was introduced so that the flow does not skip steps during operating range (Fig. 7b). With this configuration of sloping steps (Figs. 3a & b) it was found that further arrangements for energy dissipation in downstream are minimum.

Figure 7a. Initial design of steps – Flow skipping steps.

Figure 7b. Flow with steps lowered and widened.

Table 2. General comparison of cost of storage with new reservoirs and PKW in Giritale.

	Reservoir	Capacity (hm^3)	Cost (LKR* Mn)	Cost (US$ Mn)	Cost/unit capacity (LKR Mn /hm^3)	Cost/unit capacity (US$ Mn /hm^3)
1	Giritale PKW	3.5**	30	0.3	8.9	0.05
2	Kekiriobada	2.7	445	3.4	164.8	1.26
3	Moragahakande	465.0	26195	202.3***	56.3	0.44
4	Lower Malwathu Oya	197.0	7000	54.0	35.5	0.27
5	Yan Oya	169.0	15240	117.6	90.2	0.70

*US$ 1 = LKR 129.55; **Incremental capacity; ***Excluding power house; Data source: Ministry of Irrigation & Water Resources Management and Irrigation Department, Sri Lanka.

6 STRUCTURAL DESIGN

The spillway structure was designed to meet the BS 8110 requirements in general. Weak concrete in the top layer of old spillway was demolished and bond between old and new concrete was ensured with scrabbling and placing steel dowels. In order to prevent shrinkage cracking contraction joints were maintained at the same locations in the central blocks while those which were at the abutments were shifted along with the widening of the abutments to suit. Anti- crack reinforcement was provided for the surfaces. While the superstructure concrete was of Grade 30 and the foundation filling was with mass concrete of grade 15 covered with Grade 25 concrete for exposed areas. The foundation rock was fissured and seepage was noticed after exaction in downstream and hence upstream clay cut off with clay blanketing was proposed as a remedial measure.

7 ECONOMICS OF PK WEIR

A general comparison of the economy of increasing storage of existing Giritale reservoir using PKW was assessed against that by construction of new reservoirs considering some new reservoirs done during recently or being done at present in Sri Lanka (Table 2). The Kekiriobada reservoir was completed in 2009 and Moragahakande reservoir is under construction. The lower Malwathu Oya and Yan Oya reservoirs are in the final stages of feasibility.

8 CONCLUDING REMARKS

Capacity enhancement of Giritale with a PK Weir is already commenced. PK Weirs appear to be promising as a structural intervention for adapting to climate change in particular and in increasing water storage in general. Taking into account the high discharge enhancement ratio at low heads there is potential for adopting PKW in Sri Lanka's large number of medium and small reservoirs. Government of Sri Lanka has now recognized the capacity enhancement of reservoirs using PKW as a strategy for adaptation for climate change in irrigation and water resources sector (Ministry of Finance and Planning, Sri Lanka, 2013).

ACKNOWLEDGEMENTS

Authors gratefully acknowledge the policy decision taken by Eng. K.W. Ivan De Silva, Secretary, Ministry of Irrigation and Water Resources Management, Sri Lanka to implement this as a pilot initiative in Giritale.

REFERENCES

Irrigation Department, Hydraulics Research Laboratory. 2012. *Internal memo Physical model testing of PKW for Giritale*. Colombo: Irrigation Department.

Jayatillake, H.M., Chandrapala, L., Basnayake, B.R.B.S. & Dharmaratne, G.H.P. 2005. Water Resources and Climate Change. In N.T.S.Wijesekera, K.A.U.S.Imbulana and B. Neupane (eds), *Proc. Workshop on Sri Lanka National Water Development Report*. Paris: UN World Water Assessment Programme.

Leite Ribeiro, M., Boillet, J.L., Schleiss, A.J., Le Doucen, O. & Laugier, F. 2011. Experimental parametric study for hydraulic design of PKWs. [on line] Available from: *http://infoscience.epfl.ch/record/167398* [Accessed 21st August 2012].

Lemperiere, F. 2009. New Labyrinth weirs triple the spillways discharge, Data for an easy design of P.K.Weirs. [on line] Available from: *http://www.hydrocoop.org/publications/Labyrinth_%20weirs_09[1].2009.pdf* [Accessed 21st August 2012].

Lemperiere, F. & Ouamane, A. 2003. The Piano Keys weir: a new cost-effective solution for spillways. [on line] Available from: *http://www.hydrocoop.org/publications/Piano_Keys_Weir.pdf* [Accessed 21st August 2012].

Ministry of Finance and Planning. 2013. Irrigation – Adaptation to Climate Change. *Annual Report 2012*, [on line] Available from: *http://www.treasury.gov.lk/reports/annualreport/2012/1-EconomicPerspectives-Sri%20Lanka%20.pdf* [Accessed 5*th* June 2013]. 40–42.

Ministry of Irrigation and Water Resources Management, Ministry Planning Cell, 2011. *Lower Malwathu Oya Reservoir Project Pre-feasibility Report*. Colombo: Ministry of Irrigation and water Resources Management.

Ministry of Irrigation and Water Resources Management & Sino Hydro Corporation Limited, 2013. *Basic Design, Headworks Conceptual (Layout) Design*. Colombo: Ministry of Irrigation and water Resources Management.

Ministry of Irrigation and Water Resources Management & Sino Hydro Corporation Limited, 2010. *Proposal for Yan Oya Reservoir Project in Sri Lanka, Volume I, Technical Proposal*. Colombo: Ministry of Irrigation and water Resources Management.

Pralong, J., Vermeulen, J., Blancher, B., Laugier, F., Erpicum, S., Machiels, O., Pirotton, M., Boillet, J.L., Leite Ribeiro, M. & Schleiss, A. 2011. A Naming Convention for the Piano Key Weir Geometrical Parameters. [on line] Available from: *http://infoscience.epfl.ch/record/167402/files/2011-781_Pralong_Vermeulen_Blancher_Laugier_Erpicum_Machiels_Pirotton_Boillat_LeiteRibeiro_Schleiss_A_naming_convention_for_the_piano_key_weirs.pdf* [Accessed 14th August 2012].

Labyrinth and Piano Key Weirs II – PKW 2013 – Erpicum et al. (eds)
© 2014 Taylor & Francis Group, London, ISBN 978-1-138-00085-8

Piano Key Weir spillway: Upgrade option for a major dam

M.A. Phillips
URS Australia, Brisbane, Queensland, Australia

E.J. Lesleighter
Hydraulics Engineering Specialist, Cooma, New South Wales, Australia

ABSTRACT: A spillway upgrade conceptual design and selection process was undertaken to identify options for upgrading the Dartmouth Dam to pass the Probable Maximum Flood (PMF). The piano key weir was initially developed from the limited available publications on the weir design, and further developed with the use of a 1:60 scale model. The piano key weir, a variation of the labyrinth weir, is a passive spillway that utilises a total weir length several times that of the existing spillway width. For the Dartmouth Dam study, the piano key weir design that was developed consisted of a 7-cycle, 9 m high structure, with a total weir length of nearly 600 m, or more than 6 times the existing spillway width of 91 m. The spillway was designed to pass the routed PMF outflow of approximately 11,500 m^3/s with a head of approximately 11 m.

1 BACKGROUND

Dartmouth Dam is located on the Mitta Mitta River, one of the main tributaries of the Murray River in north eastern Victoria, Australia. The dam is located approximately 25 km upstream of the town of Mitta Mitta and about 70 km upstream of the Hume Dam. When full the reservoir stores 3,856 M m^3 and acts as long term carry over storage for the Murray Darling Basin.

The dam embankment, as shown in Figure 1, is a 670 m long, 180 m high central core earth and rockfill embankment. The existing spillway, located in the left dam abutment, incorporates a 3 m high, 91 m long ogee crest and a concrete lined chute which discharges into a series of rock cascades.

A spillway upgrade conceptual design and selection process was undertaken to identify options for upgrading Dartmouth Dam to pass the Probable Maximum Flood (PMF). A number of upgrade

Figure 1. Dartmouth Dam Existing Spillway and 1:60 Scale Physical Model.

options were investigated, including variations of dam raise heights and spillway modifications. One of the options, the piano key weir, was developed further for the study because:

- The weir could be constructed at the same location as the existing ogee crest spillway;
- The weir is a passive structure (i.e. no mechanically moving parts); and
- The maintenance requirement would be similar to the existing ogee crest structure.

The Dartmouth Dam piano key weir option was developed and tested in three phases:

- Testing of an initial design in a 1:60 scale physical model of the dam and spillway;
- Hydraulic optimisation of the piano key weir was conducted using Computational Fluid Dynamics (CFD) modelling; and
- Testing of the refined piano key weir design in the 1:60 scale physical model.

The 1:60 scale physical hydraulic model testing was conducted at the SMEC Fluid Mechanics laboratory in Cooma, New South Wales, Australia. The model was constructed, as shown in Figure 1, to include the approach reservoir, upstream face of the dam, spillway area, concrete chute, rock cascades and downstream river area. The model was constructed to allow physical measurements for flow rates, including reservoir level, velocity magnitude, static pressure levels, dynamic pressure measurements (transducers) and dye tracing to observe flow patterns.

2 PIANO KEY WEIR – DESCRIPTION

The piano key weir is a passive structure that, much like a labyrinth weir, utilises a longer weir length for low flows and acts similar to an ogee crest or a broad crested weir at higher discharges. The piano key weir, as shown in Figure 2, can generally be characterised as follows:

- Consisting of one or more cycles (1 cycle shown in Figure 2);
- Rectangular shaped cycles;
- Sloping apexes which overhang both upstream and downstream; and
- The width of the inlet bay is greater than the width of the outlet bay.

A piano key weir spillway is a modified labyrinth weir design, the major differences being:

- sloped apexes with overhangs;
- a rectangular plan shape; and
- varied upstream and downstream cycle widths.

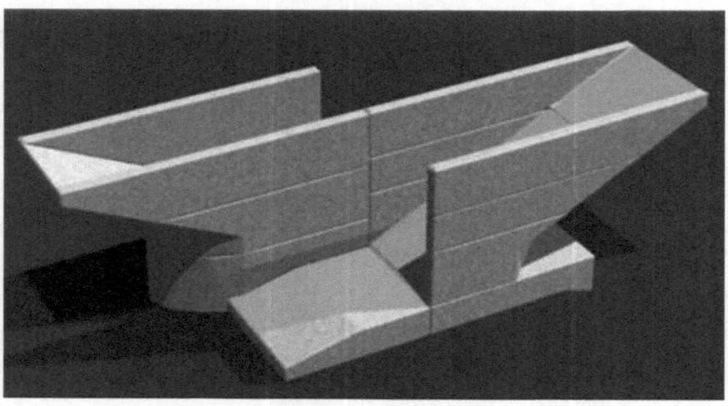

Figure 2. Single cycle Piano Key Weir.

The concept of the piano key weir spillway was developed about ten years ago where many have been constructed around on dams around the world. Model testing and research of the piano key weir structure has been conducted at about half a dozen hydraulic laboratories around the world. Several technical papers have been published by HYDROCOOP in France on the model testing and design of piano key weirs and many are publicly available. The recommendations and results from these papers formed the basis of the initial conceptual design.

3 INITIAL PIANO KEY WEIR DEVELOPMENT AND TESTING

The initial Dartmouth Dam piano key weir model was developed using the hydraulic design criteria for a "Type A" piano key weir structure from several technical publications by HYDROCOOP, primarily from Barcouda et al. (2006) and Lempérière, F. & Ouamane, A. (2003).

The initial Dartmouth Dam piano key weir concept consisted of the following:

- 91 m wide spillway;
- 9 m high weir structure;
- 36 m long upstream to downstream;
- 7 cycles with the inlet wells 7.1 wide and the outlet wells 5.9 m wide (a 1.2 ratio);
- Total weir length of 595 m;
- Overhangs at a 2H:1V slope;
- 1 m high sill;
- Maximum head of 11.4 m; and
- 1 m approach excavation (equal to sill height).

The model testing for the piano key weir generally focused on:

- observing how the structure functioned, as there was limited information available at the time about the design of piano key weir structures; and
- the efficiency of the structure and testing various methods to improve the rating curve, these included:
  - Adding a trapezoidal shaped "nose" under the upstream overhang;
  - Adding a rounded shape to the top of the upstream sill face; and
  - Adding a rounded shape to the face of the upstream overhang.

Based on the model testing of the piano key weir, the following observations and key design results are summarised as:

- The initial testing of the piano key weir showed a peak discharge capacity of 10,200 m^3/s at the maximum head of 11.4 m;
- It was observed and recorded that the addition of the shaped noses under the upstream overhang:
  - Improved the efficiency of the structure by reducing the head loss under the overhang and increasing the flow in the inlet bays;
  - The piano key weir with the pier noses increased the peak discharge capacity to 10,500 m^3/s at the maximum reservoir level with 11.4 m of head; and
  - The observable frequency of pulsing water over the weir, as a result of vortex shedding around the upstream walls under the overhang, was greatly diminished.
- It was observed and recorded that the addition of a rounded sill and rounded face on the upstream overhang (as shown in Figure 3):
  - Improved the efficiency of the structure by reducing the head loss at the inlet bays and increasing the flow at the overhang structure;
  - The piano key weir with the rounded faces further increased the peak discharge capacity to 10,900 m^3/s at the maximum reservoir level with 11.4 m of head; and
  - The observable frequency of pulsing water over the weir, as a result of vortex shedding from the sharp corner on the upstream sill face, was nearly eliminated.

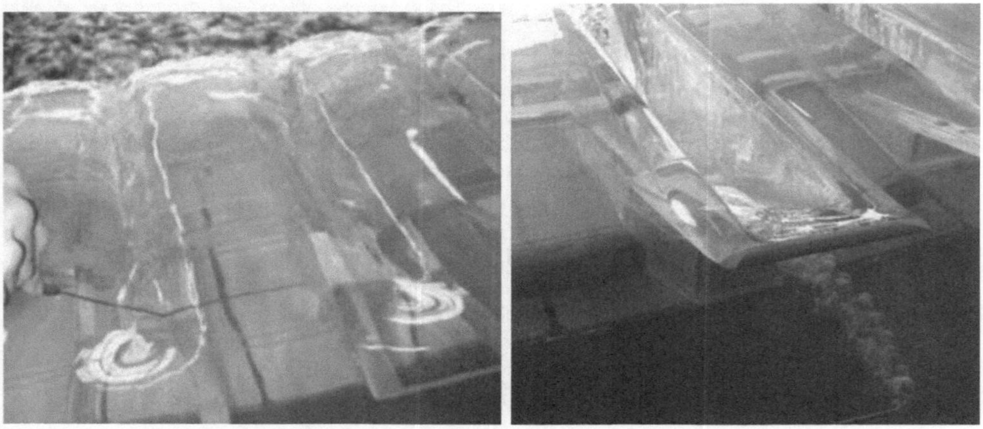

Figure 3. Vortex Shedding in Initial Model and Tapered Nose to Reduce Vortex Shedding.

- At the highest discharges, it was observed that the flow at the downstream end of the piano key weir was unstable due to negative pressures under the downstream overhang. These negative pressures resulted in violent pulsing under the downstream overhang as air was introduced and then removed.

4 HYDRAULIC OPTIMISATION OF PIANO KEY WEIR USING CFD MODELLING

Following the successful testing of the initial piano key weir arrangement in the physical model, a three-dimensional CFD hydrodynamic model was developed to optimise the shape and configuration of the piano key weir structure based on observations from the initial physical model testing and advice from HYDROCOOP in France. A single cycle CFD model was developed to:

- Estimate the capacity of the piano key weir for various design shapes in order to increase the overall hydraulic efficiency of the structure;
- Estimate the potential for jet breakup from the downstream overhang, reducing the potential for negative pressures at high heads; and
- Provide additional hydraulic information for the structural analysis of the weir structure.

The use of a CFD model was recommended for the optimisation process because of the ease with which the model geometry could be modified (particularly weir shape, bay widths and structure height), tested, and results obtained in a short timeframe, as compared to making modifications in the physical model. A CFD model was developed to simulate the final configuration of the initial piano key weir in the physical model and verify that the CFD model results were comparable to the results from the physical model. The CFD model (10,955 m^3/s) showed results within 1% of the physical model (10,900 m^3/s) results for a head of 11.4 m. Figure 4 shows the flow over the single cycle weir with stream lines to show flow direction.

Following verification of the CFD model with the physical model results a number of changes to the weir design were made and tested in the CFD model. A summary of the weir modifications and results tested in the CFD model are presented below:

- A varied approach depth of 1 m and 2 m. The 2 m approach depth resulted in an increase in flow over the spillway of 231 m^3/s or 2.2% at a head of 11.4 m (as shown in Figure 4);
- Centre wall crest shape of ½ round or ¼ round. The ½ round crest resulted in an increase in flow over the spillway of 32 m^3/s or 1.9% for 2 m of head. There was minimal difference in results for heads greater than 5 m;

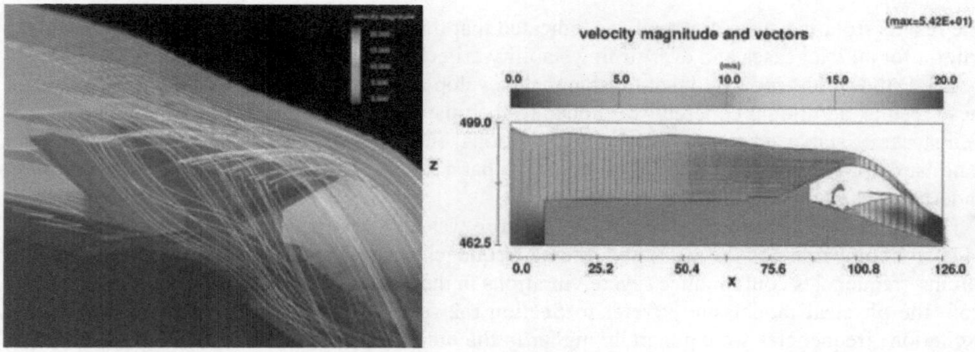

Figure 4. Streamlines and Velocity Profile from CFD model for 11.4 m Head.

- A sloped entrance slab (as opposed to a rounded corner as tested in the initial model) for the inlet bays resulted in an increase in flow over the spillway of 175 m^3/s or 1.7% for 11.4 m of head;
- The shape of the support section under the downstream cantilevered over hang was tested as a square shape and with a 45 degree angle to form a nose shape. It was observed in the initial testing that at this location there was a constriction in flow that limited the discharge of water out of the outlet bay for flows greater than approximately 6,000 m^3/s. The 45 degree angled nose shape resulted in an increase in flow over the spillway of 196 m^3/s or 1.8% for 11.4 m of head;
- A varied inlet to outlet bay ratio of 1.2 and 1.4 The increased inlet to outlet bay ratio of 1.4 (versus 1.2 for all other runs) resulted in a marginal increase in flow over the spillway of 49 m^3/s or 0.5% for 11.4 m of head. Further correspondence with HYDROCOOP (Correspondence 2009), regarding the inlet to outlet bay ratio and the CFD model results was undertaken. Their recommendation was to maintain the 1.2 ratio or possibly investigate a lower ratio toward 1.0. Based on the recommendation by HYDROCOOP, the small benefit in maximum discharge, and the fact that the 1.2 ratio had already been modelled and proven to work, the existing ratio of 1.2 was selected for further design and modelling; and
- Type "A" piano key weir or Type "B" piano key weir. It was recommended by HYDROCOOP (Correspondence 2009) that a 11.7 m high Type "B" piano key weir be tested because of the higher discharge capacities that can be realized as opposed to a 9 m high Type "A" structure. A Type "B" structure is generally characterized as a taller structure, with steeper bay slopes (1H:1V) and lower cycle length. The CFD model results showed that the Type "B" structure could increase the peak discharge by 400 m^3/s or 3.7% at 11.4 m. However, due to the lower total cycle length for the Type "B" (391 m versus 595 m), the Type "B" could not pass the PMF routed hydrograph with a maximum head of 11.4 m.

Based on the CFD modelling, the following modifications were recommended for testing in the physical hydraulic model:

- Type "A" piano key weir structure retained;
- ½ round crest of centre walls;
- Cycles with the inlet wells 7.1 wide and the outlet wells 5.9 m wide (a 1.2 ratio);
- 2.0 m high sill; and
- 2 m approach excavation (equal to sill height).

5 STRUCTURAL ANALYSIS OF REVISED PIANO KEY WEIR

A structural analysis was conducted to size the walls, slab and overall structural to meet stability criteria (sliding, overturning, and flotation). The structural analysis used both the results of the initial physical model testing and the CFD model results for estimating hydraulic input parameters.

The results from the structural analysis indicated that the piano key weir satisfied sliding stability criteria for all load cases and overturning stability criteria for all load cases except for the PMF. In the PMF, the results indicated that tension may develop at the upstream edge of the slab. Based on these results, additional concrete was added to the upstream pier supports to make the piano key weir structure stable as a passive structure. Additional cursory analyses suggested that the structure is not susceptible to flotation, even in the event that a crack allows full reservoir head to develop underneath the width of the slab.

The natural frequency of the prototype structure was also estimated to be in the range of 12 to 24 Hz for comparison to the estimated forcing frequencies from the physical model to determine if forcing frequencies could induce severe vibrations in the structure. The analysed spectral densities from the physical model study (refer to Section 6.2 of this paper) showed that the forcing (or excitation) frequencies were generally higher in the outlet bay, and between 0 to 4.5 Hz at port A and between 3 and 25 Hz in port B. While the forcing frequencies at port B were within the range of the natural frequency of the structure, the spectral densities were significantly lower and this port was located near the base of the inlet bay where the structure is expected to be stiffer.

6 DETAILED HYDRAULIC TESTING OF PIANO KEY WEIR

Based on the optimisation analysis from the CFD modelling and the structural analysis for wall sizing and stability analysis, a revised piano key weir structure was developed for testing in the physical hydraulic model that consisted of the following arrangement (refer to Figures 5 and 6):

- 9 m high weir structure;
- 36 m long upstream to downstream;
- A total weir length of 618 m or 6.73 times the length of the existing 91 m long ogee crest spillway;
- ½ round crest of centre walls (750 mm thick);
- Cycles with the inlet wells 7.05 wide and the outlet wells 5.88 m wide (a 1.2 ratio);
- Overhangs at a 2H:1V slope;
- 2.0 m thick sill;
- Maximum head of 11.4 m; and
- 2 m approach excavation (equal to sill height).

Detailed hydraulic testing of the piano key weir structure was conducted for a range of flood discharges. Testing of the weir structure focused on the following:

- Final spillway rating curve;
- Eliminating the potential for negative pressures on the underside of the structure through the use of aeration splitter piers located at the downstream end of the centre walls;
- Water surface profiles and velocity measurements along six longitudinal alignments and more than twenty sections at the following locations:
  - Spillway approach channel;
  - Three upsloping (inlet) bays;
  - Three downsloping (outlet) bays;
  - Downstream spillway chute; and
  - Along the top three rock cascades.
- Static pressure measurements (piezometers) along six longitudinal alignments at 29 locations on the structure and 36 locations in the spillway chute, including:
  - On the upsloping bays from the base slab to the downstream nose on the underside;
  - On the downsloping bays from the upstream nose to the downstream edge of the base slab; and
  - Along the downstream spillway chute slab.
- Dynamic pressure measurements (transducers):
  - On the centre wall of one bay; and
  - On the downstream spillway chute slab.

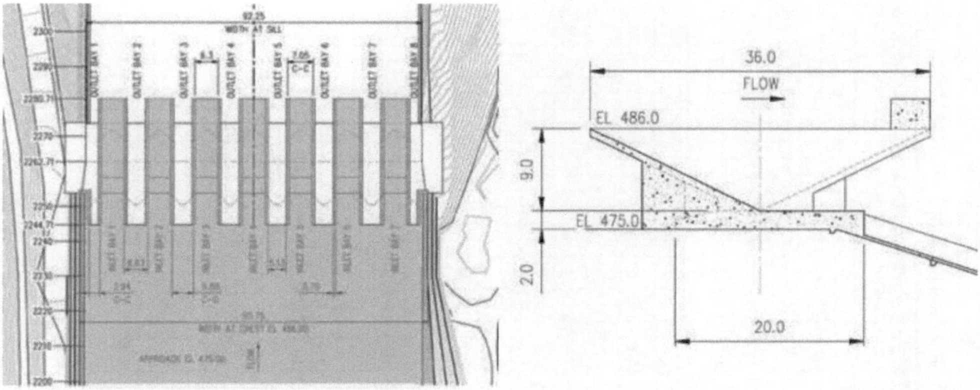

Figure 5. Final Piano Key Weir Configuration in the Physical Model.

Figure 6. Final Arrangement of Piano Key Weir and 11.4 m head, or 11,650 m^3/s.

– Dye testing, with photographic and video recording, along the structure for noting flow direction and potential vortex formation

Numerous sets of measurements were recorded over the model testing schedule, however for the purposes of this paper, the key model test results as a guide for future piano key weir designs are presented below.

6.1 *Spillway rating curve*

The spillway rating curve for the final piano key weir model configuration showed that at the maximum head of 11.4 m, a discharge of 11,650 m^3/s, as shown in Figure 7, could be passed over the structure. This is approximately a 6.9% increase over the discharge achieved in the initial physical model testing. It should be noted that there was an approximately 10% increase in discharge over the entire spillway rating curve from the initial piano key weir arrangement to the final configuration, as shown in Figure 7, largely a result of a combination of the following:

– Increased total weir length (598 m to 618 m);
– Improved flow hydraulics through the optimization process (CFD modelling); and
– Increased flow area due to sloped spillway abutment walls.

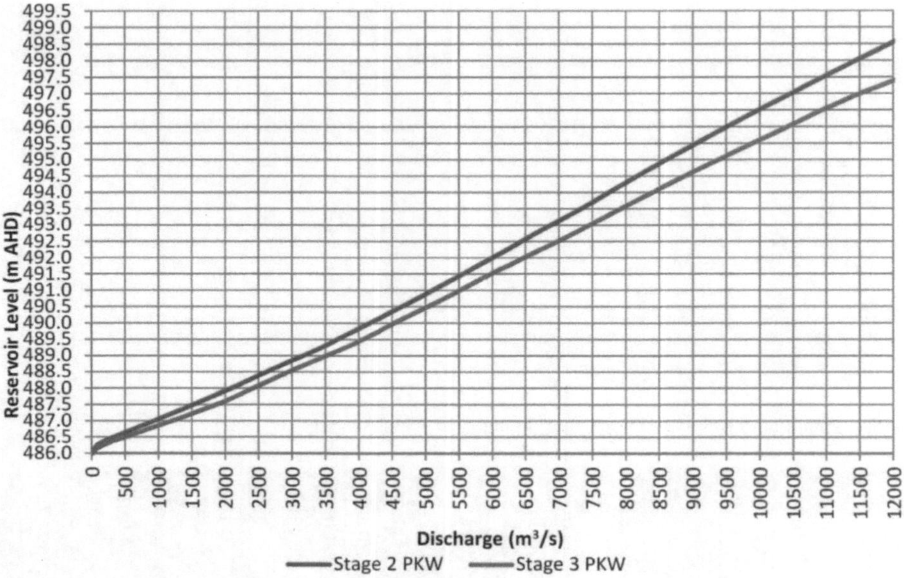

Figure 7. Spillway Rating curves for the Initial Model (Stage 2) and the Final Model (Stage 3) arrangement.

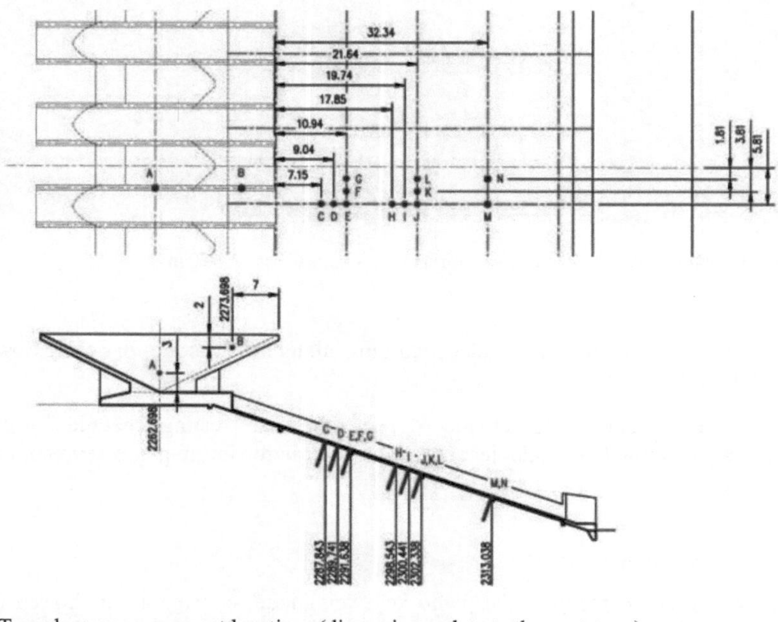

Figure 8. Transducer measurement locations (dimensions relate to the prototype).

6.2 *Dynamic pressure measurements*

During the hydraulic model testing of the PK weir, consideration was given to the possibility of flow-induced vibration of the walls within the weir structure. In order to address this issue, transducers were placed in two locations on the model weir. The locations, A and B, are shown in Figure 8. The figure shows also the location of transducer measurements on the spillway chute below the weir, areas which were significant for the spillway which was the subject of the model

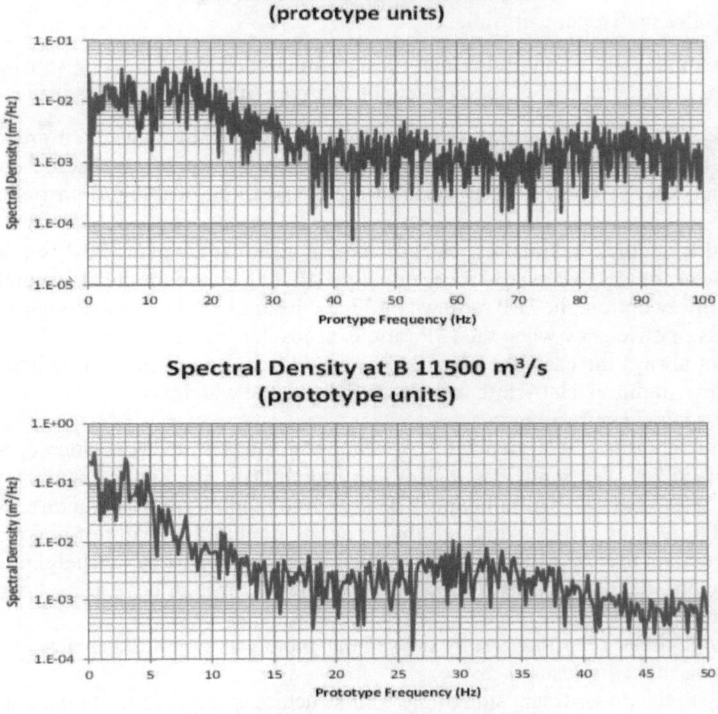

Figure 9. Spectral Density Results for 11,500 m^3/s at Locations A and B.

studies. For the present discussion the measurements at A and B only will be considered as they relate to the specific design of piano key weirs.

The dimensions of the walls in the weir have been described elsewhere. Pressure measurements were taken at four discharges, up to 11,500 m^3/s. The sampling was at a number of rates for durations between 60 seconds and 10 minutes (model time). The digitised signatures were used to carry out a Fourier analysis to compute the spectral density. Figure 9 is a sample only of the spectra in prototype units, based on conversion of model data using Froude law scaling. It is known that a direct scaling of model frequencies to the prototype in this way is subject to some scale effect due to the model's inability to reproduce the higher frequencies. The purpose of the procedures described herein is to illustrate how model or experimental data on the pressure transients are needed as part of the structural evaluation and design of the piano key weir.

7 CONCLUSIONS

The piano key weir concept for Dartmouth Dam was developed in three separate modelling phases in order to optimise the proposed design, to be hydraulically efficient, cost effective and to minimise the impact to current operations. The model study proved a success and showed that the final piano key weir configuration would operate as intended. The piano key weir was selected for model testing and design development for the following reasons:

- It is a passive concrete structure (i.e. no mechanically moving parts), thereby requiring minimal maintenance;
- The structure would be located upstream of the existing concrete spillway chute and fit within the existing 91 m spillway width; and

- The weir would be hydraulically efficient to meet the flood capacity requirements in conjunction with a dam raise with a parapet wall.

Through the model study and preliminary design process of the piano key weir for Dartmouth Dam, several suggestions are presented below for this type of structure:

- The piano key weir is a unique variation on the labyrinth spillway, where a greater total weir wall length can be achieved in the same spillway width. For spillways that are constrained by width, a piano key weir may be a superior choice over a traditional labyrinth weir for the same weir height;
- Previous model studies and this model study have shown that the piano key weir performs well for hydraulic heads (H) that exceed the weir height (P). In the case for the Dartmouth Dam piano key weir spillway option, the H/P ratio was 1.27 at a head of 11.4 m. Traditional labyrinth weirs exhibit a loss of efficiency when the H/P ratio exceeds 0.9 (Falvey 2003);
- Although not always the case, the piano key weir will require a significantly larger volume of concrete than a traditional labyrinth weir, largely due to the wider base slab and volume required for stability against overturning;
- Hydraulic model testing of large piano key weir structures should be considered. At the time of the initial piano key weir model testing for Dartmouth Dam, the highest piano key weir tested and built in the world was 5 m in height. The proposed piano key weir structure for Dartmouth Dam would be 9 m in height. Although this is nearly double the height of any previous piano key weir structure, there is precedence globally for labyrinth weirs of this height. However, this model study did identify areas that should be considered for high head, large height piano key weir structures:
  - The dynamic pressure testing results showed that there was a minimal chance of the flow over the structure inducing resonance;
  - Air supply to the downstream side of the weir structure is required for larger head structures. This was observed in the model for heads over the weir greater than 8 m and was eliminated through the introduction of 3 m high aeration (flow) splitters located on the downstream end of each centre wall.
- The combined use of physical hydraulic modelling and CFD modelling for this study provided a streamlined means of optimising the piano key weir structure arrangement and obtaining static and dynamic measurements.

ACKNOWLEDGEMENTS

The authors would like to acknowledge Goulburn-Murray Water and the Murray-Darling Basin Authority for allowing us to present this study. We would also like to acknowledge the SMEC hydraulics laboratory, Dr. Henry Falvey, Mr. Brent Mefford, and Mr. Ian Hampton for their contributions to the physical model studies.

REFERENCES

Barcouda, M., et al. 2006. Cost Effective Increase in Storage and Safety of Most Dams Using Fusegates or P.K. Weirs, ICOLD Conference 2006.

Corresponence F. Lemperiere and Dr. Henry Falvey regarding suggested modifications for the piano key weir spillway, 15 September 2009.

Falvey, H. T. 2003. *Hydraulic Design of Labyrinth Weirs*, American Society of Civil Engineers Press, 2003.

Lempérière, F. & Ouamane, A. 2003. The Piano Keys weir: A new cost-effective solution for spillways. International Journal on Hydropower and Dams, 10, 144–149.

Phillips, M. A., Maslin, K. & Reynolds, A. 2011. Increasing spillway capacity at Dartmouth Dam. International Water Power and Dam Construction, 63, 36–38.

Labyrinth and Piano Key Weirs II – PKW 2013 – Erpicum et al. (eds)

Designing a new spillway for the Rassisse Dam

A. Bail, L. Deroo & J.P. Sixdenier
ISL INGENIERIE, France

ABSTRACT: The Rassisse Dam (Tarn – France) is a 60 year old arch and gravity dam. Following previous studies made by the CEMAGREF and Coyne et Bellier, the Tarn Prefet – State representative – obliged the owner (Syndicat Intercommunal pour l'Aménagement Hydraulique du Dadou – SIAH) supplying drinking water to the nearby communities to lower the normal reservoir level by 2.9 meters and upgrade the dam (stability and flood protection). Main issues are the present spillway capacity to withstand extreme floods and the stability of the dam in case of very high water elevation level. An upgrading project was derived, that had to take into account a design flood twice (431 m^3/s) as high as the present gated spillway capacity, considering also that Max Water Level (MWL) has to be kept under elevation Normal Water Level (NWL) + 1.20 m, little space to accommodate for the new spillway, requirements of low costs and low environmental impacts and the need to dismantle the present gated spillway, 3.30 meter under NWL, without modifying the arch dam load on its abutments.

1 INTRODUCTION

1.1 *The dam*

Built from 1952 to 1954 for the Syndicat Intercommunal d'Aménagement Hydraulique du Dadou (SIAH), the Rassisse Dam is located on the Dadou River, in the French Tarn Department, 20 km south East of Albi. The reservoir is mainly used for water consumption (3 hm^3), low flow support during dry months (2.5 hm^3) and irrigation (1 hm^3).

The dam design by Coyne-et-Bellier consists of a 35 m high and 100 m long high arch dam closing the Dadou river gorge, and a 12 m high and 300 m long gravity dam closing the valley (Fig. 1). The crest is at elevation 360 NGF, the parapet wall at 360.60 NGF. The designed Normal Water Level (NWL) is at elevation 360 m, for a reservoir capacity of 13.39 hm^3.

The dam is equipped with two gated spillways each 7 m wide, at 356.10 NGF, located on the arch dam crest (Fig. 2). The discharge capacity at 360.60 NGF has been computed at 284 m^3/s. Since the flood attenuation in the reservoir is low, the discharge capacity of the spillways needs to be in accordance with the flood peak flow.

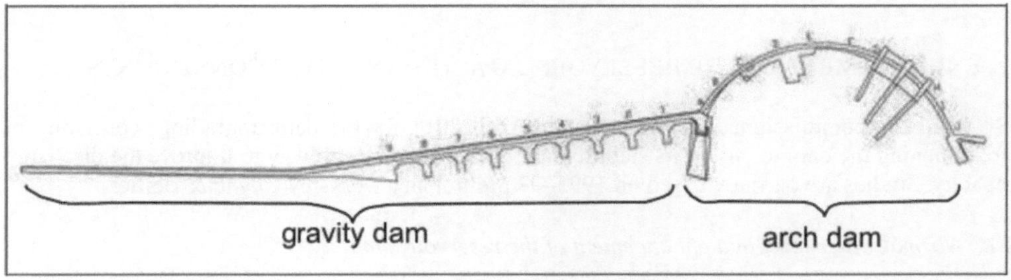

Figure 1. Dam general layout.

Figure 2. Existing gated spillway on the arch dam.

1.2 *Previous studies on the dam safety*

Previous studies on the dam showed the urgent need to improve the dam safety. Indeed, the stability of the dam lies on passive anchors in the foundations for the gravity dam and active anchors for the arch dam. Their conditions are unknown and their load capacities have to be considered as non-existent.

Furthermore, according to recent hydrological studies, the maximum discharge capacity of the existing spillway matches a 200 year return flood for the Dadou River, which is by far lower than the usual standards.

This led the prefecture of the Tarn Department to lower the normal water level in the reservoir, by letting the spillway gates open all year round. The Normal Water Level is therefore lowered at elevation 356.10 NGF, with a loss of reservoir capacity of 4.62 Mm^3.

1.3 *Previous studies: 1996 weir design*

Due to the high difference between the existing discharge capacity and the needed one, a simple upgrade of the existing could not be achieved. The 1996 study by Coyne et Bellier proposed to build a new weir on the right bank and to improve the existing weir.

This design was based on a 10 000 years return period and allows a discharge capacity of 350 m^3/s at 360.00 NGF elevation for the new weir, i-e a total discharge capacity of nearly 570 m^3/s for the two spillways. It consisted in the creation of a 17 m wide gated broad crested weir, with a sill at elevation 355 NGF and replacement of the existing spillway gates.

2 ENHANCEMENT OF THE RESERVOIR CAPACITY AND SITES CONSTRAINTS

ISL has been commissioned by the owner in April 2011 for the dam upgrading, consisting in strengthening the dam to ensure its stability and building a new spillway to improve the discharge capacity. Studies are partially based on 1995–97 preliminary works by Coyne & Bellier.

2.1 *Normal water level and enhancement of the reservoir capacity*

The project initially considered a NWL at elevation 359 NGF, one meter below the original one, and a MWL at 360 NGF, corresponding to the dam crest. With these elevations the reservoir capacity

Table 1. Enhancement of the reservoir capacity.

Solution	NWL (m.NGF)	MWL (m.NGF)	Reservoir capacity gain/ initial project (hm^3)	Raise of the upstream parapet (cm)
Low	359.20	360.50	0.33	+15
Max	359.40	360.65	0.65	+75

Table 2. Hydrology review and corresponding reservoir levels.

Flood	Return period (year)	Peak flow (m^3/s)	Reservoir level (m.NGF)
Project	3,000	431	360.60
Safety	100,000	675	361.35 – Raised Dam crest

is nearly 1.4 hm^3 lower than the initial one. Although the reservoir capacity is by far greater than the water needs, the owner took the opportunity of the upgrading works to increase the storage capacity by raising the NWL above 359 NGF. ISL proposed to increase the height between NWL and MWL at about 1.30 m instead of 1 m, thus reducing the size of the spillway. Two solutions were proposed to the owner, who chose the "max" one (Table 1).

2.2 *Hydrology review and normal water level*

Appraisal of the Dadou River floods was carried out by ISL in 2012. It was held in accordance with the latest recommendations of the French Committee on Dams and Reservoirs (CFBR). Main results are summarized in Table 2.

2.3 *Site constraints*

The available space is restricted by the purification plant downstream of the dam, the arch dam on the Dadou left bank and the steep topography on the right bank. Another restriction was the incentives to minimize earthworks for the spillway channel leading to the stilling basin in the Dadou River below. Eventually, the existing drinking water canalizations and access roads to the various parts of the dam and water treatment plant have to be kept in use.

2.4 *A spillway with zero operation and maintenance*

A key issue was that the Client has expressed interest in a spillway with little or no operation & maintenance. A solution including hydromechanical equipments (gates) requires maintenance and strict operational processes. A spillway without mechanical equipment was strongly preferred.

3 NEW SPILLWAY DESIGN

This context led ISL to seek a compact with no-equipment spillway design. A no-gated broad crested weir would be too wide, due to the low gap between NWL and MWL.

Therefore, ISL proposed in 2010 a complete PK weir solution, consisting in upgrading the existing one and building a new one on the right bank (Fig. 3).

3.1 *Main design issues*

On the arch dam, closing the existing weir would result in a large modification of its abutment loads. It was considered better to upgrade this weir to improve its discharge capacity with no gates as required by the owner. The upgrade can be achieved by building two PK.weirs on the ogee

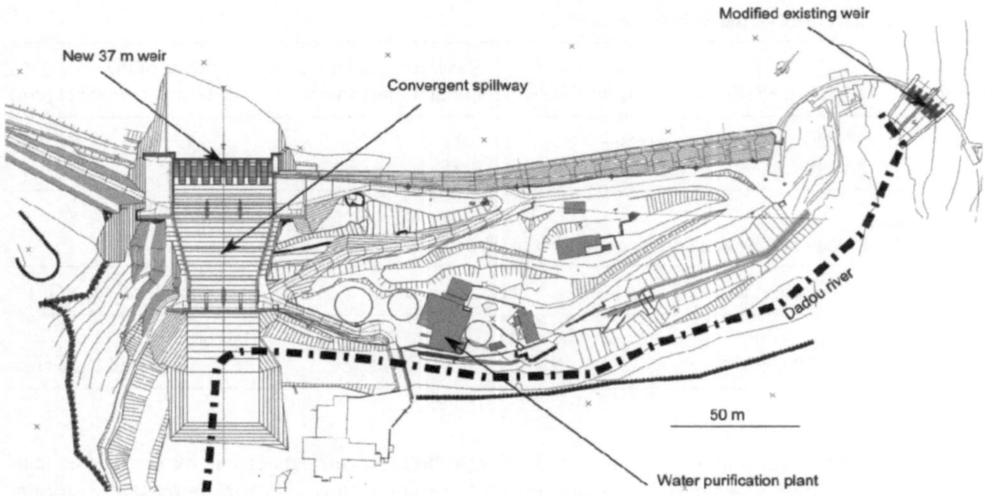

Figure 3. Plan view of the proposed works.

crest sill. Main design issues are related to: (1) how avoiding any change in the arch dam stresses repartition and (2) how to ensure the stability of the PW weir on a low-width arch dam crest.

On the right bank the little space available led to a 37 m PK.weir, and a 55 m long and 20 m wide channel with a convergence leading to the Dadou river below. This weir is build at the end of the gravity dam and is spanned by the access road to the water purification plant.

Main design issues derive from the little space available to accommodate the spillway: (1) not enough space to fit a standard stilling basin downstream the PK weir and (2) need for a rather sharp convergence: reduction of the spillway width.

3.2 *Spillways data*

The PK-Weirs geometrical data are based on the PK.weir model A proposed by Lempérière (Lempérière et al., 2006) (Table 3). With this design, well studied and known, no model study was required to derive its discharge rating curve.

4 HYDRAULIC CHARACTERISTICS AND STUDIES

4.1 *Discharge rating curves and flood attenuation*

The discharge rating curves used in the study were derived from those defined for the PK. weir model A (Lempérière et al., 2006). For very low head less than $0.3P_m$ (i.e. 0.75 m for the new PK.weir and 0.5 m for the modified existing weir), the discharge capacity was estimated considering a classic crested spillway, with a specific discharge coefficient of 0.36 on the developed PK.weir length.

For these low heads, the discharge reduction induces by wall thickness of the PK weir can not be neglected.

Under MWL (360.65 NGF elevation), the rebuilt existing spillway has a discharge capacity of 97 m^3/s and the new built spillway 318 m^3/s, i.e. a global 415 m^3/s discharge capacity at MWL (Fig. 4).

Flood routing in the reservoir was computed accounting for these rating curves and led to the following results (Table 4).

Table 3. Main characteristic of Rassisse PKW Project.

	Dam crest – Modified existing PK.weir	Right bank – New PK.weir
Total width of PK.weir (W)	2 × 7 m	37 m
Number of upstream alveoli	2 × 3	9
Number of downstream alveoli	2 × 2	4
Height of the inlet and outlet entrance ($P_i = P_o$)	2.7 m	3.75
Difference between P_i and P_o (P_m)	1.8 m	2.5 m
Upstream-downstream length of element (B)	6.5 m	9 m
Upstream and downstream overhang crest length ($B_i = B_o$)	1.65 m	2.25 m
Base length (B_b)	3.25 m	4.5 m
Width of a unit (W_u)	3.25 m	4.5 m
Inlet key width (W_i)	1.5 m	2.2 m
Outlet key width (W_o)	1.15 m	1.7 m
Crest and wall thickness (T)	0.3 m	0.3 m
Total developed length of a unit (L_u)	15.6 m	21.9 m
Total developed length (L)	63 m	180 m

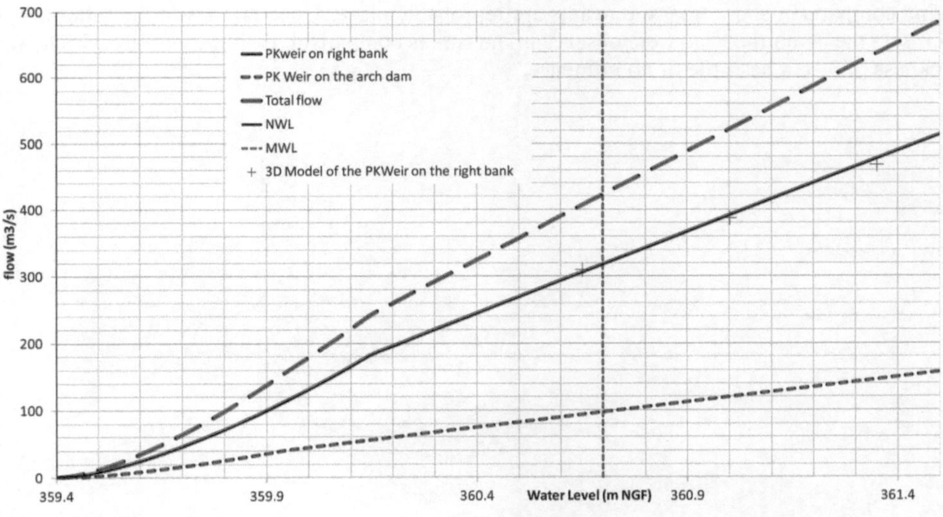

Figure 4. Discharge rating curves.

Table 4. Flood attenuation.

Flood	Peak flow (m^3/s)	Reached reservoir level (m.NGF)	Flow trough the new weir (m^3/s)	Flow trough the modified existing weir (m^3/s)
Project	431	360.65	306	101
Safety	675	361.35	477	157

Several hydraulic issues were nevertheless raised and led to check the main spillway hydraulics with a 3D-hydraulic model. These issues were as follows:

– Influence of the shorter-than-normal stilling basin downstream the PKweir and influence of the spillway strong convergence on the hydraulics within the spillway: secondary waves cannot be avoided, but are they acceptable or not?

– The PKweir to be constructed as a reinforced concrete structure has wall thicknesses thicker than what is used in physical mode. It was deemed necessary to investigate whether this might have influence on the PK-weir discharge capacity.

4.2 *Hydraulic study of the new spillway*

A 3D computer fluid dynamic model was set up using Computational Fluid Dynamics (CFD). This model (Fig. 5) allowed designing the convergent sidewalls and the water elevation in the channel, initially based on USBR recommendations.

It also allowed to compare, for the new weir, the computed elevation-discharge of the PK.weir with the one proposed by Lempérière et al. The influence of the concrete wall thickness was therefore investigated.

These computations were performed using CFD commercial code CFX. Volume Of Fluid method, an Eulerian-Eulerian multi-phases model, was used for free surface tracking. Turbulence was modeled with the shear stress transport (SST) based k-ω turbulence model. Solid walls were treated with the scalable wall functions.

The water elevation was computed only for the project flood, the safety flood and an intermediate flood between these two.

The comparison shows that the results are reasonably close. Uncertainties in the order of 1 to 3.75% on the hydraulic head were observed. The curves comparison is shown on Table 5. The wall thickness of 30 cm has little or no influence.

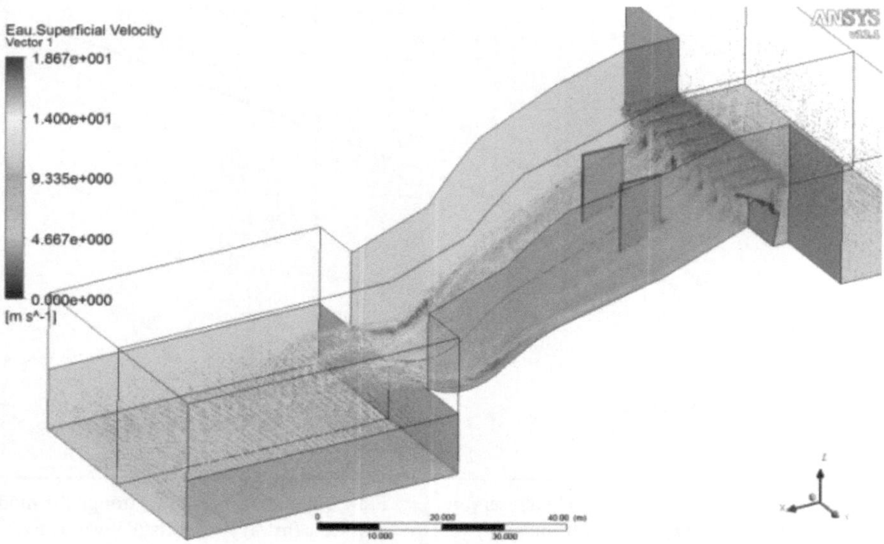

Figure 5. Speed water profiles under the design flood (reservoir level at 360.65 NGF).

Table 5. Comparison between 3D model and design formula.

Flood	Reservoir Level (m NGF)	Discharge capacity according to Model A formula (m^3/s)	Discharge capacity according to 3D model (m^3/s)	Uncertainties on the hydraulic head
Project	360.65	310.3	309.9	0.14% = 0 cm
Intermediate	361	398	387	−2.70% = 4.3 cm
Safety	361.35	485.8	467.6	−3.75% = 7.3 cm

5 SCHEME AND MAIN WORKS DIFFICULTIES

The design being approved by the owner in July 2012, the project is currently under examination by the State dam safety authority CTPBOH (Comité Technique Permanent des Barrages et des Ouvrages Hydrauliques). Works are to begin by November 2013 for 15 months duration.

5.1 *Main works difficulties*

During the works, drinking water supply and support of low water during dry months have to be maintained. For theses reasons, the reservoir can not be lowered and all works are to be done without altering the dam safety during works. The existing discharge capacity shall therefore be maintained. The works have also to take into account that the dam structural strengthening works are to be done before raising the normal water level to its future level. Furthermore, the existing weir is not accessible via the right bank, due to the thin arch dam.

The low flow in the Dadou River has 4 months duration, and all theses works can not achieved within this period.

5.2 *Construction schedule*

All theses difficulties led to propose two construction stages.

Construction stage I begins with the excavation of the new convergent spillway on the right bank. The excavated materials are partially reused to build an earthfill cofferdam in the reservoir, upstream of the new PK.Weir to be built. During this first phase, the existing gated spillway remains available with its gates open. The new PKweir spillway is constructed including free outlets within the PKweir structure. These sluices will control the reservoir elevation during stage II; they are shown on the drawing of Figure 6.

Once the new spillway put into operation, stage II includes the construction of the new PK.weir on the existing arch dam. This stage II is performed with a reservoir elevation not much below the working platform: floods have to be carefully anticipated.

One debatable issue deals with the interaction between the PK.weir structure and the arch dam on which it rests. Because of the rather thin support, it is not possible to make sure the PKweir would be stable on it under all circumstances. Anchoring or keying the PK.weir to the arch dam is necessary, at least for stability under earthquakes, and also to guarantee the PK.weir would still be stable at MWL in case the underdrainage were deficient. The proposed design includes the following:

- The pkweir is stable by its own weight (with no anchoring) at NWL and MWL.
- Stability under accidental conditions depends on anchors drilled into the arch dam.

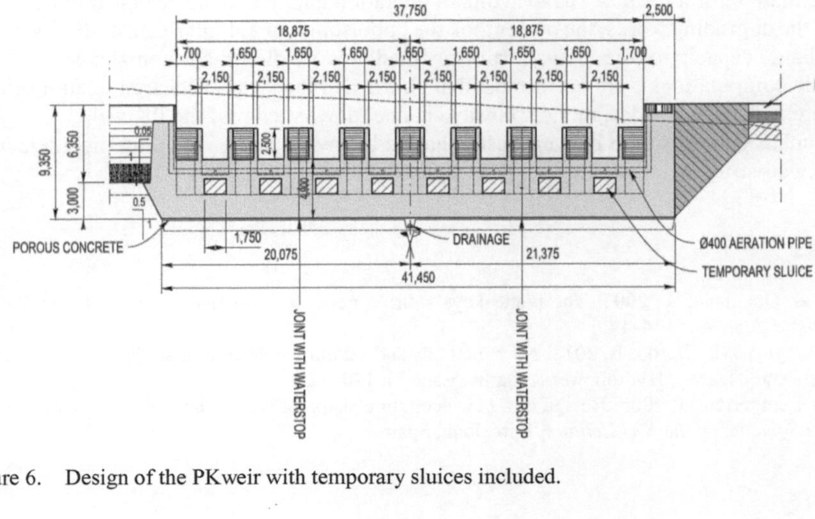

Figure 6. Design of the PKweir with temporary sluices included.

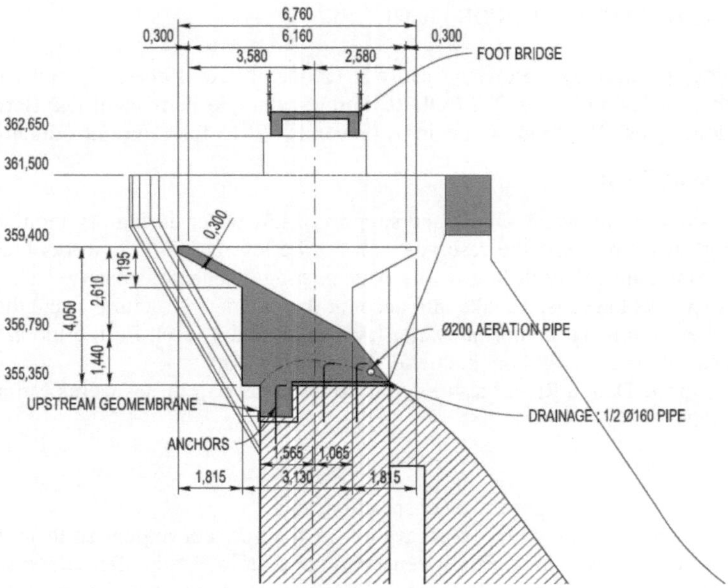

Figure 7. Design of the PKweir on top of the arch dam.

It shall be underlined that the arch dam concrete is reinforced there, so as to compensate for the thrust by the existing gates. The anchors would then be drilled within an already reinforced concrete structure.

The design is shown on the drawing of Figure 7.

6 CONCLUSION

Rassisse dam, built in the 60's, has to be thoroughly upgraded to meet present standards of flood safety.

The initial upgrading design included refurbishment of the existing spillway and an additional new gated spillway. An alternative design was proposed with two PK.weirs instead of the two gated spillways. The main reason for this was to avoid all operation and maintenance constraints.

Along with the upgrading works, the owner took the opportunity of a slight rising of the reservoir. The high discharge capacity of the PK.weir spillway made it possible for a minimal cost.

3D hydraulic computations were performed, that confirmed the stage-discharge relationship of Mr. Lempérière and helped to design the spillway channel downstream of the PK.weir.

Some particular features of the PKweir design had to be investigated, including the interaction of the new PKweir structures with the arch dam on which it rests.

REFERENCES

Lempérière, F. & Ouamane, A. 2003. The piano keys weir: a new cost-effective solutions for spillways. *Hydropower and dams (5)*: 144–149.

Lempérière, F., Vigny, J.-P., Deroo, L. 2012. New methods and critera for designing spillways could reduce risks and costs significantly, Hydropower & Dams (issue 3): 120–128.

Ouamane, A. & Lempérière, F. 2006. Design of a new economic shape of weir. *International Symposium on Dames in the Societies of the XXI Century,* Barcelona, Spain.

Labyrinth and Piano Key Weirs II – PKW 2013 – Erpicum et al. (eds)
© 2014 Taylor & Francis Group, London, ISBN 978-1-138-00085-8

Hydraulic scale model of the daisy-shape spillway on the Causse Corrézien dam

P.E. Loisel
Artelia Eau & Environnement, France

F. Duval
EDF – CIH, France

S. Chanourdie
Syndicat Mixte d'Aménagement du Causse Corrézien, France

ABSTRACT: The Causse Corrézien dam is an earthfill structure built on the River Couze in the Corrèze department of France for tourism purposes in 1978. The dam is equipped with a daisy-shape spillway (morning glory spillway with labyrinth crest), which discharges into a low-head 6 m diameter shaft and a 5 m diameter tunnel through the dam. The spillway was originally designed in order to pass the 1,000-year flood (160 m^3/s). However, current French practice stipulates that earthfill dams must be designed for the 10,000-year flood (232 m^3/s). The owner of the dam (SMACC), assisted by EDF-CIH, appointed Artelia's Water and Environment sector (formerly Sogreah Consultants) to perform a study to confirm the present discharge capacity of the spillway and the corresponding shortfall compared to the objective. Artelia Water & Environment built and studied the hydraulic scale model of the dam in its hydraulic laboratory in Grenoble (France) in 2012, in order to determine the actual discharge capacity of the present spillway, identify the reasons for capacity limitations and evaluate the effectiveness of easy structural solutions.

1 INTRODUCTION

The Causse Corrézien dam is a gravity earthfill structure built in 1978. According to the French legislation, it is a class B dam ($h > 10$ m and $h^2 \times V^{0.5} > 200$, with h (m) = height above ground, and V (Hm^3) = reservoir volume). Current best engineering practice in France recommends that its spillway should be sized to pass the 10,000-year flood while remaining below the highest water level elevation in satisfactory conditions of safety. It is also anticipated that a new French practice to come recommends to check that the spillway is able to pass a "safety" flood, with peak discharge of 1.15 × the 10,000-year flood without destroying the dam (CFBR, 2013).

When it was designed, the spillway was sized for the thousand-year flood flow. It should hence now be recalibrated, in regard to these two new criteria.

With this aim in mind, the Causse Corrézien development agency (Syndicat Mixte d'Aménagement du Causse Corrézien), with technical assistance from the Brive branch office of EDF's hydraulic engineering centre (Centre d'Ingénierie Hydraulique), appointed Artelia Water and Environment to conduct a hydraulic study to determine the maximum capacity of the spillway in its current condition (EDF, 2012).

2 PRESENTATION OF THE DAM AND ITS DAISY-SHAPE SPILLWAY

The Causse Corrézien dam key characteristics are as depicted in Table 1.

Table 1. Dam key characteristics.

Dam type	Earthfill dam with upstream asphaltic facing and alluvial clay cover
Maximum height above ground level	10.45 m
Crest length	214 m
Spillway crest elevation	122.55 NGF (ordnance survey datum) according to the 2011 topographical survey (122.50 NGF according to the original drawings)
Highest water level (HWL) elevation	123.46 NGF
Dam crest elevation	125 NGF according to the 2011 topographical survey (124.75 NGF according to the original drawings)
Reservoir surface area at FSL	0.755 km^2
Reservoir active storage at FSL	2.564 million m^3
Catchment area	73.8 km^2
1000 years flood peak discharge	160 m^3/s
10,000 years flood peak discharge	232 m^3/s

Figure 1. General plan view of the dam and spillway (with scale model extension).

The spillway is a free-surface, eight-lobed (daisy-shape) shaft spillway structure located very close to the right bank a few metres upstream from the dam (Fig. 1 and 2). It has a developed length of 100 m and is linked to the outlet tunnel via a circular shaft with a diameter of 6 m. The total height of the structure is 9.00 m. The outlet tunnel (Fig. 3) has a horseshoe-shaped section measuring 20.6 m^2, and a straight length of 100 m.

The downstream end of this tunnel leads to a free-flow channel about 50 m long. Downstream of this channel, the flow becomes pressurised again as it passes through two conduits with almost semi-circular sections about 25 m in length which empty into the natural river bed (Fig. 4). These conduits, which were built several years after the dam to protect the outlet channel from landslide from the right bank, can be bypassed via the left bank, from relatively low flows.

The dam was originally sized to handle the 1000-year flood ($Q1000 = 160\ m^3/s$), considering a negligible damping effect. No scale model study was performed at that time. According to the original hydraulic calculation note from the Syndicat Intercommunal d'aménagement du Causse Corrézien, at peak flow during the 1000-year flood:

- The upstream water level elevation is equal to the HWL elevation (123.46 NGF).
- "The spillway is not submerged; the structure hence discharges the 1000-year flood in normal conditions".
- The water level downstream of the outlet structures reaches an elevation of 115.90 NGF.
- The tunnel enables the flow to be discharged without excessively raising the water level in the shaft, and this water level in the work does not influence flows over its crest.

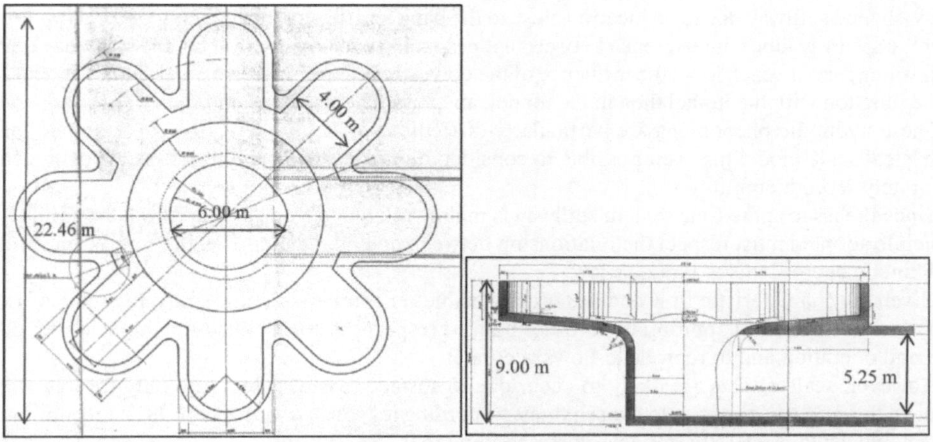

Figure 2. Plan view (left) and cross-section (right) of the daisy shape spillway.

Figure 3. Picture of the daisy shape spillway (left) and outlet tunnel (right).

Figure 4. Picture of the tunnel outlet (left) and downstream area (right).

3 PRESENTATION OF THE SCALE MODEL

The decision to use a physical scale model for this hydraulic study was taken in particular due to the complex nature of the expected flows on the structure, such as asymmetric flows into the daisy-shaped spillway due to its location close to the bank and the upstream facing of the dam, flows over the spillway lobes, interference between nappes facing each other in a lobe, the influence of the backwater curve in each lobe, the influence of the downstream level in the vertical shaft, head losses at the junction with the tunnel then in the tunnel, and parameters influencing the downstream level.

These hydraulic phenomena are particularly difficult to model numerically. On the other hand, a physical scale model makes it possible to consider all the decisive parameters intrinsically while accurately reproducing flows.

Since the issue at the Causse dam spillway is mainly related to free-surface flows, the similitude principle adopted must respect the relationship between potential energy, i.e. the surface elevation, and kinetic energy: this is Froude's law.

Given the characteristic flow and water depth values encountered in this study, the scale of 1:50 appears to be the most appropriate in order both to respect the local constraints related to model size and operation and to reproduce flows accurately.

However, scale effects are likely to occur due to surface tension effect, but only for low water head on the spillway crest and low discharges. According to Novak & Cabelka (1981), the minimum acceptable depth is 0.015 m (model), corresponding to 0.75 m (prototype). All measures above this value can be trusted, measures below this value are to be considered with precaution, however the scale effect acts in the direction of safety.

The physical scale model represents the following aspects (extension displayed on Figure 1):

- A part of the Causse dam reservoir measuring approx. 150 m by 150 m around the spillway structure, with the bathymetry in this area. Given the approach flow velocity (lower than 0.15 m/s on prototype), an extension greater than 6 times the spillway width is sufficient to ensure that approach currents around the structure are well reproduced,
- The weir, in the form of an eight-lobed daisy shaped spillway, including the concrete platform for the bottom outlet operation,
- The vertical shaft, the outlet tunnel and its junction with the shaft,
- The downstream area comprising the free-flow channel as well as its overflow on the left bank and the conduits,
- A 50 m reach of the river downstream of the mouth of the conduits, including the presence of a bridge. Reproducing the profile of this bridge at the downstream end of the model ensures that the downstream stage-discharge relation is reproduced faithfully, since this section acts as the control section.

The test programme was designed so as to be able to identify the hydraulic capacity of each component of the spillway structure: the weir, shaft inlet, the tunnel and the downstream area.

4 TESTS RESULTS

4.1 *Hydraulic capacity of the daisy crested spillway and shaft*

In the first model configuration (Fig. 5), the flow passing over the eight-lobed weir is discharged via a vertical pipe that short-circuits the outlet tunnel, from the bottom of the shaft. Diameter and height of this vertical outlet pipe ensures no hydraulic influence on the flow passing over the eight-lobed weir and into the vertical shaft, and the intrinsic hydraulic behaviour of these two elements is characterised in this first configuration. The Figure 6 illustrates the rating curve obtained during these tests.

The shape of the curve obtained reflects the typical characteristics of both a "labyrinth" weir and a "morning-glory" weir:

- For the flow range between 0 and 160 m^3/s: the eight-lobed weir operates like a 100 m long circular weir (purple dotted line). The weir nappe is thin enough for the shape of the weir not to influence the discharge capacity (no interference between two nappes facing each other) (Fig. 7 left).

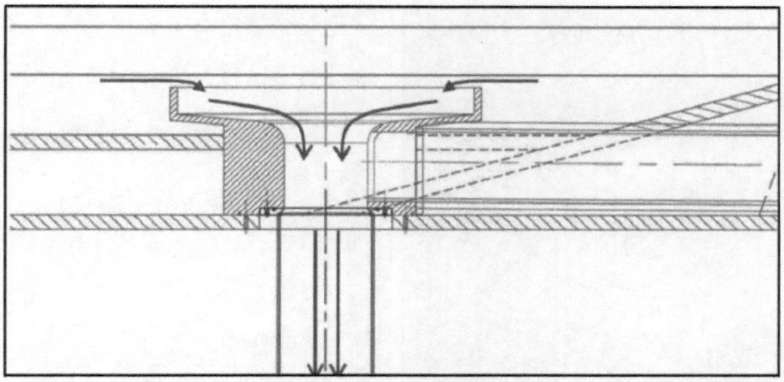

Figure 5. Sketch of the first model configuration: daisy shape spillway + vertical shaft.

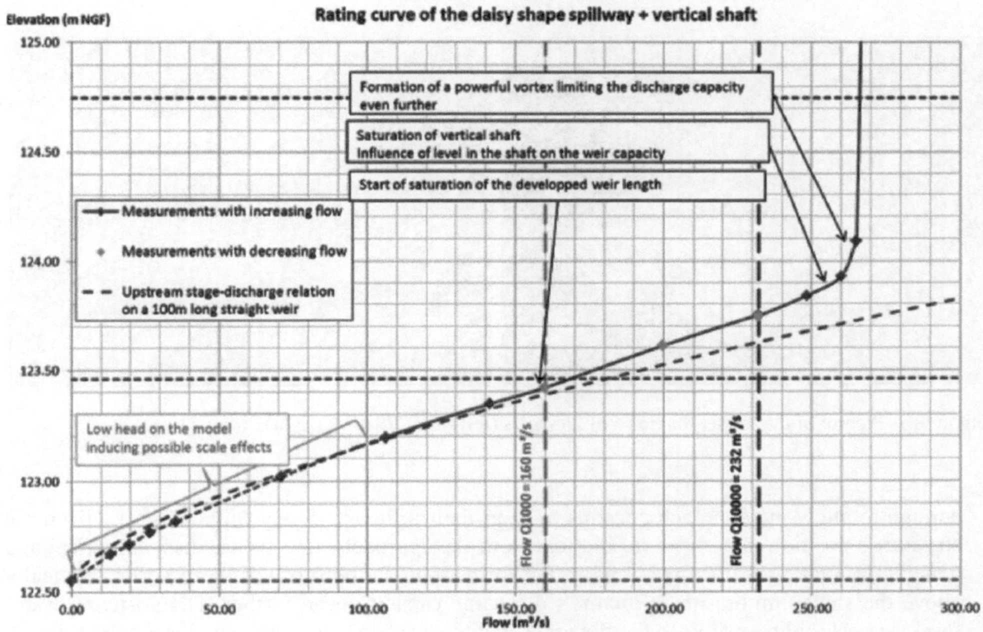

Figure 6. Scale model result: rating curve of the first model configuration (daisy shape spillway + vertical shaft).

- For the flow range between 160 m^3/s and 255 m^3/s: the "labyrinth" shape of the weir starts to have an influence and the developed length of the weir can no longer be exploited to the full. This is because the nappe is such that the effective overspill length of the weir gradually decreases as the flow increases (Fig. 7 right). The lobe shapes gradually disappear as nappe thickness increases. the upstream stage-discharge relation then ceases to follow that of a fictitious straight weir (see Fig. 6).
- For the flow range between 255 m^3/s and 265 m^3/s: the 6 m diameter vertical shaft is no longer capable of discharging the flow and the water level at the centre of the eight-lobed spillway starts to influence flow over the weir Fig. 8 left). The flow rate is now controlled at the level of the vertical shaft and no longer at the weir.
- Above 265 m^3/s: the spillway now behaves like a submerged orifice. In the absence of water-guiding walls arranged radially around the weir crest, the water above the weir is now deep

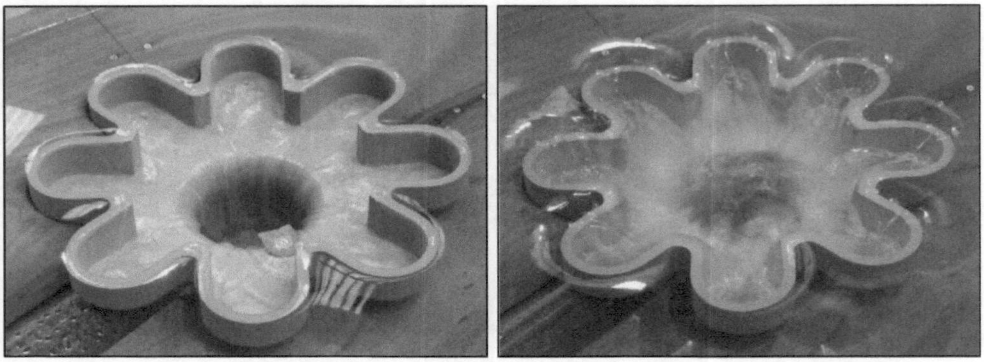

Figure 7. Picture of the model for flows of 13 m³/s (left) and 200 m³/s (right).

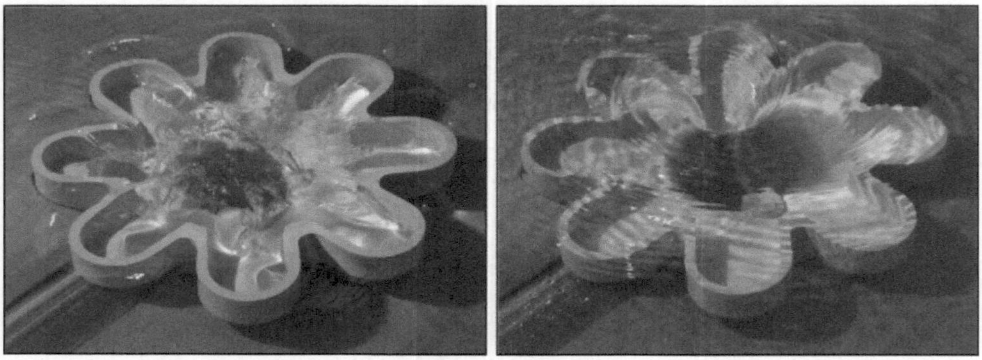

Figure 8. Picture of the model for flows of 265 m³/s (left) and 270 m³/s (right).

enough to allow the approach currents to exert their influence freely (Fig. 8 right). Given the structure's position in a corner of the reservoir, the approach currents are dissymmetrical and generate a rotating inflow, resulting in the formation of a powerful air-core vortex vertically above the shaft, limiting the structure's discharge capacity even further. The upstream water level is now highly sensitive to the flow rate, and the discharge capacity curve is almost vertical.

To conclude, the study shows that the inlet to the 6 m diameter vertical shaft becomes saturated when the water flow reaches approx. 255 m³/s, which is higher than the objective (Q10,000 = 232 m³/s).

However, the eight-lobed spillway itself is an obstacle to achieving the objective, since the flow rate passing below the HWL elevation (123.46 NGF) is approx. 168 m³/s. Moreover, the absence of water-guiding walls arranged radially around the weir crest leads to the formation of a powerful vortex vertically above the shaft.

4.2 *Hydraulic capacity of the complete spillway structure*

The last configuration tested reproduces the present state of the structure and the topography downstream. The water thus flows through the eight-lobed weir, the shaft, the tunnel and then the land downstream, including the two conduits and the bridge.

The Figure 9 illustrates the rating curve obtained during the tests, added to the one obtained during the first test series.

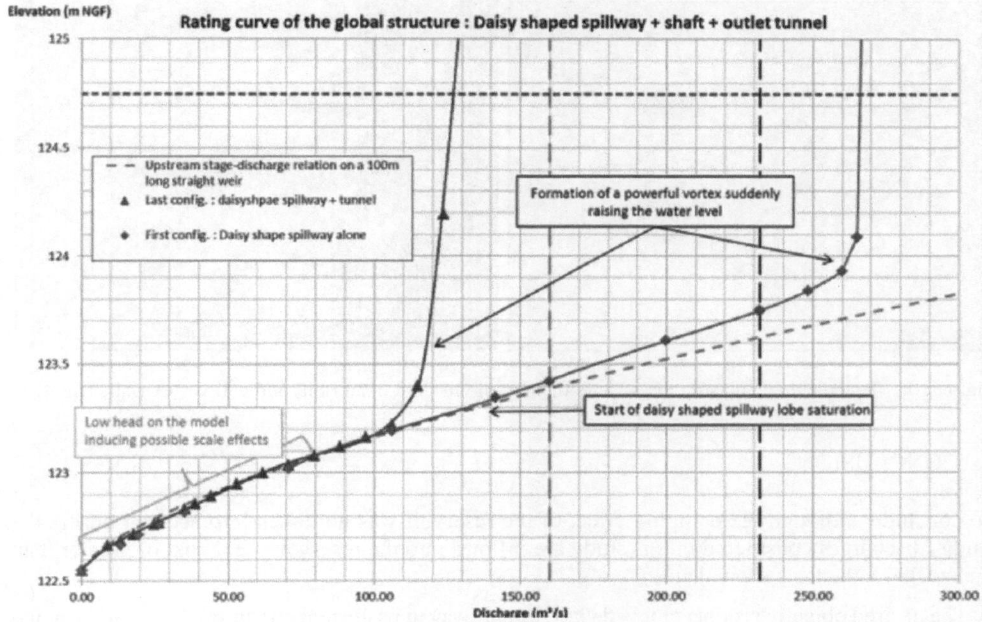

Figure 9. Scale model result: Rating curve of the global structure.

To conclude, the series of tests performed on the present spillway configuration produced the following results:

- The junction between the shaft and the tunnel becomes a limiting factor once the flow rate exceeds 100 m^3/s. Above this rate, the water level at the centre of the structure submerges the weir nappe and the structure quickly becomes saturated.
- The HWL elevation is reached with a flow rate of 115 m^3/s.
- With flows up to at least 125 m^3/s, the downstream flow conditions have no influence on the flow and water level in the structure.
- In a similar manner to the observations made on both tested configurations, a powerful air core vortex forms above the structure.

4.3 *Tests of small enhancing structures*

In order to attempt to reduce the problem of the structure's limited capacity, two separate developments were tested on the model (Fig. 10):

- Changing the profile of the bend between the shaft and the outlet tunnel: this is likely to improve the passage of water from the shaft to the tunnel, while limiting head losses. The radius of curvature adopted for the modified profile is 10 metres, along the tunnel axis.
- Installing 4 anti-vortex walls around the rim of the eight-lobed weir: this is likely to delay the formation of the vortex observed on the previous configurations shaft vertically above the shaft, and hence gain additional capacity.

These developments, tested on the structure, correspond to simple modifications and minor works, but unfortunately do not improve its discharge capacity.

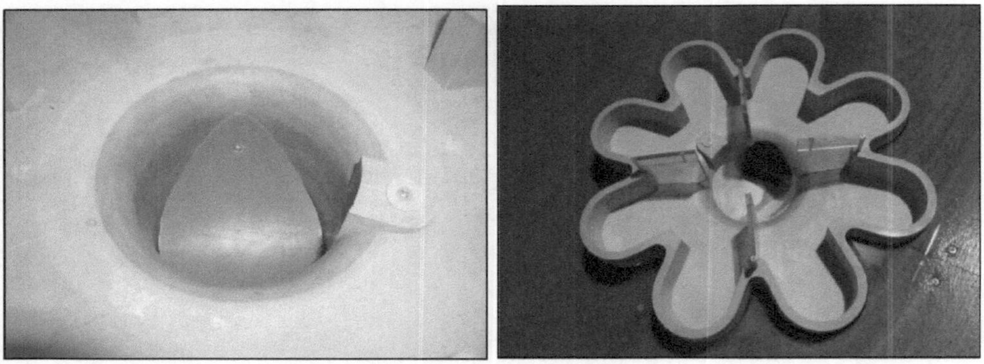

Figure 10. Tested enhancing structures: smoothening of the shaft bend (left) and anti vortex walls (right).

5 CONCLUSIONS

To conclude, although the damping effect of the reservoir was not taken into account during the study (discharges considered in this study are inflows into the reservoir, i.e. equal or greater than the outflow), the tests show that:

- The desired objective is not achieved with the spillway in its current condition. It should be noted that merely removing the passage through conduits would have no effect.
- The original dimensional design, allowing for the passage of a 160 m^3/s flow rate below the HWL elevation, is incorrect since it does not fully take into consideration the submerged operation of the daisy-shape weir.
- Changing the profile of the bend or building four anti-vortex walls does not bring about any significant improvements.

The only remaining solution is to design an additional spillway, free flow type, on the right abutment of the dam, which could be a labyrinth weir in order to limit its width.

This physical scale model shows that development works on a weir crest to boost its specific capacity cannot be designed without considering the structure's environment, in particular the downstream hydraulic conditions:

- firstly, the submersion of the labyrinth weir from its downstream side significantly reduces its capacity,
- and, in the present case, because the flow is forced into a shaft of low height, this submersion comes quickly and generates a circular current that forms a strong vortex, further reducing the capacity.

This study provides a good illustration of the submersion risk due to the fitting of an earthfill dam with a shaft spillway (morning glory or, in this case, daisy-shape) with very low height and complex hydraulic behaviour due to the eight-lobed spillway crest.

This option would not be chosen today when drawing up the dimensional design for a new dam, as simpler solutions can be identified.

REFERENCES

CFBR, 2013, Recommandations pour le dimensionnement des évacuateurs de crues de barrages, http://www.barrages-cfbr.eu/Recommandations.html.

EDF-CIH, 2012, Recalibrage de l'évacuateur de crues, Détermination de la débitance maximale de l'EVC, Cahier des Clauses Techniques Particulières, IH CAUS-EVC CCTP 00002 A BPE.

Novak and Cabelka, 1981, *Models in hydraulic Engineering – Physical Principles and Design Applications*, Pitman, London.

Labyrinth and Piano Key Weirs II – PKW 2013 – Erpicum et al. (eds)
© 2014 Taylor & Francis Group, London, ISBN 978-1-138-00085-8

Application of the Piano Key Weir spillway in the Republic of South Africa

A.J. Botha & I.P. Fitz
Ingerop, Johannesburg, South Africa

A.J. Moore, F.E. Mulder & N.J. Van Deventer
Department of Water Affairs, Pretoria, South Africa

ABSTRACT: The original design for Hazelmere Dam provided for the use of radial gates to raise the full supply level by 7 m to meet increasing water demands. The hydrological characteristics of the catchment area are however such that the time of concentration of only 6 hours could result in operating staff being unable to open the gates timeously in the event of a flood. This risk led to the designers considering the use of an uncontrolled piano key weir spillway, which was found to accommodate flood waters extremely efficiently and consequently impose lower hydraulic forces on the existing structure, thus reducing the required stability improvements. The raising of Tzaneen Dam does not require the same extreme discharge efficiency. Hydraulic modeling indicated that a PKW, modified for simpler construction, would nevertheless result in a much lower headwater level than a conventional labyrinth. This in turn dictated that the level of the existing 1.5 km long non overspill crest would require less raising.

1 INTRODUCTION AND USE OF PIANO KEY WEIRS IN RSA

The use of labyrinth spillways is not a new concept in South Africa and they have already been utilized to raise dams like Midmar and Bospoort. However, with the need to meet specific requirements, the "new" concept of PKW structures is presently under consideration for a number of dams in South Africa.

Notwithstanding the fact that a number of labyrinth type spillway structures have already been constructed in South Africa, the first piano key weir (PKW) spillway structures will be constructed at the Hazelmere and Tzaneen dams. The main reasons for the choice of a PKW is that additional spillway capacity can be obtained and that the full supply level (FSL) of an existing dam can be raised using this type of structure, particularly when the length of an existing spillway is limited and a larger flow capacity is required with a restriction in the maximum allowed water level.

In this paper the authors describe some of the aspects considered to decide on the use of PKWs at Hazelmere and Tzaneen Dams to raise the FSL, including confirmation of theoretical flow calculations through model tests in a hydraulics laboratory.

2 GENERAL

During the latter half of 2012 various layouts of PKWs were tested in the hydraulic laboratory of the Department of Water Affairs (DWA), South Africa.

These models were for Pm-values of 4.8 m to 7 m. The higher wall heights were for the Hazelmere Dam raising and the lower heights for the Tzaneen Dam raising. The 6 m to 7 m high walls were associated with a B-value of approximately 25 m and W-values of 7.5 m to 13 m, while the 4.8 m high walls were for a B-value of 20 m and a W-value of 15.24 m. The linear scales used varied between 1:21 and 1:44.

The data is summarized as follows:

The discharge coefficient is defined by equation 1 below:

$$q = CH\sqrt{Pm} \tag{1}$$

The C-values reported are determined using the head (H) for $q = 40\ m^3/s/m$ for all layouts. The H-values varied between 3.7 m and 5.2 m. The C-values were in the range of 3.5 to 4.4. The best fit equation indicates a C-value of 4.0 for L/W-ratio of 5 and that C varies according to L/W to the power 0.25.

The single data point residing below the minus 5% curve is for a layout of equal incoming and outgoing channels. To minimize the required head, the incoming channel should be approximately 55% of the cycle width, which means that the outgoing channel is only 80% as wide as the incoming channel.

Other features like rounding of crest and inlet-walls seem useful to reduce the required head, but remain within the ±5% range.

For an L/W ratio of less than 4.0 it was found that non-parallel walls (in plan) performed significantly better (+5%) compared to the standard parallel PKW wall layout (−5%), with no rounding of the inlets in both cases.

3 HAZELMERE DAM: SUMMARY OF CONTENTS FOR MAIN PAPER

3.1 *Basic relevant characteristics of Hazelmere Dam*

Hazelmere Dam was designed and built in 1975 to 1976 with provision for the future raising of the water storage level. The intention was for the raising of the dam to be accomplished by the installation of radial gates for which provision was made in the existing structural concrete arrangement.

The crest level of the uncontrolled spillway (present FSL) is at 86 m and the planned 7 m raising of the spillway will increase the FSL to 93 m. The non overspill crest (NOC) level is 94 m. The dam is slightly curved in plan with a radius of 750 m and comprises 32 concrete monoliths each approximately 15 m wide and ranging from 15 m to 45 m in height. The spillway is located on 7 of these blocks above the river channel and is 103 m long. The spillway center block is only 13 m wide.

Existing vertical joints between the blocks have been sealed with rubber waterstops and during construction the contractor will have to experiment with the joining of new waterstop with the old.

The planned construction process, which will include the improvement of dam stability by installation of multiple strand rock anchors, is planned to start in October 2013 with a two year contract period. The PKW will constitute about 4500 m^3 of concrete and will comprise about 40% of the construction costs which will be predominantly split between the PKW and rock anchors.

The PKW has been positioned on the upstream side of the wall to maximize contribution to the stability of the wall and to ensure that most of the discharged water impacts the ogee spillway where the existing splitters are positioned. It has been visually assessed that this will facilitate adequate energy dissipation.

Increased water demand in certain areas of Durban North, exclusively supplied with water from Hazelmere Dam, have now dictated that the FSL of the dam be raised by the planned 7 m in order to improve the yield so that demand can be met by a supply that is within acceptable safety margins.

Although provision was made in the original design for a future raising of the FSL by means of radial gates, this was considered by the current owners and operators to be inappropriately expensive to install, operate and maintain. As a result the designers were required to investigate alternatives. Apart from the normal financial restrictions for this kind of operation, the two major considerations in determining the eventual preferred option were, firstly, the extremely short time of concentration (ToC) of 6 hours for the dam catchment area and, secondly, the limited height difference of 1 m between the proposed new FSL and the NOC of the dam wall.

Figure 1. Artist's impression of Hazelmere Dam wall with a PKW.

Figure 2. View of Hazelmere Dam.

The ToC restriction demanded that an uncontrolled spillway be preferred but the physical constraint of the limited freeboard between FSL and NOC militated towards mechanically operated gates in order to cope with the sheer volume of the safety evaluation flood (SEF).

In the search for alternatives to mechanical gates the designers examined the possibility of using labyrinth fusegates but the same freeboard constraints meant that the likelihood of the regular loss of fusegates was very high. This was unacceptable as it would result in the regular compromise of storage capacity and increase in operation cost as well.

This dilemma was resolved by the introduction of the PKW, as one of the main features of the PKW is to minimize the head required to pass flood waters through a given spillway. The ingenious design of the PKW offered a perfect solution to the imperatives faced in managing flood waters through the Hazelmere Dam spillway.

Four basic parameters were then assumed to define the design criteria for the Hazelmere Dam PKW structure:

- The 1:100 year flood must pass through the spillway section without overtopping the existing NOC section of the dam.
- Floods with a return period as large as possible, preferably the regional maximum flood, should be passed over the PKW structure without overtopping the existing 1.5 m parapet wall on the NOC section.

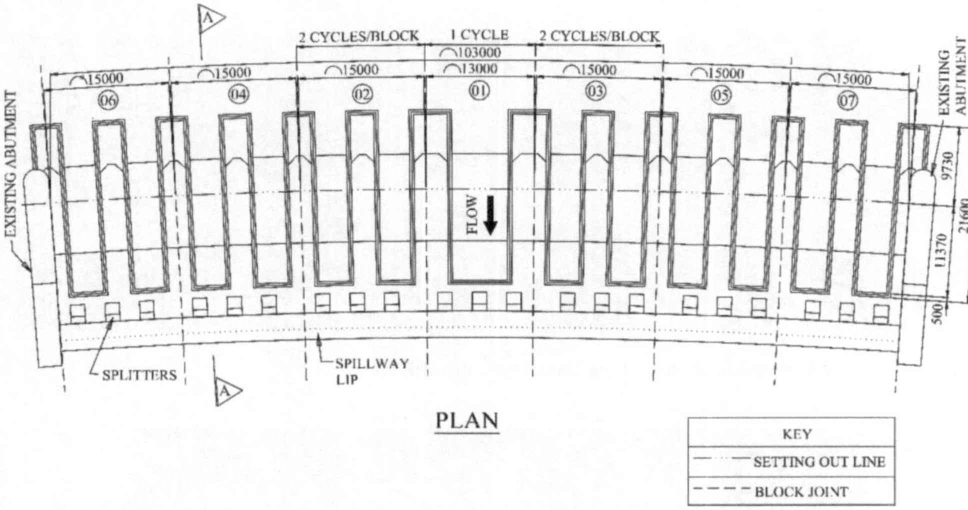

Figure 3. Sketch: Plan of Hazelmere Dam structure.

- The SEF must be safely routed over the spillway with limited overtopping of the NOC and the parapets on the NOC.
- Under the above circumstances, the dam wall stability criteria must not be compromised.

3.2 *Theoretical considerations*

It is normally more advantageous to make use of a larger flow depth to optimize the design of a spillway because flow through a spillway is defined by flow depth to the power of 3/2. This is especially so when the spillway length is limited as is the case at Hazelmere Dam.

When the spillway length is limited and the maximum allowable flow depth is likewise restricted, the only alternative is to increase the effective length of the crest. A PKW is particularly effective in this regard as it is possible to increase the overflow crest length by up to 5 or 6 times the effective length of the spillway. Hazelmere Dam is an existing gravity concrete dam with both a fixed spillway width and restrictions regarding the temporary storage depth of water above the fixed FSL.

Using the available theoretical equations published on the internet in various papers by researchers from Algeria, France and the USA who have sought to optimize the hydraulic efficiency and structural stability of the PKW, a configuration for the use of a PKW at Hazelmere Dam could be defined.

The main relevant characteristics of the proposed PKW are as set out below and shown in Figure 3 and 4:

- The existing spillway length is 103 m consisting of 6 blocks 15 m wide, and a spillway center block 13 m wide, separated by vertical construction joints.
- The raising of the FSL above the existing ogee crest is 7 m.
- The 1:100 year flood to be passed without overtopping the NOC is 1360 m^3/s.
- P was determined as 9 m.
- Pm was determined as 6 m.
- W should have been 10.8 m but was fixed as 13 m for the center block and 7.5 m for the remaining 6 blocks in order to accommodate the fixed block widths and their associated vertical construction joints.
- W_i/W_o was determined as 1.315 where W_i and W_o are the internal widths of the inlet and outlet key respectively.
- PKW top length B is 21.6 m.

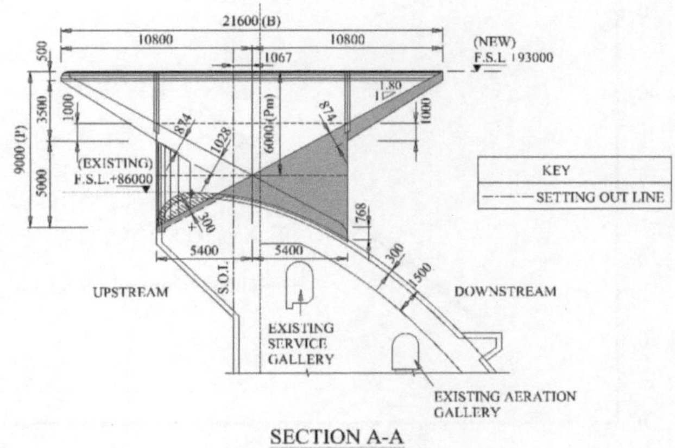

Figure 4. Sketch: Section of Hazelmere Dam structure.

- L/W was determined as 6.46 where W is the width of the spillway and L the effective length of the top of the PKW walls on their center lines.
- H/P was determined as 0.356 where H is the height of the dam water level above the PKW spillway level.
- The wall thickness at the top of the wall is 0.5 m.

A hydraulic model of the proposed PKW structure was then constructed in DWA's water research laboratory in Pretoria and used to test several configurations to determine the optimum unit flow through the weir.

After the initial results were obtained, the hydraulic model was adjusted several times to optimize the required structural layout. The model also contained a study of the energy dissipation downstream and on the existing ogee to determine minimum modification (breaking out of concrete etc.) of the structure. Finally a configuration with a 7.5 m cycle width for the 15 m wide blocks and a single cycle width of 13 m on the center block, was selected and arranged over the full width of the existing spillway. This configuration was selected to optimize flood passage whilst ensuring that existing construction joint positions were accommodated in the new PKW. The result as determined in the laboratory modeling for the final configuration is shown below with unit discharge of the overall project flow against the actual project hydraulic head (Fig. 5).

The discharge rating curve has then been developed to determine the flow depth (water level in the dam) discharge relationship. This was very important to determine the flow depth for extreme flood flow conditions as the raising of the existing structure will be done by ensuring stability of the structure against overturning and sliding by installation of rock anchor cables. The discharge rating curve is shown below (Fig. 6).

3.3 *Structural requirements*

After the flow rating curve had been determined, the structural requirements to optimally position the new PKW on top of the existing ogee crest was necessary. As far as we know the proposed PKW height, P, of 9 m is presently the highest in the world and required special attention.

The following aspects had to be considered for the structural placing of the PKW on the existing ogee concrete structure:

- Positioning of the PKW on top of the existing ogee to ensure adequate energy dissipation during flow.
- Configuration and placement in an upstream-downstream direction to optimize the effect it would have on the stability of the structure.

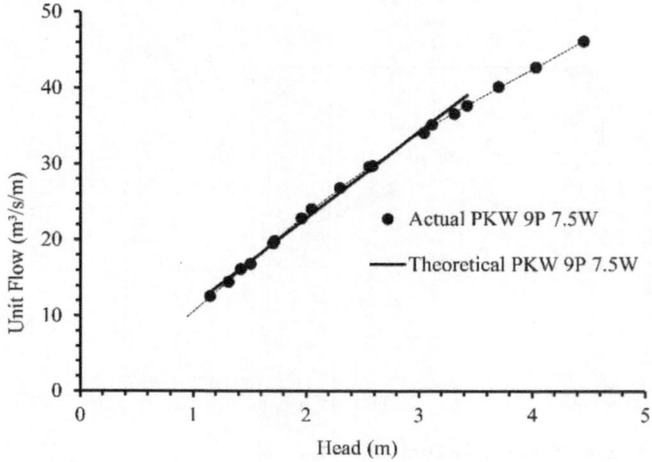

Figure 5. 7.5 m cycle width all data.

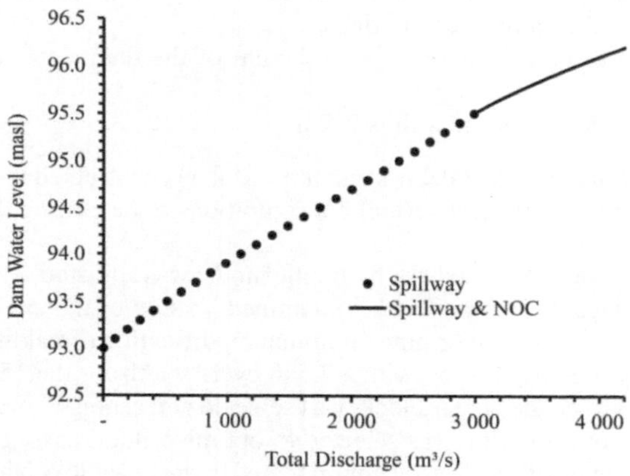

Figure 6. Discharge rating curve.

- Proper anchorage of the PKW to the existing ogee.
- Only 1 m freeboard is available between the new FSL and NOC to pass the 1:100 year flood.
- Proper placing of construction joints to coincide with the existing construction joints on the ogee.
- Minimize the removal of concrete from the existing ogee for fitting the PKW, to save costs.

3.4 *Final design*

The final design, after optimization, has the following characteristics which are depicted in the above sketches (Fig. 3 and 4):

- FSL is 93 meters above sea level (masl).
- NOC concrete level is 94 masl.
- NOC with existing parapet left in place is 95.5 masl.
- FSL top level length of each PKW is 21.6 m.
- 13 full standard cycle 7.59 m wide PKWs with a center one 6.9 m wide.
- Total effective spillway length of PKW is 665 m.

- Unit discharge capacity at 95.5 m water level is 29 $m^3/s/m$ length.
- Total discharge capacity with water level at 95.5 m is 3000 m^3/s.
- The safety evaluation flood (SEF) is 4300 m^3/s.
- Maximum water level at SEF is 96.23 m which determined the design water level for all structural considerations.

3.5 *Challenges and lessons learnt*

The main design considerations were to:

- Configure a modified PKW to conform with the geometry of the existing ogee structure.
- To ensure efficient contribution to the stability of the raised structure.
- To provide sufficient anchorage to the existing concrete. The mass concrete ogee structure was designed with a strength of 18 MPa, but concrete core tests indicated a strength of more than 34 MPa.

By carefully reviewing design criteria, it was possible to obtain an optimized design and the positioning of the PKW structure is shown in the sketch (Fig. 3 and 4).

4 TZANEEN DAM: SUMMARY OF CONTENTS FOR MAIN PAPER

4.1 *Basic relevant characteristics of Tzaneen Dam*

Additional dam yield for mostly domestic use in the Tzaneen district was required. The existing safe yield from the dam is already fully utilized and increased demand requires the raising of the dam FSL by 3 m to increase the storage in the dam basin.

Similarly to Hazelmere Dam, raising the existing dam wall structure requires implementation of an economically viable solution by the design of an efficient system with a limited spillway length.

Upstream water storage limitations in the basin define the maximum allowable water level during floods and that in turn dictates the allowable maximum water level at the dam structure thus defining the spillway requirements. Also, the overall stability of the existing gravity structure must be improved to cater for the increased loads. Minimizing the water level for the different loading conditions will reduce the loads acting on the structure.

The main characteristics of Tzaneen Dam are:

- Height above the lowest foundation level is 54.9 m.
- Total crest length is 1140 m.
- Catchment size is 652 km^2.
- Current Ogee Level (FSL) is 723.97 masl.
- Current volume at FSL is 153.870×10^6 m^3.
- Current NOC level is 730.67 masl.
- Spillway length is 91.44 m.
- Current spillway configuration is uncontrolled ogee.

4.2 *Consideration of alternatives*

The 3 m raising of Tzaneen Dam by means of a labyrinth spillway had been recommended by the feasibility study. Subsequently, a PKW spillway has also been investigated by the Directorate: Civil Engineering. This ultimately led to a modified PKW solution which comprises the distinctive overhang of the PKW theory in combination with the characteristic non-parallel walls of the labyrinth spillway structure.

Under SEF conditions the required embankment raising for the labyrinth alternative was calculated to be approximately 1.7 m whereas the modified PKW spillway resulted in an approximate 1 m raising of the embankment NOC.

Figure 7. Satellite image of Tzaneen Dam, Limpopo. 23 47′58.35″S, 30 09′58.37″E. Image copyright Google Earth.

Other alternatives such as the utilization of patented fusegates were investigated but the preferred option is a permanent arrangement due to the prolonged procurement process for the DWA to replace fusegates in the event that fusegates are lost.

4.3 *Challenges and solutions*

The initial labyrinth structure proposed by the feasibility investigation would ensure a simplified construction process, due to the absence of the upstream and downstream overhangs as well as the inclined floors for both the incoming and outgoing channels, but the removal of 7.5 m of the existing ogee crest to construct a labyrinth that would pass the floods, was considered unacceptable in regard to cost of construction and water storage/yield loss of the dam when it would need to be lowered for construction purposes.

The PKW alternative is favored therefore because it only requires a 2.8 m removal of the current ogee crest. Note that the programming of the construction process should prioritize the raising of the embankments during peak irrigation demand, before the lowering of the ogee crest commences.

The hydraulic capacity of the modified PKW allows a safe passing of the SEF (4120 m^3/s) in conjunction with a 1 m raising of the earth-fill embankments. This design meets the requirements of all the constraints such as limited spillway length as well as the limited high flood line constraints which are part of the design criteria.

With regard to the stability of the spillway, a downstream stepped roller compacted concrete arrangement is proposed. This offers adequate resistance to failure against sliding as a result of the additional weight and increased footprint area.

Floating debris is a concern at Tzaneen Dam due to the surrounding forestry environment. The tapered arrangement at the spillway was proposed as a means of avoiding debris congestion.

4.4 *Construction phase*

Although the project is still at an early stage, the estimated time for construction is two years because the objective is to construct the PKW cycles during two dry seasons in order to minimize the impact on the yield of the current dam. The works are due to start towards the latter half of 2015.

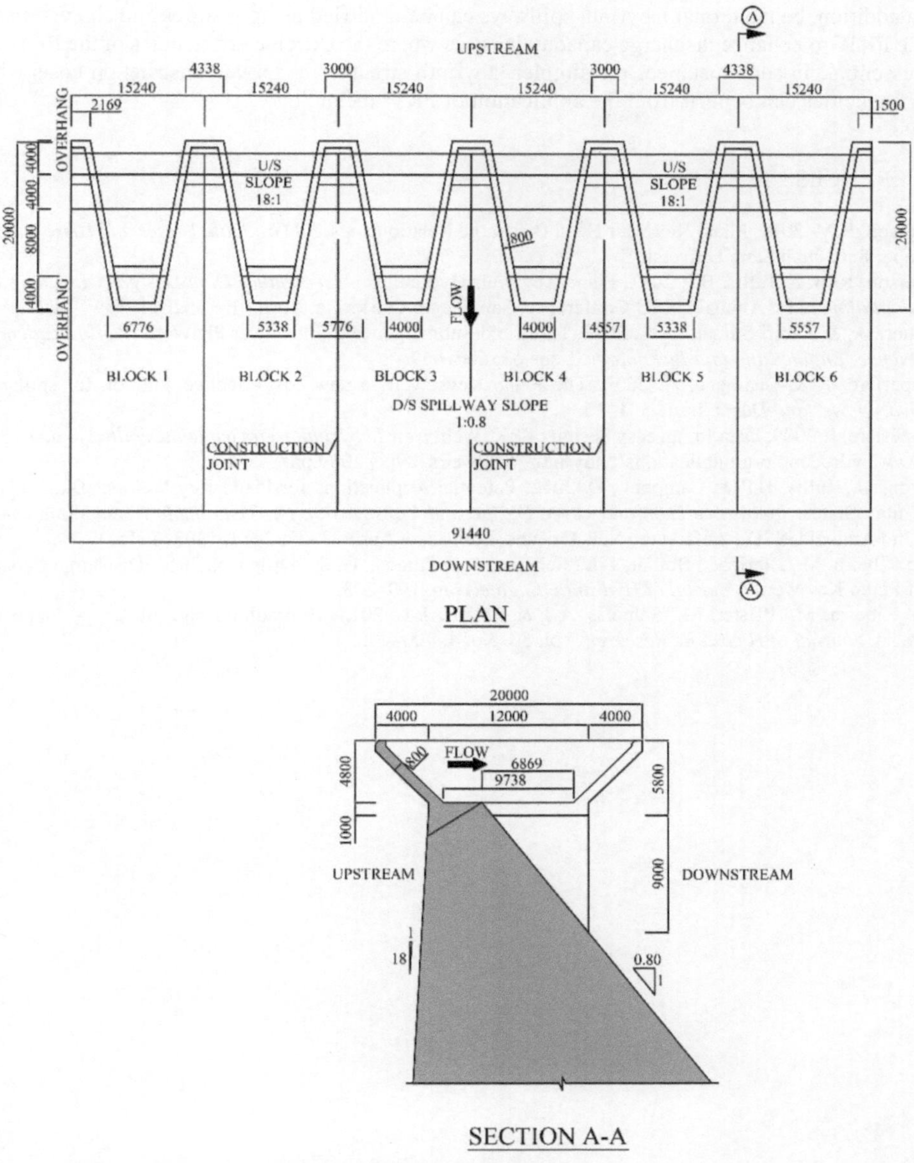

Figure 8. Typical plan view and outflow section for Tzaneen Dam.

The expected project cost has not yet been determined but the estimated quantity of PKW concrete is 3000 m^3.

5 CONCLUSION

PKW structures can be very efficiently used as permanent facilities to raise existing dams in a manner that reduces operational risks and cost of stability enhancing measures that may be necessary.

In addition, conventional labyrinth spillways can be modified using features and characteristics of the PKW to enhance discharge capacity in cases where the extreme efficiencies of the PKW are not essential. In such instances the simpler labyrinth structure can save construction costs while flow capacities can benefit from the application of PKW technology.

REFERENCES

Anderson, R.M. 2011. Piano Key Weir Head Discharge Relationships. *All Graduate Theses and Dissertations*. Paper 880. Utah State University.

Anderson, R.M. & Tullis, B.P. 2011. Piano Key Weir Hydraulics. *21st Century Dam Design – Advances and Adaptations*. 31st Annual USSD Conference. San Diego, California. April 11–15, 2011. 1697–1707.

Javaheri, A. & Kabiri-Samani, A.R. 2012. Threshold Submergence of Flow over PK Weirs. *World Academy of Science, Engineering and Technology*. Issue 61: 666–669.

Lempérière, F. & Ouamane, A. 2003. The Piano Keys Weir: a new cost-effective solution for spillways. *Hydropower and Dams*. Issue 5: 1–15.

Lempérière, F. 2009. Data for an easy design of P.K. Weir. *New Labyrinth weirs triple the spillways discharge*. www.hydrocoop.org/publications/Labyrinth_%20weirs_09[1].2009.pdf.

Paxson, G., Tullis, B.P. & Campbell, D. 2012. Potential Applications for Piano Key Weirs at Dams in The United States. *Innovative Dam and Levee Design and Construction for Sustainable Water Management*. 32nd Annual USSD Conference. New Orleans, Louisiana. April 23–27, 2012. 1303–1315.

Leite Ribeiro, M., Bieri, M., Boillat, J.-L., Schleiss, A.J., Singhal, G. & Sharma, N. 2012. Discharge Capacity of Piano Key Weirs. *Journal of Hydraulic Engineering.* 199–203.

Leite Ribeiro, M., Pfister, M., Schleiss, A.J. & Boillat, J.-L. 2012. Hydraulic design of A-type Piano Key Weirs. *Journal of Hydraulic Research.* Vol. 50, No. 4: 400–408.

Labyrinth and Piano Key Weirs II – PKW 2013 – Erpicum et al. (eds)
© 2014 Taylor & Francis Group, London, ISBN 978-1-138-00085-8

Hydraulic physical model of Piano Key Weirs as additional flood spillways on the Charmine dam

P.E. Loisel
Artelia Water & Environment, France

P. Valley & F. Laugier
Hydro Engineering Center, EDF-CIH, Le Bourget du Lac, France

ABSTRACT: Charmine dam is a concrete gravity dam built in 1950 on a tributary of the Ain river (France) for hydropower purpose. It is currently operated by EDF. Further to an update of the hydrological study, the Charmine dam flood spillway revealed itself to be unable to pass safely the design flood under the maximum reservoir level. Therefore, an additional spillway was designed by EDF and the local constraints led to the choice of two piano key weirs replacing the two existing free surface spillways on each side of both central gated weirs. ARTELIA Water & Environment built and studied in 2011–2012 for EDF a hydraulic model of the Charmine dam in its hydraulic laboratory in Grenoble (France), in order to determine the real discharge capacity of the existing configuration of the spillways, evaluate the hydraulic design of the Piano Key Weirs and optimize the downstream energy dissipation area in order to ensure satisfactory hydraulic conditions. In addition to the results needed for the project, additional tests were carried out in order to determine the discharge capacity of the Piano Key Weir by itself and to compare it to the empirical results or formulae published by international researchers.

1 PRESENTATION OF THE CHARMINE DAM AND ITS FLOOD SPILLWAY

Charmine dam (Fig. 1) is a 17 meter high concrete gravity dam built in 1950 on a tributary of the river Ain (France) for hydropower purposes. It is currently operated by EDF. It is composed of both a 150 meter long concrete dam and a 30 meter long earthfill dam on the left bank. The concrete part is divided into 11 blocks.

The structure currently possesses the following flood spillways (Fig. 2):

- A central spillway consisting of two 8 m wide surface weirs with radial gates,
- Lateral spillways located on either side of the central spillway and each consisting of six 3.4 m wide free surface weirs separated by concrete pillars supporting a pedestrian walkway.

With regard to the dam height and its earthfill section on the left bank, the design flood is the flood discharge with a return period of 10,000 years, according to French standards.

A recent update of the hydrological study estimates this design flood discharge at 685 m^3/s. The Charmine dam flood spillway has been shown to be unable to safely release this discharge under the maximum reservoir level.

2 DESIGN OF THE PIANO KEY WEIR

Therefore, EDF launched the design of an additional spillway in order to ensure the safe operation of the dam during a flood event. The local constraints, such as the unavailability of room on the dam abutments and local environmental constraints, led to the choice of two piano key weirs replacing

Figure 1. Pictures of the Charmine dam in the existing configuration.

Figure 2. Upstream view of the Charmine dam and spillways in the existing configuration.

Table 1. Main characteristics of the Charmine dam and its reservoir.

Height above natural ground level:	17 m
Height above foundations:	22.3 m
Crest thickness:	2.5 m
Base thickness:	19.8 m
Radius of crest curvature:	Left part is rectilinear, Right part is curved with R = 210 m
Batter of the upstream facing:	0.025H/1V (concrete) 3/2 (earthfill)
Batter of the downstream facing:	0.78H/1V (concrete) 3/2 and 2/1 (earthfill)
Crest length:	180 m including 30 m of earthfill dam
Present normal reservoir level:	381.35 NGF
Maximum reservoir level:	382.05 NGF
Safe limit (crest):	383.35 NGF
Usable capacity of the reservoir:	4.3 million m^3
Flood spillway:	1 automatic radial gate, 1 electrical radial gate and 12 lateral spillways
Bottom outlet:	2 sluices each controlled by a sliding gate

Table 2. Geometrical dimensions of each Piano Key Weir.

Parameter	Notation/Value		Parameter	Notation/Value	
Crest elevation	381.10 NGF		Cycle width	W_u	4.70 m
Piano Key Weir height	P	5.00 m	Developed length	L	119.5 m
Inlet width	W_i	2.75 m	Inlet number	4	
Outlet width	W_o	1.95 m	Outlet number	3 central + 2 lateral outlets	
Upstream length	B_o	3.79 m	Wall height inside the outlet		1.02 m
Downstream length	B_i	3.71 m	Wall thickness inside the outlet		0.35 m
Base length (U/S – D/S)	B_b	5.37 m	Upstream deflector	Cylindrical, under the inlet	
Total length (U/S – D/S)	B	12.87 m	Deflector diameter		1.15 m
Wall thickness	T_s	0.35 m	Wall crest shape	Round (r = 0.175 m)	

the two existing free surface spillways on each side of the two central gated weirs. This technical solution allows for:

- a structure of reduced size,
- replacing the existing lateral spillways, which are ineffective for flood evacuation and can easily retain floating debris,
- a limitation of the lowering of the operational level, which means decreasing the hydropower production.

The objective of such additional spillway is to ensure that the maximal reservoir water level is not exceeded:

- with the design flood (discharge $Q_{10,000} = 685\,m^3/s$) with both gates opened,
- with the flood discharge $Q_{1000} = 510\,m^3/s$ with one gate inoperative,
- with the flood discharge $Q_{100} = 340\,m^3/s$ with both gates inoperative.

The geometrical design of the Piano Key Weirs was carried out by the Hydraulics Engineering Center (CIH) of EDF. The initial symmetry of the dam and its spillways allows for two identical Piano Key Weirs. The height of the new structures was set in order to limit the lowering of the water level in the reservoir during construction, in order to keep it operational. The total width of each Piano Key Weir was set in order to modify only 2 blocks on each side and avoid having an additional joint on the downstream chute. The length and width of the inlets and the outlets were chosen further to a parametrical study carried out with the Wolf 1D Piano Key Weir solver developed by the University of Liege.

On each outer side, the Piano Key Weir is connected to the lateral wall through a straight profiled weir about 2 meters wide, in order to ensure proper connection to the non-overflowing part of the dam and the lateral wall.

The proposed design of each Piano Key Weir is summarized Table 2.

3 DESIGN OF THE OUTFLOW CHUTES

Downstream of the additional flood spillway, Artelia proposed two solutions for chutes to be positioned on the existing downstream facing of the dam discharging into the low water bed:

- A smooth concrete chute parallel to the existing downstream facing, ending in an inclined section in order to better guide flow,
- A chute with 0.6 m steps parallel to the existing downstream facing.

At the toe of these chutes, two side walls have been tested on the model for the left and right banks in order to protect the wooded banks and the toe of the earthfill closure structures from erosion.

These two solutions were dimensionally designed to ensure good discharge conditions in the event of the design flood, whilst being adapted to the existing natural terrain to limit the need for excavation and backfilling.

The design of the new structure is displayed on the drawings of Figure 3.

4 PRESENTATION OF THE HYDRAULIC MODEL

In order to validate the design of the structure, EDF asked ARTELIA Eau & Environnement to build and operate a hydraulic model of the Charmine dam in its hydraulic laboratory in Grenoble (France).

The model was built with a scale factor of 1/35, without geometrical distortion and run according to the Froude similitude, which allowed for good representation of free surface hydraulics phenomena. It represents a significant stretch of the reservoir (approximately 525 m long – Fig. 4), allowing the supply currents to the flood spillway to be accurately represented. The upstream limit of the model is immediately upstream of the RD18 bridge, which acts as a control section and which ensures proper water supply to the model reservoir. Downstream of the dam, an area of 100 m is represented by the model, including the first of the Charmine falls, ensuring that the real downstream rating curve is respected. The model allowed to determine the different rating curve for the flood spillway, to measure the water lines on the side walls and the flow velocities in the reservoir.

The model is equipped with two electromagnetic flowmeters and three level gauges. Spillway gates can be operated at partial or total opening rate. The model in its initial and future configurations is displayed on the Figure 5.

5 RESULTS OF THE MODEL STUDY

5.1 *Results of the project study*

The tests conducted on the existing flood spillway configuration demonstrated that this configuration allows a flow rate of 440 m^3/s to be released at high water level, i.e. a deficit of 36% in discharge capacity. In addition, the evacuation of debris such as tree trunks between 5 and 10 m long is relatively poor: although smaller debris succeeds in passing down the central spillway, debris does accumulate, with jams forming on the lateral spillway on the right bank and the central spillway (right bank sluice).

The main conclusions of these tests on the existing configuration are as follows:

- According to the geometry specified by EDF-CIH, the additional double Piano Key Weir will allow the dam to achieve the target release capacity and accommodate the design flow rate at the high water level limit ($Z_{high\ water\ level} = 382.05$ NGF), as shown on Figure 6.
- The spillway rating curves for the current configuration for different degrees of opening of one or two gates have been determined with an aim to assisting the dam's operator in a flood event. In light of the relatively low flow velocities in the reservoir, these upstream rating curves can be used as upstream head/discharge curve.
- The addition of a stabilising beam on the upstream face of the Piano Key Weirs at the foot of the inlets and under the outlet overhang (Fig. 7) does not have a significant impact on the release capacity of the Piano Key Weir on its own, or the Piano Key Weir with the gates fully open. This result shows that the approach flow velocities upstream of the Piano Key Weirs are mainly horizontal, although the Piano Key Weir is built about 9 meters above ground level.
- Tests were also carried out with the introduction of debris in the form of uprooted trees (at scale 1/35, model diameters between 0.01 and 0.03 m – Fig. 7). These tests showed that the Piano Key Weir does not allow for the successful evacuation of this debris given the relative shallowness of the water stream (less than 1 m). However, the accumulation of a large quantity of wood (with branches and foliage) on the Piano Key Weirs only led to an increase of 10 cm in the water height for the project flow rate (high water level limit).

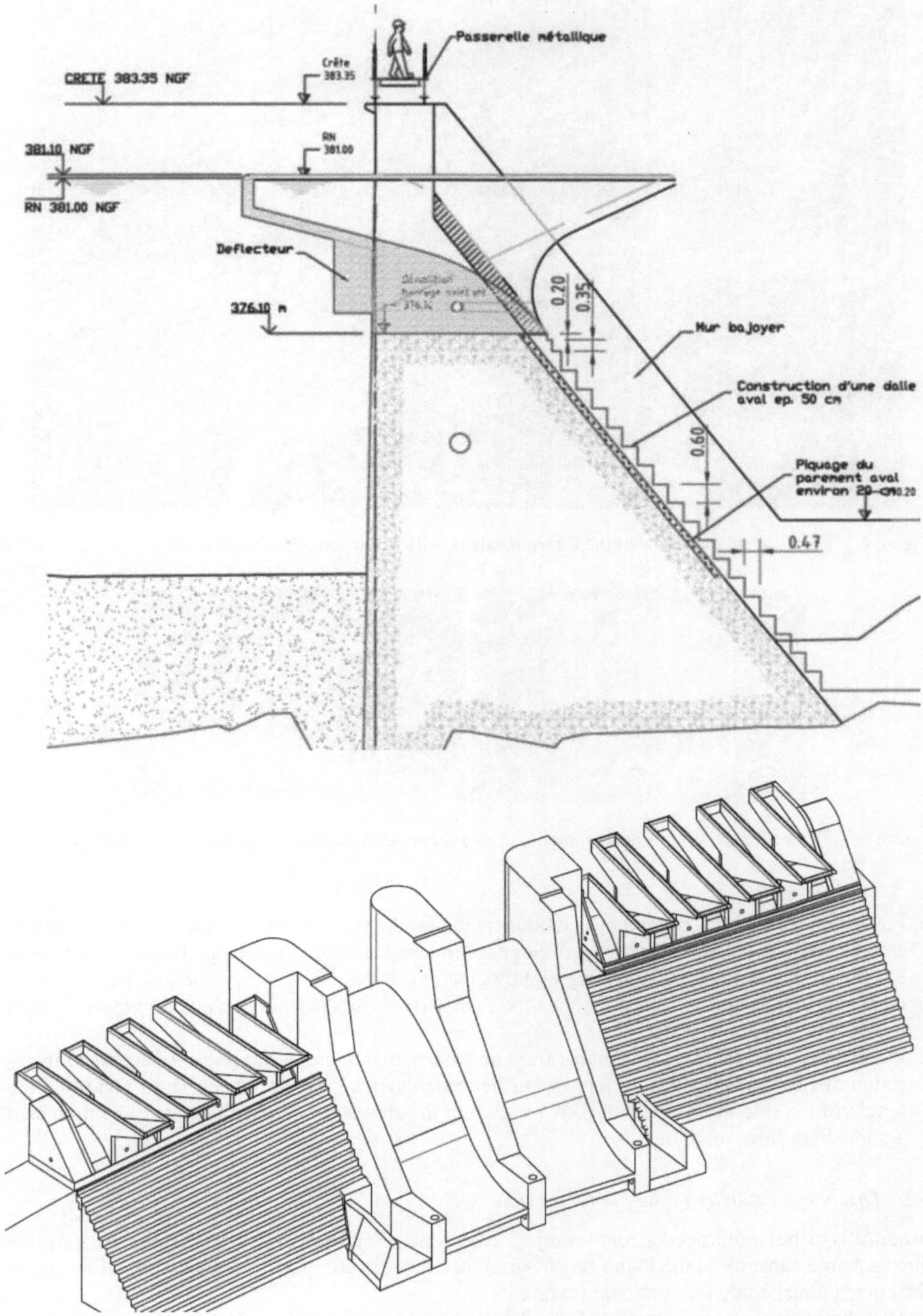

Figure 3. Drawings of the future configuration of the spillway with stepped chute and lateral walls: first configuration.

Figure 4. Localization plan view of the Charmine dam, with extension of the scale model.

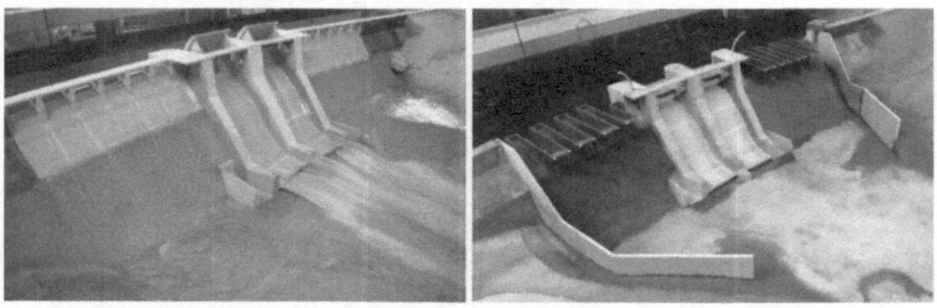

Figure 5. Pictures of the Charmine dam model in its existing (left) and future (right) configurations.

- Furthermore, the tests carried out demonstrated that the two proposed chute solutions function satisfactorily and present very few differences in terms of flow. The stepped chute nonetheless has the slight advantage of allowing slightly greater dissipation of energy and lower water levels on the side walls downstream. Despite this, EDF-CIH chose the smooth chute for reasons linked to construction constraints.
- Finally, provided that it is acceptable from an environmental point of view, the second configuration of side walls is to be preferred. Further recommendations have been given regarding the height of the side walls along and downstream of the chutes, as well as the areas to be protected against high flow velocities.

5.2 *Discharge capacity of the piano key weirs*

In addition to the results needed for the project, additional tests were carried out in order to determine the discharge capacity of the Piano Key Weir by itself and to compare it to the empirical results or formulae published by international researchers.

The dimensionless parameters of the Charmine dam Piano Key Weirs are the following: $W_i/W_o = 1.41$; $B_i/B_o = 0.98$; $W_u/P = 0.94$; $L/W = 5.87$. For such values, two empirical laws are found in technical papers, published by Machiels et al. (2010) (valid for type B PK Weirs only) on one hand and by Ouamane and Lempérière (2003) on the other hand, in which the authors have

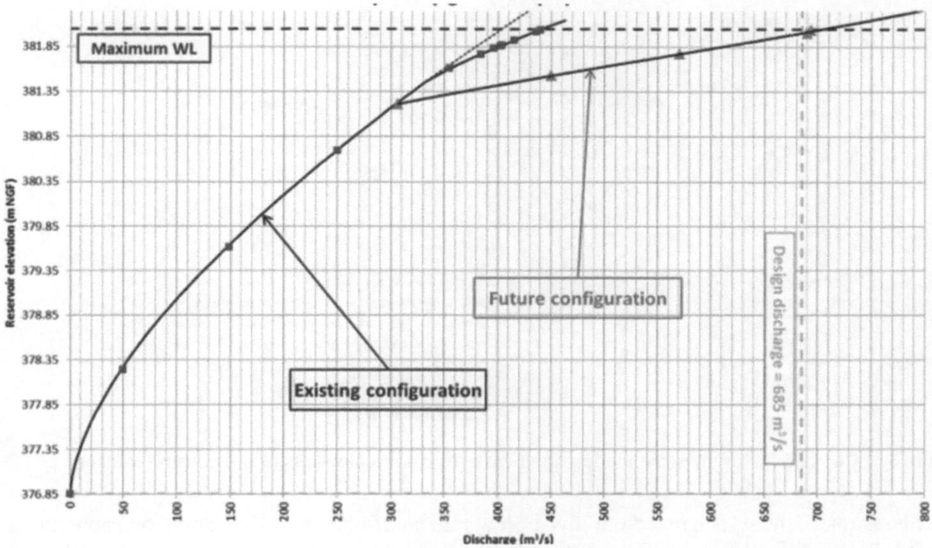

Figure 6. Rating curve of the flood spillway in its existing and future configurations.

Figure 7. Pictures of the model Piano Key Weir with upstream stabilization bar (left) and with floating debris (right).

carried out parametrical studies in order to evaluate the influence of Piano Key Weir dimensions on their release capacity. The discharge coefficient is defined as follows:

$$C_{dW} = \frac{Q}{[W-4.K_a.(Z_{upstream}-381.10)].\sqrt{2g}.(Z_{upstream}-381.10)^{1.5}} \quad (1)$$

where W is the total width of the two Piano Key Weirs, K_a is the coefficient of lateral contraction, equal to 0.1 and $Z_{upstream}$ is the upstream pool level.

The rating curve of the Piano Key Weirs was measured on the model by closing and obstructing the central spillway gates. The results and the empirical values drawn from both above-mentioned publications are plotted on the graph of Figure 8.

It is to be noted that the release capacity of the Piano Key Weir is relatively close to the empirical law proposed by Ouamane and Lempérière (2003) for geometrical characteristics similar to those of the Charmine dam. On the other hand, considering the approach proposed by O. Machiels et al. (2010), which concerns only B-type Piano Key Weirs (without downstream overhang), leads surprisingly to an overestimation of the structure's capacity, despite similarities with the key width W_i/W_o and spillway length L/W ratios.

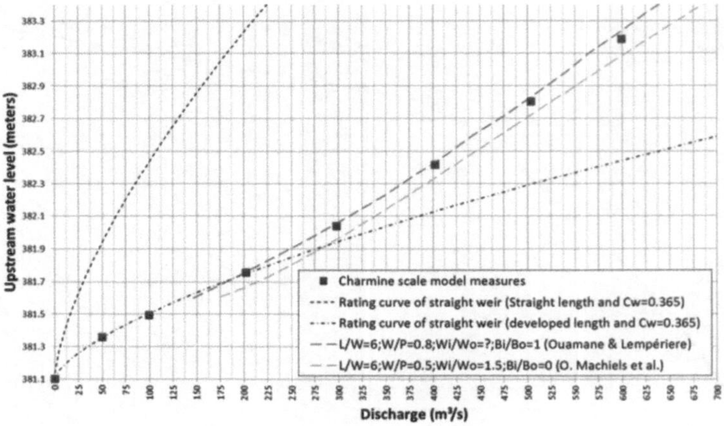

Figure 8. Comparison of the scale model rating curve to empirical results of previous researchers.

The results of the tests carried out give a flow rate coefficient and an evacuation capacity value slightly higher than those predicted by the Ouamane and Lempérière law. It must nonetheless be noted that the Charmine Piano Key Weir configuration tested on the model included joints at each end of both Piano Key Weirs which were slightly different to the systematic geometry of the central inlets and outlets, in particular the two straight profiled weirs, each of which was approximately 2 m long. For a relatively shallow water stream ($H/P < 0.1$), the flow rate coefficient stays below the empirical curve mentioned above. This difference can be attributed to scale effects (inopportune effects of surface tension) as well as to the form of the crest of the weir which is relatively thick in comparison to the water stream.

6 CONCLUSIONS

The model allowed EDF-CIH to confirm that the Piano Key Weirs to be added to the existing Charmine dam are hydraulically well designed in terms of release capacity, but it was also a very good tool to check both the hydraulic behavior of the upper pool (flow velocimetry, risk of obstruction by floating debris) and the energy dissipation conditions along the downstream chute for flood conditions, up to the design flood (10,000 year return period).

The model results were compared to previously published laws. In our case, it appears that the empirical law proposed by Ouamane and Lempérière are very close to the model results, which is due to very proper upstream hydraulic conditions (limited flow velocity) and geometrical parameters similar to those considered by these authors.

REFERENCES

Machiels O., Erpicum, S., Archambeau P., Dewals B. and Pirotton M. 2010. Experimental study of the alveoli widths influence on the release capacity of Piano Key Weirs. *La Houille Blanche*, 2, 22–28.

Ouamane A. and Lempérière F., 2003. Conception d'une nouvelle forme économique de déversoir. http://www.hydrocoop.org/fr/publications/Conceptiond_unenouvelleformeeconomiquededeversoir.pdf

Pralong, J., Vermeulen, J., Blancher, B., Laugier, F., Erpicum, S., Machiels, O., Pirotton, M., Boillat, J.-L., Leite Ribeiro, M. & Schleiss, A.J. 2011b. A naming convention for the piano key weirs geometrical parameters. In Erpicum, S., Laugier, F., Boillat, J.-L., Pirotton, M., Reverchon, B. and Schleiss, A.J. (Eds), *Labyrinth and piano key weirs-PKW 2011*, London: CRC Press, 271–278.

Labyrinth and Piano Key Weirs II – PKW 2013 – Erpicum et al. (eds)

Quipolly dam: A case study on customizing Fusegates® design for adverse spillway conditions

M. Beretta & A.G. Merino
Hydroplus, Rueil-Malmaison, France

O. Cazaillet
Artelia, Grenoble, France

R. Batterham
Liverpool Plain Shire Council, Quirindi, NSW, Australia

ABSTRACT: The feasibility and performance of the Fusegate® system were tested at LPSC's Quipolly dam; a 21 m high earthfill dam located on the Werris Creek in the State of New South Wales, Australia. The primary purpose of the rehabilitation works included increasing the water storage capacity of the reservoir by 55% and improving spillway discharge potential to meet the new 1:100,000 Annual Exceedance Probability (AEP) flood. The challenges of the project, mainly due to the presence of high tailwater levels and highly perturbed approach channel flow conditions at the spillway, required a more detailed analysis in order to optimize the design, reduce the cost of rehabilitation and minimize the risks. A large scale physical model was performed in collaboration with Artelia laboratories (France) to find a customized solution with a non-mechanical spillway control system. The model studies helped to optimize and calibrate the input for the physical model tests. These studies not only benefited the current research and development of the Fusegate® system but also supported the scientific understanding of labyrinth crested weirs including their hydraulic behaviour. This paper discusses the details of these studies including the tailored Fusegate design adopted to minimize risks and costs.

1 INTRODUCTION

Quipolly dam is a 21 m high and 200 m long earthfill embankment with a 35 m wide free overflow spillway on its left bank. Completed in 1955, the dam is located in New South Wales (NSW), Australia, approximately 300 km from Sydney and 11 km from the village of Quirindi (see Fig. 1).

Owned by the Liverpool Plains Shire Council (LPSC), its purpose is mainly irrigation and primary water supply. In order to meet the current Australian dam safety standards, a major spillway upgrade was required to enable discharging the revised Design Flood (1,617 m^3/s).

Upon request of LPSC and in cooperation with their consulting engineers (GHD Pty Ltd), Hydroplus has developed a solution to improve the flood discharge potential and the reservoir storage capacity of Quipolly Dam by using the Fusegate® system. The solution adopted involved lowering the existing spillway sill level by 0.60 m (which corresponds to the existing parapet wall on the sill) and installing eight 2.60 m high labyrinth type Fusegates®.

The increase in the Full Supply Level by 2.00 m provides an additional storage capacity of approximately 2.9 Mm^3 (+55%). The adverse hydraulic conditions of the project required physical model studies to accurately assess the hydraulic parameters and develop a tailored solution using Fusegates®.

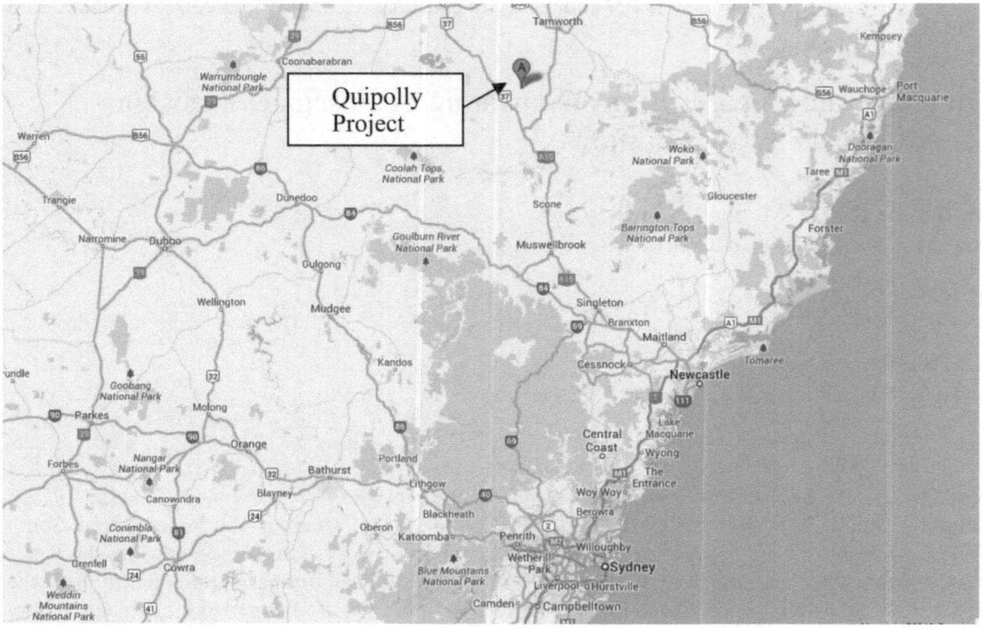

Figure 1. Quipolly Project location.

Figure 2. Upstream spillway approach.

2 HYDRAULIC BOUNDARY CONDITIONS

At Quipolly dam, the spillway is a free overflow type and is located on the left bank. A pedestrian bridge is spanned across the control section with five piers (see Figs 2 & 3). The spillway has a very short approach channel ending at a small vertical slab, which comprises the broad crested weir at the control section. The tailrace channel is concrete lined on the first 30 m section and has a moderate slope. The rest of the channel is the natural rocky river bed with a moderately increased slope.

One of the main reasons for the inefficiency of the spillway stems from the fact that the flows have to make a 90 degree turn to pass through the spillway, causing turbulence and random velocity gradients, particularly on the right side of the spillway, and perturbations in water profile.

The next issue is related to the high tailwater levels in the tailrace channel during major floods, which created a downstream hydrostatic pressure on the Fusegates® able to impact their stability and triggering. The figures below provide photos of the spillway during dry and wet seasons.

Figure 3. Downstream spillway channel.

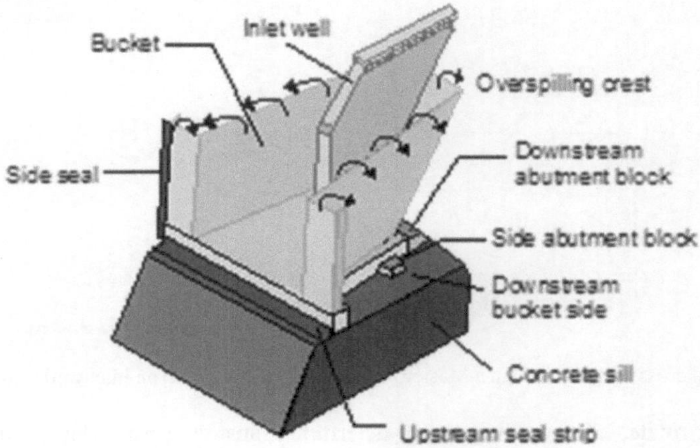

Figure 4. Typical 3D view of a Fusegate®.

3 FUSEGATE® SYSTEM: A TAILORED SOLUTION

The Fusegate® system is based on the following concept: Fusegates® are free-standing units installed side by side on a spillway sill to form a watertight barrier. The Fusegate® system operates as a weir divided into segments and each Fusegate® sits on the concrete sill and remains in place only by its own weight.

They bear against small abutment blocks set in the sill to prevent them from sliding before they are required to rotate (under extreme flood conditions). There is a chamber in the base of each Fusegate®, with drain holes to discharge incidental inflow (due to leaking seals for example). An inlet well on the upstream side of the Fusegate® crest discharges water into the chamber when the headwater reaches a predetermined level.

Under normal operating conditions, the Fusegates® act as a watertight barrier. Medium to moderate floods are simply discharged above the Fusegates® crest as they would do over a free weir. If the reservoir level exceeds a predetermined value, water will flow into the inlet well and cause an uplift pressure to develop in the chamber. The uplift pressure, combined with the hydrostatic pressure (acting from left to right on the diagrams below) is sufficient to overcome the restraining forces and the imbalance causes rotation of the unit off the spillway. The Fusegate® is then washed away from the spillway by the flood.

Each Fusegate® has its own overturning level, precisely determined by the height of the water inlet and its own unique stability. If the water level continues to raise after the first breach, more

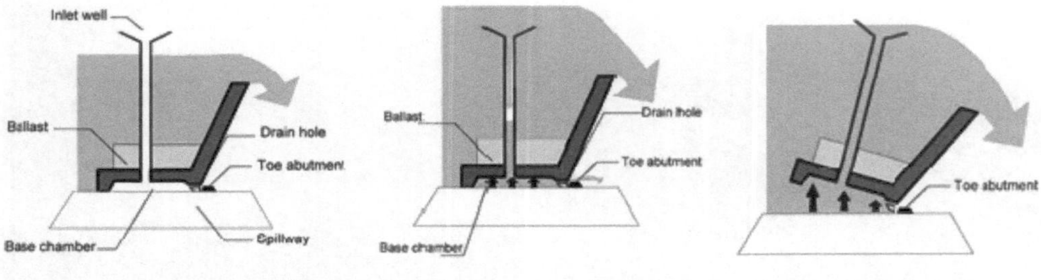

Figure 5. Fusegate spilling over the crest (left), Fusegate well being fed (center), Fusegate tipping (right).

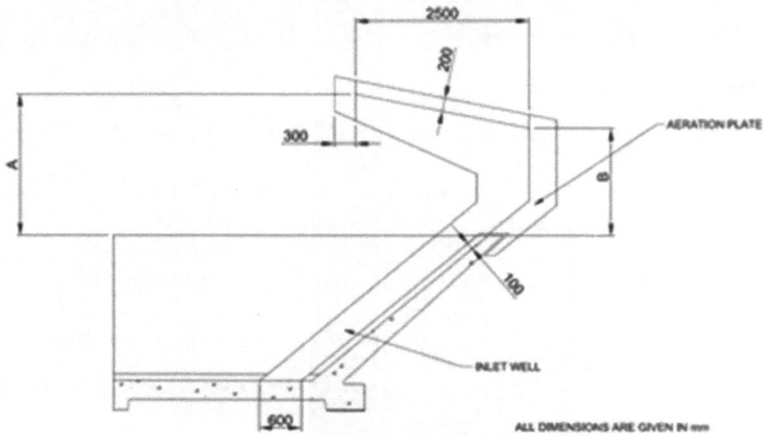

Figure 6. Typical cross section through a Master Fusegate® equipped with an inlet well (dimensions in mm).

Fusegates® can rotate, all according to pre-determined upstream water levels, until eventually there are no more units remaining. In such a case, the spillway is free to pass the original maximum design flood.

4 FUSEGATE® SYSTEM APPLIED TO QUIPOLLY DAM

The installation at Quipolly involves a total of eight Fusegates® each 4.40 m wide and 2.60 m high. Labyrinth crest type Fusegates® have been selected for this application, which allows for maximum spilling capacity before a Fusegate is required to overturn. At Quipolly, the first Fusegate is designed to tip in excess of the 1:1,000 annual exceedance probability (AEP).

Four Fusegates® are equipped with conventional inlet wells (located on their bucket) which are referred as "master" Fusegates® (see Fig. 6).

The other four Fusegates®, referred to as "secondary" Fusegates®, are not equipped with inlet wells. Each base chamber is fed by water flowing through a pipe cast into the sill and which daylights in the base chamber.

The pipe opening consists of a deflector facing the upstream reservoir. Such deflector is maintained watertight by a cover plate which is fixed to each "master" Fusegate® belonging to the same tipping sequence. When the "master" Fusegate® tips off, the cover plate is removed and the deflector diverts and concentrates the flow into the embedded pipe connected to the base chamber of the "secondary" Fusegate® which tips instantaneously (see Figs 7 and 8).

Such an arrangement has been designed through model tests for other projects in South Africa, Zimbabwe and Australia.

Figure 8 highlights the configuration of the sill with deflectors and embedded pipes.

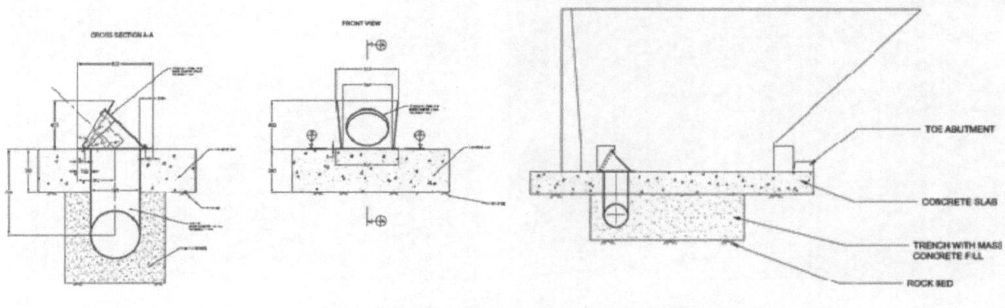

Figure 7. Typical views of a deflector and lateral view of a "master" Fusegate®.

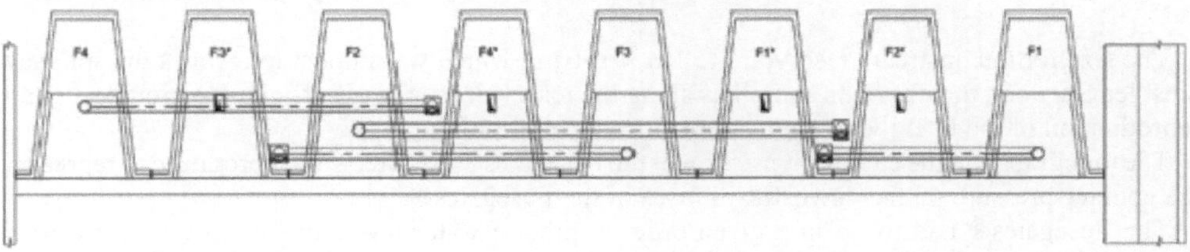

Figure 8. Plan view of sill with deflectors, embedded pipes and Fusegates® installed.

Figure 9. Fusegates® installed at Quipolly dam.

The main structure of the Fusegates® were constructed in pre-cast concrete and all steel components are made in steel (LDX 2101 type). The units were manufactured off-site and transported and assembled on site. The entire assembly and installation works including installation of the watertightness sealing system, setting of the concrete ballast blocks and calibration of the inlet wells, took 10 working days. The project was completed in December 2012 (see Fig. 9).

5 HYDRAULIC MODEL: OVERVIEW AND TARGETS

The challenges associated to the hydraulic conditions of the spillway justify the use of scaled physical model studies. The scale of the model is 1/20 which allows a good representation of the eight 2.60 m high labyrinth Fusegates® (see Fig. 10).

Figure 10. General view of the model and Fusegates®.

The size for the upstream reservoir (150 m × 150 m), which was almost four times the spillway crest length along the dam axis as well as along the reservoir axis, was sufficient to provide a good reproduction of the lateral contraction and current pattern of the flow.

The topography of the chute downstream of the Fusegates® was precisely reproduced to represent the counter-pressure on the downstream faces of the Fusegates®.

The Fusegates® had to tip in a given order at precise water levels of the upstream reservoir corresponding to given flood return periods. To meet these requirements, the two most important phenomena to be precisely known and controlled are:

- upstream flow characteristics: the way the inlet wells are filled,
- downstream flow characteristics: the counter-pressure exerted against the Fusegates® by the tailwater level.

These phenomena depend on the three dimensional hydraulics of the local flows which were largely unknown, unsteady and fluctuant. The main advantage of the scaled model was its ability to reproduce these non-computable phenomena.

The scope of model tests focused on determining:

- the impact of the lateral spillway walls on the approach flow;
- the proper impact of the well design on the flow and on its feeding;
- the impact of the global approach flow on the feeding of the inlet wells: the selection of the best tipping order;
- the fluctuant parameters controlling the tipping: the water pressure above the Fusegate® just before the tipping, the uplift pressure at the tipping, the downstream pressure just before the tipping.

Many static pressure taps were equipped on the model and the Fusegates®. There were also two dynamic pressure sensors in the base chamber of Fusegate® and one optical displacement sensor linked to a computer to record precisely the tipping conditions. The water head above the deflectors (alveolus water level) was measured after tipping of each "master" Fusegate® and stabilization of the upstream flows.

6 HYDRAULIC MODE: FLOW CONDITIONS IMPROVEMENT

On the right bank of the spillway, the approach channel wall and the abutment wall generated strong local perturbations which prevented the proper control of the tipping of the right bank Fusegates®. It was possible to improve these conditions at a reasonable cost by the implementation of an optimized 7 m long wall as an upstream extension of the abutment wall in the reservoir (Fig. 11).

The upstream nappe breaker (Fusegate® nose) that was initially considered semi-circular did not give satisfactory results as the rising of the water level in front of the well was too high and the Fusegate® was not properly fed. A triangular nappe breaker proved to be more effective (Fig. 12).

Figure 11. Strong lateral perturbation and wall abutment extension.

Figure 12. Semi-circular nappe breaker and triangular nappe breaker.

7 HYDRAULIC MODEL: TIPPING TESTS AND MEASUREMENTS

The water level in the reservoir just prior to the tipping should be as close as possible to those desired for the prototype. The first unit was required not to tip before the 1:1000 flood. In addition, the reservoir offered attenuation to the incoming Design Flood so it was not possible to delay the last tipping too much without causing the Maximum Water Level to rise.

Many measurements were carried out to collect data for the calibration of the mathematical model developed by Hydroplus, including water profiles along both the left and right sides of each inlet well, pressures immediately downstream of the Fusegates® measured by 14 taps before each tipping, pressure acting on the bucket of some of the Fusegates®, and water height in the base chamber during the tipping of two Fusegates®.

The graph of Figure 13 presents the water pressure evolution recorded in the chamber and the moment of the tipping for the third tipping unit.

The fluctuation of the pressure observed in the chamber was due to the strong downstream water level agitation caused by the spilling and the aeration, and transmitted through the drain hole to the chamber. The tipping occurred followed by the rising of the water pressure in the chamber which was not compensated by the discharge through the drain due to the difference of water pressures between the chamber and the downstream part of the Fusegate®. These tests were repeated to measure the variance in results, but it was observed that the tipping always occurred at the same water elevation. It proves that the Fusegates® will tip very precisely for a known upstream water level with an uncertainty of 1 cm, even with the highly fluctuating counter-pressure observed downstream of the Fusegate®.

8 HYDRAULIC MODEL: DESIGN OPTIMIZATION

Model results were instrumental to optimize the design in two different ways. In fact, part of the data were used as an input for the design, in particular the measurements that characterized the

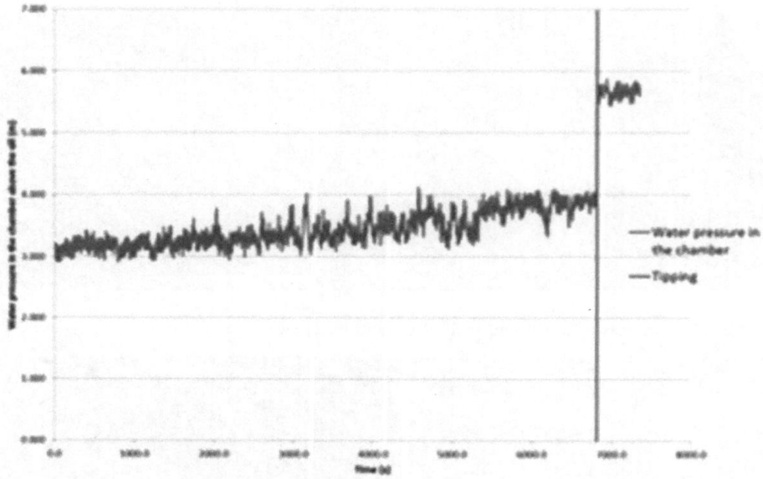

Figure 13. Base chamber pressure at tipping.

particular hydraulic conditions of the site (discharge coefficient, downstream water levels, alveolus water levels and water profile along the wells).

Downstream and alveolus water levels have helped to evaluate water pressure on the downstream faces of the Fusegate® and also related water heads used for the pressure chamber balance equations. Alveolus water level has also been used to calculate the deflector hydraulic efficiency and head losses between deflectors and "secondary" Fusegates® pressure chamber balance equations. Furthermore, water profile measurements were used to optimize the inlet wells' opening geometrical angle (upstream to downstream slope) and to calculate the inlet discharge during the incremental feeding of the wells.

The upstream curved flows resulted in an asymmetry in the feeding of the wells and were accentuated mainly on the right bank. This fact was taken into account to accurately size the wells' length and geometry in order to assure the proper feeding of the wells.

On the other hand, the measurement of the water head in the pressure chamber was reviewed to cross check the design accuracy. The theoretical pressure chamber, partially calculated by using independent experimental test results, was compared with proper measurements.

The test results allowed for improvement of the design accuracy, maximizing the savings on construction works and minimizing the risks.

9 CONCLUSIONS

The application at Quipolly dam has shown that the Fusegates® system can be implemented in a wide range of spillway sill configurations with severe upstream and downstream adverse hydraulic conditions (high tailwater levels, curved flows, etc.).

The numerical and physical model tests helped to optimize the design and to develop a tailored Fusegate® solution to upgrade the dam. In particular, Quipolly project confirmed the feasibility of the Fusegate® System and the deflector technology for applications involving very high tailwater levels that even exceed the Fusegate® crest level.

Compared to a traditional labyrinth spillway solution with the Fusegate® system, it was not required to increase the width of the existing spillway. As a consequence, the Fusegate® solution allowed reducing the excavation and concrete works and consequently the project investment cost.

In the same time the, Site Works schedule could be strongly reduced, since Fusegates® modules have been composed by precast concrete units and steelworks items produced off-site.

E. Existing projects

Labyrinth and Piano Key Weirs II – PKW 2013 – Erpicum et al. (eds)
© 2014 Taylor & Francis Group, London, ISBN 978-1-138-00085-8

Overview of Piano Key Weirs experience developed at EDF during the past few years

F. Laugier, J. Vermeulen & V. Lefebvre
EDF – CIH Hydro Engineering Center, France

ABSTRACT: PKW spillways appear to be a performing solution in order to improve discharge capacity from both technical and economic aspects. In addition to the fact that they can be installed on top of most existing concrete dams, their great specific discharge, even for high heads, is their main advantage, as well as their operating free flow spillway reliability. PKW solutions have revealed to be reasonably cheap solutions thanks to a very short time of construction and low operating losses. Electricité de France (EDF) decided to build the first worldwide PKW spillway at Goulours dam in 2006. Because of the very good feedback of Goulours PKW project, EDF actually decided to launch other PKW projects to rehabilitate dams having an insufficient discharge. Five other PKW units were actually built from 2008 to 2012. Four other PKW projects units are currently under construction by EDF and a couple of other projects are in the design phase. Some current PKW projects tend to be designed higher and wider with significantly increased specific discharge. Future developments might lead to even more cost-effective structures or at least structures easier to build (steel, sandwich structures mixing steel and concrete). The paper will also present existing hydraulic and structural design tools and methodology.

1 INTRODUCTION

Piano Key Weir (PKW) labyrinth spillways are an innovative technical-economical optimization of traditional labyrinth weirs. They were introduced by Lemperiere and Ouamane (2003) ten years ago and they are now well known in the sphere of free-flow spillways. PKW appear to be an interesting option to increase spillway discharge capacity by providing high specific flow rates. They can also be used to increase the storage of reservoirs while keeping the maximum water level constant. Unlikely traditional labyrinth weirs, PKW can be installed on the top of most existing concrete dams. As free-flow spillways they also provide an excellent reliability and would generally require less maintenance than active controle spillways.

After building the first worldwide PKW spillway at Goulours, France (2006), EDF actually decided to launch other PKW projects to rehabilitate dams having an insufficient discharge. Five other PKW units were built from 2008 to 2012. Four additional PKW units are under construction by EDF. A couple of other units are planned in the near future thanks to the great increase in design and knowhow developed by EDF.

2 PKW BASICS

The PKW spillways concept is based on labyrinth weirs technology. PKWs are innovative thanks to their overhangs that enable to keep a very long developed crest length without requiring a large footprint on a foundation. This point is one of its more favorable characteristics as it can be installed on top of most existing concrete gravity (or arch) dams.

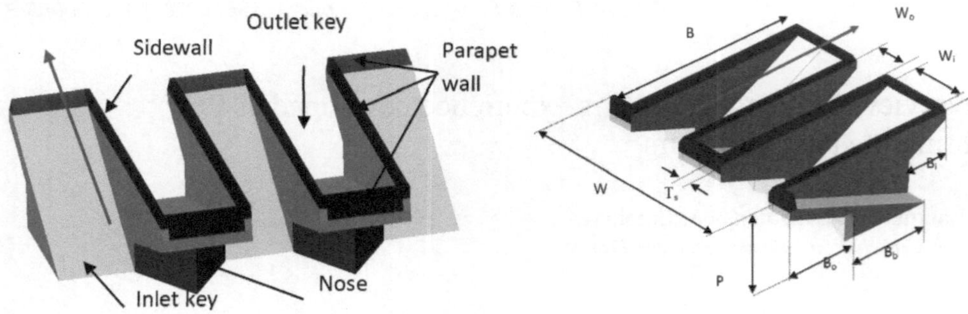

Figure 1. Main elements of principal parameters defined by the PKW naming convention.

PKW geometry involves many parameters making its configuration description difficult. The international scientific interest about PKW requires a unity in terminology. A workgroup involving the Laboratory of Hydraulic Constructions of Lausanne University, the Hydraulic Laboratory of the University of Liège and EDF – Hydro Engineering Center developed a specific nomenclature in Pralong et al. (2011). This naming convention was presented during the first workshop on PKW and labyrinth spillways held in Liege in 2011.

The main geometrical parameters presented in Figure 1 are:

- The PKW height P;
- The PKW total width W;
- The PKW upstream-downstream length B;
- The keys widths (W_i for inlet key and W_o for outlet key);
- The overhang lengths (B_i for downstream overhang and B_o for upstream overhang); and
- The sidewall thickness T_s

The main geometrical parameters governing PKW discharge capacity are:

- Ratio "L/W". This "labyrinth" ratio represents the developed length. It is a major factor for small heads.
- PKW height "P". High P value improves the flow approach and distribution within PKWs inlet keys. It also reduces submergence effects in outlet keys.
- Key width ratio "W_i/W_o". High W_i/W_o value improves flow approach and distribution within PKWs inlet keys. On the other hand, high W_i/W_o value increases submergence effects in outlet keys.
- Sidewall thickness T_s. Thicker sidewalls significantly reduce the available space on the dam crest. This structural date can affect up to 10–15% PKW specific discharge capacity (for example the reduction of Malarce PKW sidewall thickness from 40 cm to 20 cm on top of wall increased PKW specific discharge capacity by 10%).

3 OVERVIEW OF CONSTRUCTED PKW BY EDF

Six PKW have been built by EDF since 2006, almost one per year. This structure is particularly interesting for rehabilitation projects with respect to discharge capacity since it can be placed on top of existing dam. Numerous dams have been equipped with PKWs: Goulours (Pyrénées), St Marc and l'Etroit (center of France), Gloriettes (Pyrénées) and the largest and newest one, Malarce (Ardèche). Escouloubre PKW was built for an operation optimization purpose on an expansion basin. Table 1 presents the specific parameters of each constructed PKW according to the naming convention.

Table 1. Achieved PKW (characteristics).

	Goulours	St Marc	Etroit	Gloriettes	Escouloubre	Malarce
Dam type	Concrete gravity	Concrete gravity	Concrete gravity	Slim arched	Concrete gravity	Concrete gravity
Height on ground level (m)	20	40	25	43	13.4	28.4
Dam's date of construction	1946	1932	1933	1951	1971	1968
PKW's date of construction	2006	2008	2009	2010	2011	2012
PKW localization	Bank	Dam crest	Dam crest	Bank	Bank	Dam crest
Dam design flowed (m^3/s)	162	660	500	150	13	4600
PKW Discharge capacity at MWL (m^3/s)	68	138	82	90	13	570
q_{sW} at H = 1 m ($m^3/s/m$)	5	6.5	9	6.8	~4 (2)	9.4
Total project cost (k€)	400	1800	1500	1600	615	4500
PKW Cost (k€)	300	400	400	400	230	800
Head on PKW at MWL (3) (m)	0.95	1.35	0.95	0.8	0.65	1.5
Time construction (month)	3	6	6	5 (1)	6	12
B (m)	9.3	12.7	12.2	10	5.1	13.46
B_o (m)	3.35	4	3.2	3.5	1.2	6.63
B_i (m)	1.5	4	2	2.6	1.2	2.03
B_b (m)	4.4	4.7	7	3.6	2.7	4.8
P (m)	3.1	4.2	5.3	3	1.8	4.4
W (m)	12	18	18.7	18.5	4.8	42.5
W_i (m)	2.7	3.1	2.45~2.75	2.3~2.5	1.3	1.25~1.65
W_o (m)	1.5~1.8	2.2	1.5	1.5	0.9	1.58
T_s (m)	0.2	0.25	0.35	0.25~0.35	0.3	0.2~0.4
L (m)	59	77	78	86.8	22	350
P_{px} (m)	0	0	0.5	0	0	1.65
B_n (m)	1	1.5	1	1.5	1	1.5
Nose form	Triangular	Triangular	Triangular	Rounded	Triangular extruded	Triangular

(1) 5 months for the dissipation channel and 5 months for the PKW itself in two seasons
(2) Extrapolation from smaller values of hydraulic heads
(3) MWL: Maximum Water Level

Most of the built PKWs present a small number of alveoli (One to three inlet keys) and relatively balanced overhangs. The PKW of Malarce is by far the largest one ever with 12 units and an upstream overhang three times longer than the downstream one.

The maximum hydraulic head on these spillways is in the range between 0.5 m and 1.5 m and they can pass discharge up to 570 m^3/s which represents specific discharges from 2 to 4 times better than classic linear weirs.

Figure 2 shows the plan view, typical cross section and general picture of each PKW built by EDF so far.

4 PKW PROJECTS BY EDF

EDF is currently working on several other PKW projects in France mainly for discharge capacity improvement. The knowledge about hydraulic optimization and structural stability developed at the Hydro Engineering Center is combined with the building experience drawn from the previous projects to design large and efficient PKW. Dams of Campauleil and Charmines are

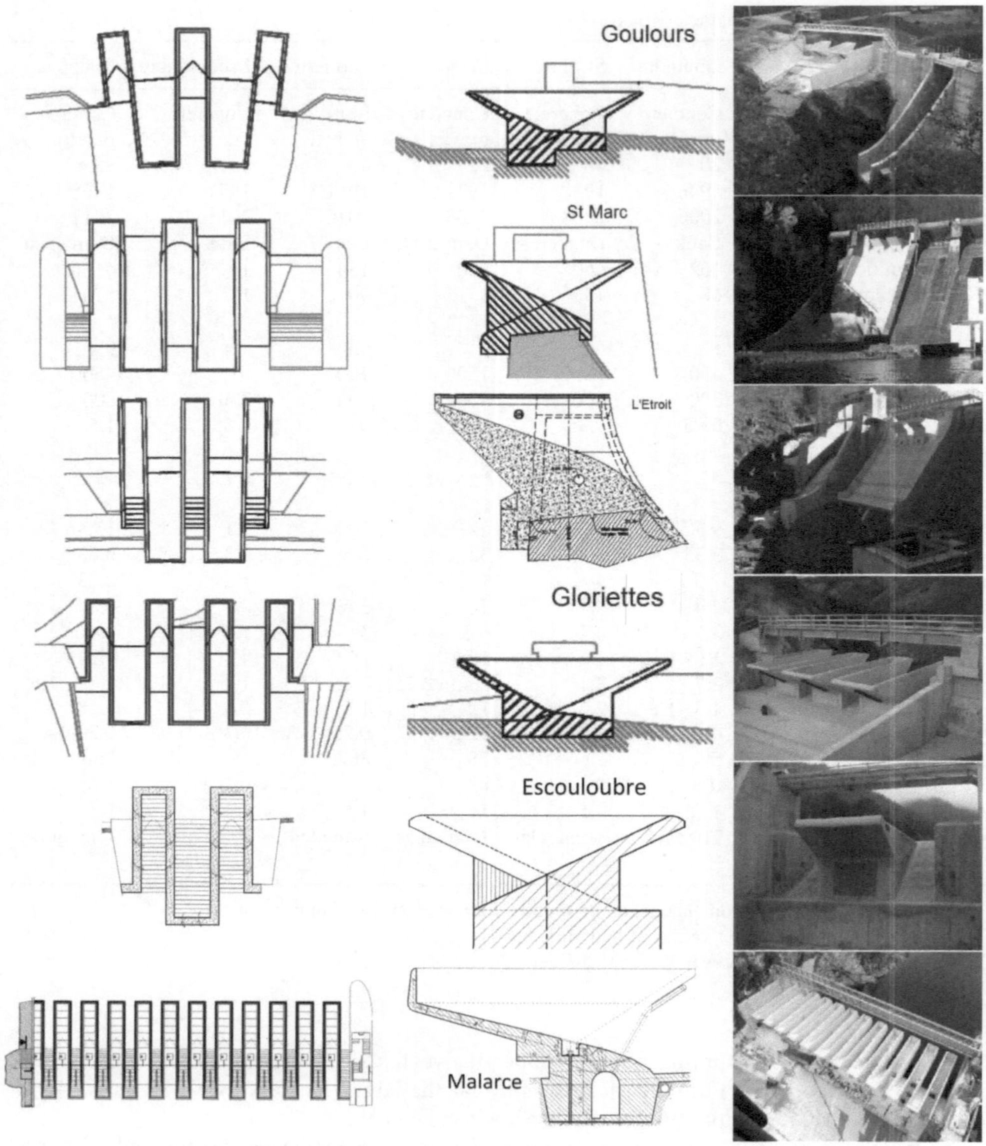

Figure 2. Plans and cross-sections of the six first PKW built by EDF.

under construction and should be completed within the next two years. They include one 16 m PKW unit and two 23 m PKW units respectively. The detailed design of two other projects aiming to increase the discharge capacity by 300 m^3/s and 455 m^3/s is completed for the dams of La Raviège and Gage. They are on call for tender phase and construction should start in 2014.

Table 2 and Figure 3 summarize all the characteristics of the PKW projects lead by EDF in the short term.

Table 2. Characteristics of planned PKW.

	Campauleil	Charmines	Raviège	Gage
Dam type	Concrete gravity	Concrete gravity and earthfill	Concrete gravity	Slim arched
Height on ground level (m)	15.8	17	37	39.6
Dam's date of construction	1939	1948	1958	1954
PKW's date of construction	2014	2015	–	–
PKW location	Dam crest	Dam crest	Dam crest	Bank
Dam design flowed (m^3/s)	192	685	1700	675
PKW Discharge capacity at MWL (m^3/s)	120	300	300	455
q_{sW} at H = 1 m ($m^3/s/m$)	8.3	8	8.5	11
Total project cost (k€)	2800	6500	4000~6000	15000~25000
PKW Cost (k€)	450	1800	700~1500	700~1500
Head on PKW at MWL (m)	0.9	1	1.4	1.75
B (m)	13.1	13.24	13.24	13
B_o (m)	4.9	3.97	4	4
B_i (m)	2.8	4.41	3.33	3
B_b (m)	3.5	4.87	5.91	6
P (m)	5.35	4.38	4.67	6
W (m)	16.55	2*23	25.8	26.6
W_i (m)	1.55	2.4	2.4	1.6
W_o (m)	1.4	1.6	1.65	1.3
T_s (m)	0.35	0.35	0.25~0.4	0.25~0.4
L (m)	115	2*120	177	208
P_{px} (m)	0.7	1	1	0.8
B_n (m)	1	1.15	2	2
Nose form	Triangular	Rounded	Triangular	Triangular

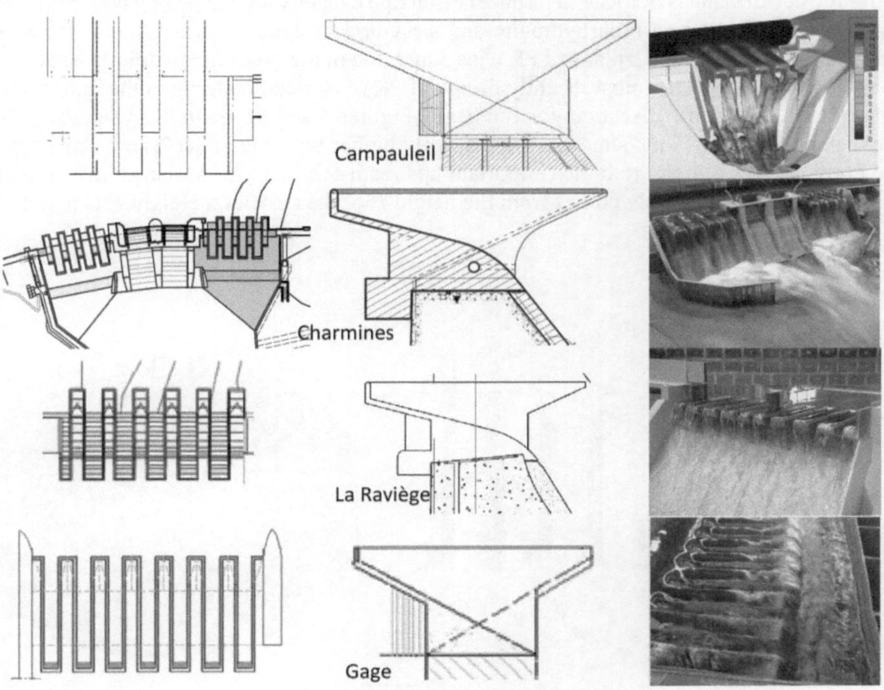

Figure 3. Plans and cross-sections of the next PKW to be built by EDF.

5 LESSONS LEARNT

5.1 *Design for existing dams*

Design constraints and considerations will be quite different whether the project concerns a new dam or an existing one. This point was emphasized by the Machiel (2012) who distinguished purely hydraulic optimization and technico-economical optimization.

In the case of existing dams, many parameters are imposed by the existing structures: for example the total available width on a dam crest, the difference between maximum water level (MWL) and normal water level (NWL), the base length (B_b). PKW height parameter P is also quite dependant on the capacity to keep a lowered reservoir level during works and the corresponding cost in term of operation loss. All these crucial parameters change for each existing dam and will lead each time to a specific design. However, in most cases, the overall cost of the project will be governed by the specific PKW discharge capacity. In most cases, it will be more interesting to optimize the hydraulic PKW specific discharge capacity. This point is illustrated by Figure 4 below. The cost of the whole PKW project (including dissipation structures, site installations, etc.) is compared to the cost of the lone PKW structure. For most PKW projects, the cost of the PKW represents less that 25% of the total project cost. This has to be seriously considered when designing a new spillway. Other cost matters more than the actual spillway. For examples, Gloriette and Gage PKW projects have very important downstream structures and highlight the aforementioned point.

Most EDF dams are equipped with gates in order to optimize the available reservoir volume for hydroelectric purposes. Technico-economical comparisons are performed for new spillway projets. They often show that the installation of free flow spillway is more suitable for cost, maintainance and operation reasons rather than the installation of an additional mechanical gate. However, a minimum water head between NWL and MWL is required so that free flow spillway can properly operate. The analysis of EDF major dams characteristic shows that the available water head for free flow spillway (Maximum Water Level minus Normal Water Level) is less than 1 m for 40% of them. This small water head points out the relevancy of labyrinth or PKW for these applications.

Hydraulic performances of PKW in terms of discharge capacity are increasing constantly thanks to common optimization on both hydraulic and structural designs. Figure 5 shows the specific discharge of each achieved and planned PKW as a function of the upstream hydraulic head. We can see that the efficiency of the most recently designed PKW is much better than the initial designs. Considering PKW specific discharge for 1 m head, (Figures 5 and 6), Gage PKW appear to be the most efficient one with 11 $m^3/s/m$. It is significantly higher than Malarce (9.4 $m^3/s/ml$) although the structure is much simpler to build (important upstream overhang for Malarce close to type B). Gage PKW efficiency partially comes from the height P of the structure. Height P is major factor of PKW efficiency.

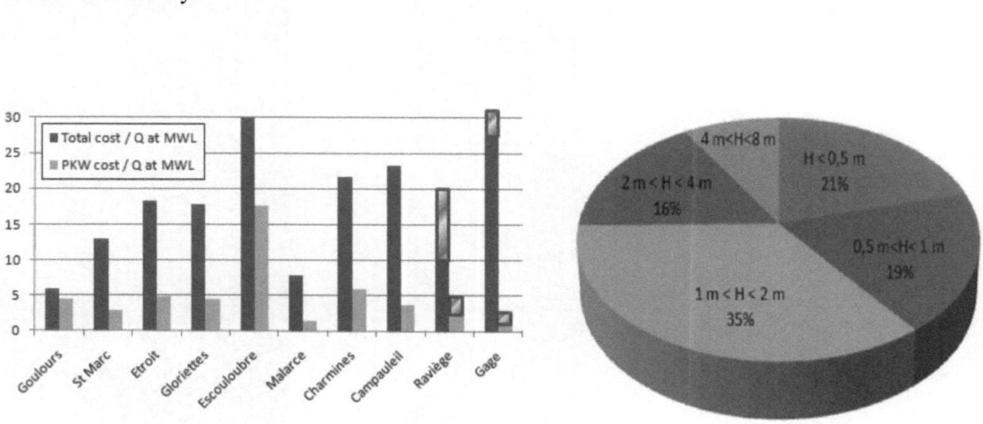

Figure 4. Economic ratios and potential hydraulic head repartition on EDF dams.

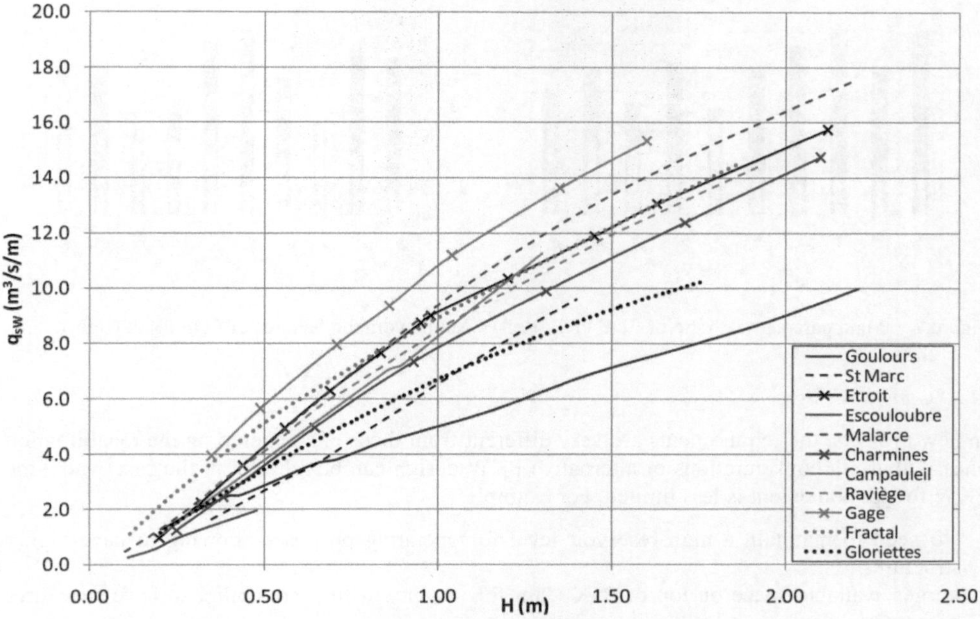

Figure 5. Specific discharge versus upstream head for constructed and planned PKW.

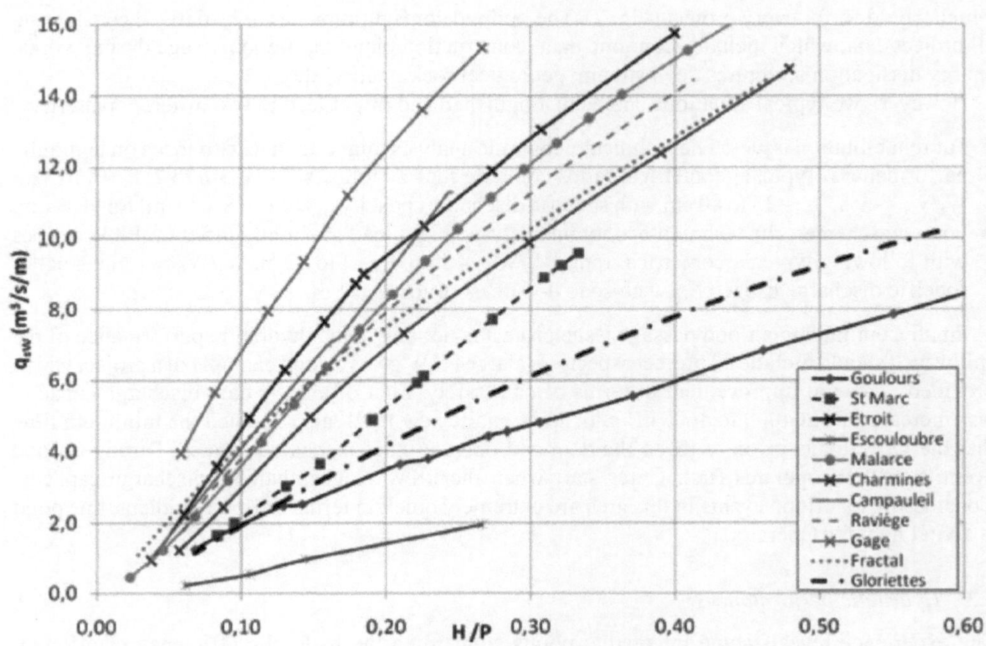

Figure 6. Specific discharge versus H/P for constructed and planned PKW.

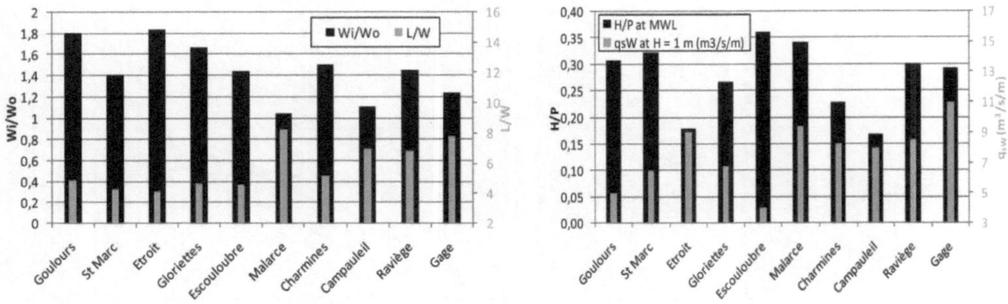

Figure 7. Main parametrical ratio of EDF PKW (left) – Main hydraulic features of EDF PKW (right).

5.2 *General design*

In new projects, the requirements are very different from those encountered on the rehabilitation ones. Labyrinth configurations or alternative PKW design can be relevant in the case where the PKW foundation extent is less limited. For example:

- No need to maintain a high reservoir level for operating purposes allowing to have higher structures.
- Larger available space on top of RCC dam for instance with a possibility to have structures without overhangs.
- Available space on the banks for spillway foundation.

In this case, Labyrinth weirs should be closely compared to PKW on a technical-economical point of view.

After 10 years of design experience, it is still difficult to propose a typical set of parameters which would fit for every project (Fig. 7). The optimal configuration has to be derived considering all project data, which include economic data, construction planning, flood passage during works, energy dissipation structures downstream, geological rock quality, etc.

However, two typical situations can be distinguished and might lead to two different patterns:

- For rehabilitation issues: The technical-economic analysis might often lead to focus on hydraulic performances. Typical geometrical ratios could be such as follows : $L/W \sim 6$ to 7, $P \sim 4$ to 5 m, $W_i/W_o > 1.3$, $T_s \sim 35$ to 40 cm with specific discharge capacity $q_s w$ close 8 $m^3/s/ml$ for $H = 1$ m.
- For new schemes, the technico-economic analysis might lead to slightly different PKW shapes with following typical geometrical ratios: $L/W \sim 4$ to 6, $P \sim 4$ to 10 m, $W_i/W_o > 1$ and smaller specific discharge capacity $q_s w$ close to 4–6 $m^3/s/ml$ for $H = 1$ m.

Finally, optimizing a flood passage issue project is not only about hydraulic performance of the spillway. It should include all project aspects. Malarce PKW gives a good example of a project which provided a general improvement in terms of dam safety. First of all, the dam discharge capacity was increased from 4000 to 4600 m^3/s. In addition the new PKW also doubled the minimum time that the Operator is given to reach the dam and operate other mechanical gates. During a flood event, the PKW operates first. Gates start when the PKW reaches half its discharge capacity. Considering that flood events in this area are extremely quick in terms of flood gradient, this point is a vital help for Operators.

5.3 *Hydraulic performances*

The experience reveals some interesting points concerning the hydraulic efficiency of different devices and configurations.

The nose built under the upstream overhangs improves the inlet key entrance and consequently the PKW discharge capacity. Various shapes can be designed: triangular, rounded, or triangular

Figure 8. Nose at Goulours, parapet wall at l'Etroit and inclined sidewalls at Malarce PKW.

extruded in the base under the outlet key. This last solution has been chosen at Escouloubre and seems to be the most efficient if the stability is not affected by such modification. A sensitivity analysis has been performed on CFD software and gains in specific discharge are about 5%.

Parapets walls have been installed on some PKW built by EDF, all along the crest or only on the outlet key. The idea was to increase the outlet key volume for a better evacuation and improve the stream lines shape upstream of the PKW. Simulations reveal that the global gain is limited to 5% for a parapet wall localized on the outlet key crest. For some configuration, this gain might be quite small. The gain will be very small if the outlet key slope is high.

Crest shape design is linked to the constructive dispositions executed on the structure. Quarter rounded, half rounded, flat and chamfered crest have been realized on EDF PKW and a numerical analysis showed that the quarter rounded and half rounded are the most efficient (Lefebvre, 2013). However, the influence is almost insignificant. Construction and durability constraints are more critical. Considering the difficulty of formwork assembly and long-lasting considerations, the quarter or half-rounded shapes should be preferred to respect both hydraulic and building imperatives.

It has been shown that the sidewall thickness has a major influence on the discharge capacity of the PKW, especially for narrow keys. Yet, it is rather hard to build thin concrete walls respecting the structural constraints of the project. In order to improve the flow conditions and to keep a good internal resistance, battered or inclined sidewalls have been imagined, large at the PKW base and getting thinner close to the top. This kind of geometry is used on Malarce PKW (40 cm to 20 cm) and gains about 5% to 10% are expected compared to thick sidewalls. On a practical point of view, it is easier to place the inclination of the wall only on one side to keep the other face of the formwork vertical.

5.4 *Aeration*

In order to aerate the flow on the PKW, air intakes (>400 mm diameter) have been implemented for all PKW projects. Both structures start below downstream alveoli and come out to the dam crest on both sides of the PKW. They are protected by stainless steel covers and connected to each other in order to allow doubled air supply in case of blockage of one pipe. Although physical models show that PKW flow appear to be naturally aerated, additional aeration devices were added as their additional cost is very small compared to the PKW project. They can only improve flow characteristics and stability.

5.5 *Structural design*

PKW happen to be very hyperstatic and therefore stiff structures. Various loads such as gravity, water pressure and seismic accelerations are very well supported by the reinforced concrete with low stresses. As per all thin-walled hyperstatic concrete structures, attention must be paid to the thermal loads and the stresses induced. Thermal calculations showed that high temperature gradients can establish during summer when the sunlight heats up the downstream face and the water keeps the

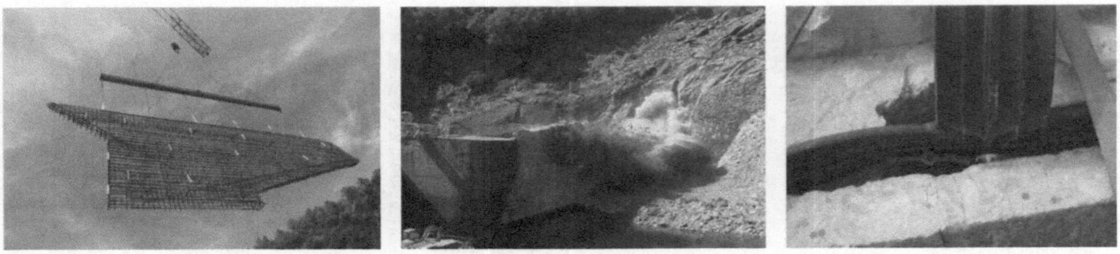

Figure 9. Sidewall armature, dam demolition and waterstop junction at Malarce.

upstream face cool. Because PKW are thin and stiff, more steel reinforcement is required for the thermal loads than for classic structural loads.

PKW overall stability can be calculated according to usual gravity dams rules. At first, PKW seemed to be very unbalanced structures, hard to stabilize without anchoring. But, calculations generally show that the PKW center of mass at NWL is located above the base ensuring the stability of the spillway. At low water level, especially when the PKW is totally out of water, the stability is also easily reach without anchor or specific devices. PKW self-stability allowed installing two of them on top of existing dams without anchoring them to the existing structure for Saint-Marc and l'Etroit projects. The purpose was to take into account alkali-aggregate reaction affecting dam concrete. Swelling concrete could impact new structures if anchored by bringing imposed displacements and additional stresses. However, prestressed anchors could be used, as in Malarce for example, in order to reinforce the overall stability of the dam and to ensure the stability of the very unbalanced PKW.

Demolition aspects are also an important part of the rehabilitation project. It is necessary to ensure all structures close to the new location of the PKW are not damaged by any explosion or vibrations. Diamond-cable saws can be used to cut a part of the dam before using explosives.

To achieve the goal of no leakage at the PKW-dam interface, a flat waterstop is embedded in the new concrete and sealed in the existing dam. In addition, a resin injection device is installed in case of waterstop failure. These devices are still reachable and could be injected if leakage was discovered during the PKW's life.

5.6 *Tools for design*

Recent performance increase in CFD has enabled numerical modeling to become competitive in the frame of hydraulic engineering studies. 3D numerical modeling turns out to be an interesting opportunity as it is possible to model various geometries in a short timeframe. CFD modelling has demonstrated a good correlation with the experimental results. The accuracy of the CFD-model developed is about 5% on average and does not exceed 10%. Many sensitivity studies were performed on geometrical parameters (Pralong et al. 2011) and improved the hydraulic understanding of PKW. Nevertheless, physical models remain necessary for many projects (in addition to numerical models) to better understand 3D complex phenomena such as downstream energy dissipation features.

PKW stability is easy to perform. It can be studied on a spreadsheet program according to usual gravity dam methodology and design criteria. Stability calculations were based on several loading cases: empty dam, dam at NWL, design flow and seismic load. The estimation of foundation and existing concrete parameters was made through several geotechnical campaigns. Structural design softwares codes were also used for structural calculations on PKW and thermal effects.

5.7 *Lessons learnt from PKW actual flood events*

Although they were recently build, Gloriettes and Malarce PKW already experienced significant flood events (Fig. 11). A serious flood hit Gloriettes dam in October 2012 (estimation of the return

Figure 10. Water pressure analysis on CFD software and thermal load used for reinforced concrete calculation on structural design software.

Figure 11. Overflow at Gloriettes and Malarce PKW.

period close to 50 years). Without new PKW, the MWL would have been reached on the dam. Visual inspection after the flood showed that PKW structures were not damaged. In addition, the downstream dissipation structure experienced minor damages which will not require much repair.

Malarce dam experienced two flood less than 6 months after completion in May 2013 accounting 100 and 200 m^3/s respectively for a 600 m^3/s PKW total discharge. As per Gloriettes, visual inspections after flood revealed that PKW structures did not suffer any damages. Slight damages can be seen at the downstream dissipation structures slab joints. Superficial soft filling Thioflex joints were partially removed by flow energy. However, they do not have any major action. Waterstop joints between slabs are located 30 cm beneath slab top and are not affected by the flow.

5.8 *Passing debris*

The floating debris issue is a crucial one with regard to the actual and safe operation of spillways.

In most dam rehabilitation projects, the dams are equipped with existing mechanical gates. Specific discharge of mechanical gates is generally much higher than PKW. For instance a moderate 4-m high gate will discharge almost 20 $m^3/s/m$ when the PKW is initially engaged. Therefore, during the first flood hours, most debris will be naturally conveyed by existing gates.

On the other hand, even if debris reach PKW spillway (no gate or peculiar configuration leading them towards PKW), PKW remain a safe solution as free flow spillway. Indeed, when the water level increases, debris will be naturally discharged over PKW free flow crest.

In addition hydraulic tests were performed on physical models to assess PKW residual performance regarding floating debris. Goulours PKW was tested in relation to this topic. Tests were carried out with equivalent 4 to 10 m long trees with branches.

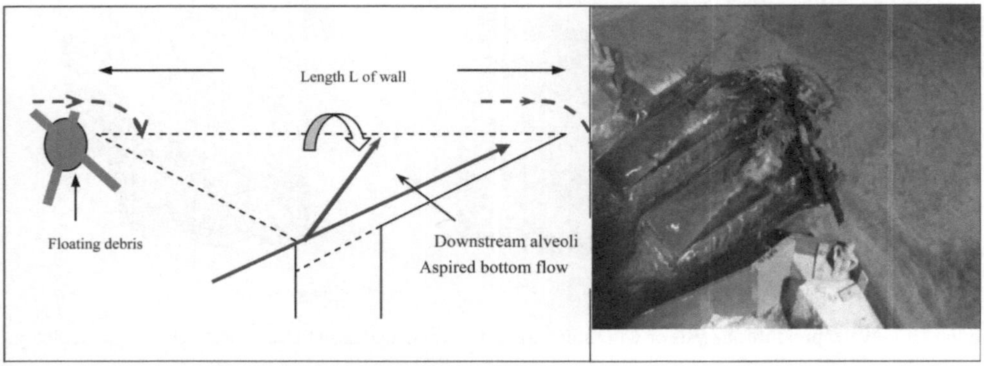

Figure 12. Residual discharge of PKW with debris.

Notwithstanding the aforementioned points, it is nevertheless possible that large floating debris might block the PKW. In this case, the residual PKW discharge capacity was simulated. Experimental results showed that:

Blocking of PKW potentially affects only a small part of the overall developed crest wall length (5 to 10%) corresponding to the witdth of the outlet key which are the part of the PKW that are mostly affected by debris).

Tests performed with dye and additional 3D numerical simulation highlighted the path actually followed by the water flow. Tests showed that the water actually flow beneath floating debris at the bottom of the PK Weir and then rises upwards to the sidewall crest.

As a result, with floating debris blocking the PKW, the residual discharge capacity remains very high. It is more than 80% of the normal one without debris.

6 PERSPECTIVES OF DEVELOPMENT

6.1 *Trapezoidal PKW*

An attempt to increase PKW discharge capacity was made by designing a trapezoidal PKW rather than a rectangular one. This idea was developed thanks to the study of Falvey (2003) and more recent publications, concerning triangular and trapezoidal labyrinth weirs. Decreasing the sidewall angle decreases the discharge coefficient but increases the crest length (in a given footprint). There is a balance between sidewall angle and crest length and construction cost – many labyrinth weirs are between 8 and 12 degrees.

Experiments have been performed both at EDF Hydro Engineering Center and National Hydraulic and Environmental Laboratory on this new geometry. They revealed interesting outcomes; in the position where the inlet is larger than the outlet, trapezoidal PKWs are more efficient than the rectangular one, with a gain in discharge that should be greater than 5%. However, this gain is not as promising as expected when compared to usual vertical labyrinth weirs.

6.2 *Fractal PKW*

In order to keep increasing PKW discharge capacity, a first option consists of increasing the size of PKW units. This generally means significant increase in civil engineering cost and building complexity. Another way to increase PKW discharge capacity is proposed. A new concept of PKW has been developed, called fractal PKW. It consists of adding a smaller PKW on the crest of a bigger PKW spillway.

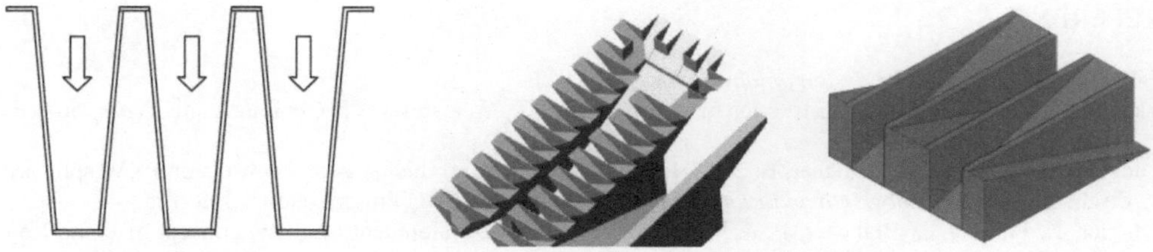

Figure 13. Innovative schemes: Trapezoidal, Fractal and Type D Piano Key Weirs.

The concept was first tested through 3D numerical simulations with Flow3d software. Several different configurations were tested before arriving at an improved design. Finally this configuration was tested on a physical model, which confirms results given by the 3D simulation. It appears that the fractal PKW is quite efficient for low head smaller than 1 m, especially lower than 50 cm. Above this point, the flow over the fractal PKW tends to saturate and this option is not very interesting any more. Fractal PKW appears to be very efficient for low heads and an interesting option for reservoir level control to replace gates or other mechanical components.

In practice, a fractal PKW would be made of prefabricated steel elements that would be assembled on site on the top of a traditional concrete PKW with slightly enlarged sidewalls (50 cm). In such a way, the overcost of the fractal PKW compared to usual concrete PKW should be small.

6.3 *Type D (without overhangs)*

PKW without overhang is a hybrid solution between PKW type A and labyrinth weirs. PKW Type D offers a very interesting stability behavior which can represent a massive advantage for low dams founded on weak rocks or a thick layer of alluviums.

The concrete amount required for this solution is significantly higher than for PKW Type A or Rectangular labyrinth weir, but the reinforcement ratio is much less, which can balance the cost comparison between the structures depending on the local material cost of the project. PKW Type D seems to be easier to build, requiring less complex formworks and lower mass concrete to fill the keys.

Being less sensitive to submergence effect, its discharge capacity almost matches the PKW Type A performances under very submerged conditions.

7 CONCLUSION

PKW recent achievements in France confirm the strong interest in this innovative spillway solution for discharge capacity improvement. Six PKW have been realized by EDF and 4 are under construction for existing dams. They reveal that this structure perfectly matches the requirements of many rehabilitation projects.

Lessons learnt from the design and construction of 10 PKW lead to significant improvements of various PKW parameters, including hydraulic design, structural design, cost and schedule and project management. This global experience leads to optimized PKW structures, mainly designed for rehabilitations projects with heads generally between 1 and 1.5 m. PKW spillway discharge capacity can be higher than 10 m^3/s for 1 m head for PKW with reasonable dimensions that can be built on dam crest with usual site installation and scaffoldings. Thanks to their geometrical simplicity, PKW projects were generally kept on control in terms of cost and schedule. First overflow events affecting PKW in France in 2012 and 2013 gave satisfaction.

REFERENCES

Falvey, H. 2003. *Hydraulic design of labyrinth weirs*. ASCE.

Laugier, F. 2007. Design and construction of the first Piano Key Weir spillway at Goulours dam. *Hydropower & Dams* 14(5): 94–100.

Laugier, F., Pralong, J. & Blancher, B. 2011. Influence of structural thickness of sidewalls on PKW spillway discharge capacity. *Labyrinth and Piano Key Weirs-PKW2011*, CRC Press, London. 159–165.

Laugier, F., Pralong, J., Blancher, B. & Montarros, F. 2011. "Development of a new concept of Piano Key Weir spillway to increase low head hydraulic efficiency: Fractal PKW", *Labyrinth and Piano Key Weirs-PKW2011*, CRC Press, London, 281–288, 2011

Leite Ribeiro, M., Bieri, M., Boillat, J.L., Schleiss, A., Delorme, F. & Laugier, F. 2009. Hydraulic capacity improvement of existing spillways – Design of Piano Key Weirs. *23ième Congrès de Grands Barrages – CIGB-ICOLD*, Brasilia, Q. 90 R. 43.

Leite Ribeiro, M., Pfister, M., Schleiss, A.J. & and Boillat, J.-L. 2012. Hydraulic design of A-Types Piano Key Weirs. *Journal of Hydraulic Research*. 50(4):400–408.

Lefebvre, V., Vermeulen, J., Blancher. 2013. Influence of geometrical parameters on PKW discharge with 3D numerical analysis. *Labyrinth and Piano Key Weirs-PKW2013*, CRC Press, London

Lempérière, F. & Ouamane, A. 2003. The piano key weir: a new cost-effective solutions for spillways. *Hydropower & Dams* (5): 144–149.

Machiels, 2012. Experimental study of the hydraulic behaviour of Piano Key Weirs. *Phd Thesis*, University of Liège, Belgium.

Pralong, J., Vermeulen, J., Blancher, B., Laugier, F., Machiels, O., Erpicum, S., Pirotton, M., Leite Ribeiro, M., Boillat, J.L. & Schleiss, A.J. 2011. A naming convention for the piano key weirs and geometrical parameters. *Labyrinth and Piano Key Weirs-PKW2011*, CRC Press, London, 271–278.

Pralong, J., Montarros, F., Blancher, B. & Laugier, F. 2011. A sensitivity analysis of Piano Key Weirs geometrical parameters based on 3D numerical modeling. *Labyrinth and Piano Key Weirs-PKW2011*, CRC Press, London, 133–139.

Labyrinth and Piano Key Weirs II – PKW 2013 – Erpicum et al. (eds)
 ISBN 978-1-138-00085-8

The first commercial Piano Key Weir in Switzerland

P. Eichenberger
Hydro Engineering GmbH, Andelfingen, Switzerland

ABSTRACT: The spillway of an existing small hydropower scheme on the River Emme in central Switzerland has been equipped with a piano key weir (PKW). Initially, a second flap gate had been proposed to increase the spillway capacity at the powerhouse. However, the cost of this additional flap gate and the risks involved in reliably controlling the gate obliged the design team to look for alternatives.

The proposed PKW at the place of the flap gate was readily accepted by the plant owner as the simplicity of the solution and the low cost proved to be quite substantial. The PKW was commissioned in November 2012 and the operational experience so far has been good. This PKW application – the first of its kind in a commercial setting in Switzerland – must be considered as a niche market not normally associated with the PKW concept.

1 INTRODUCTION

1.1 *Project overview*

The "Emmenau" small hydro scheme was established in the 1940s; it consists of a canal development with two low-head power plants in serial operation. Water is diverted from the River Emme, a torrential river in central Switzerland with a catchment area at the intake of some 640 km^2. The old turbine-generator units were at the end of their service lives and the owner initiated a refurbishment including an upgrade to a higher design flow. The refurbishment project for the second hydro plant along the canal, the so-called "Emmenau II" plant, was challenging due to the very limited space available at the site. The plant is squeezed into a plot of land surrounded by the River Emme, the railway track and another water course, a tributary to the main river. In addition, a new cycling path was to be established over the old powerhouse.

As the plant was no longer used in stand-alone operation for an old disused factory but was solely run in grid connection mode, the design flow of the plant could be increased from 3.75 m^3/s to 4.65 m^3/s. The existing gated spillway at the powerhouse was insufficient to evacuate this new design flow in the case of a sudden load rejection followed by a turbine shutdown. Previously, canal overflow had occurred and authorities as well as land owners upstream of the plant insisted on the establishment of a fail-safe spillway system. In addition, the damage potential due to flooding went up due to new commercial premises which were established along the canal in 2012.

1.2 *Spillway design and alternatives*

The first idea to enlarge the existing flap gate had to be dropped due to insufficient space between the abutments of the railway and cycling path bridges. The second approach, a new spillway gate under the cycling path, turned out to be associated with high costs due to the required fast opening action and high safety requirements (redundant control system). In addition, the oil hydraulic cylinder and the associated pressure system required for fast gate opening are an environmental threat that was unwanted by the owner of the plant.

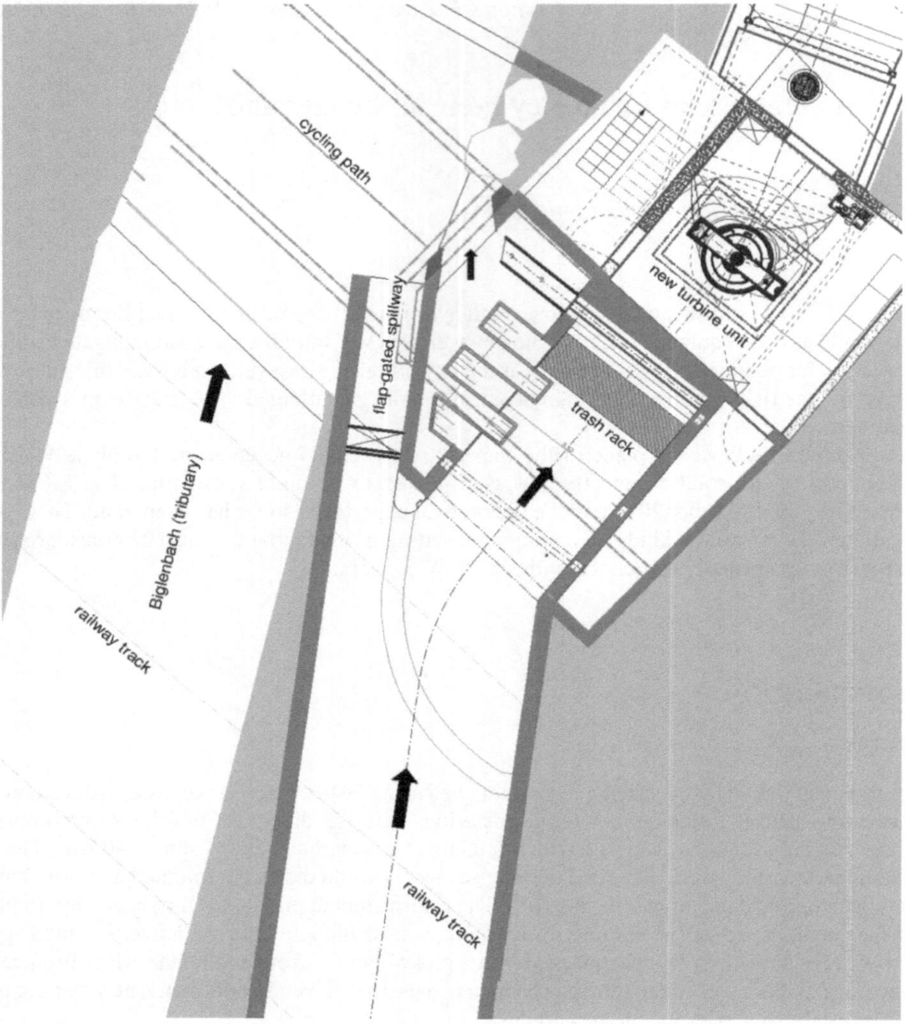

Figure 1. Layout of the PKW at Emmenau II.

In this situation, the idea of an ungated, free overfall spillway was considered. As the length for a lateral spillway available at the approaching canal to the powerhouse was limited to some 3.20 m, a non-linear, labyrinth-type weir had to be selected. In 2010, when the preliminary design of the scheme was taken up, information on PKW design was still rather limited. A first concept was proposed based on PKW configurations reported by Lempérière & Ouamane (2003). The head-discharge relationship given was not accurate enough for the very low water heads considered for this small-scale application at the Emmenau II hydro plant. Nevertheless, this conceptual design was used to obtain the building permission from the authorities. They agreed in principle with this new spillway type even though it involved a temporary exceedance of the full-supply level of the diversion canal whenever the PKW is in operation.

The PKW at Emmenau II is located just downstream of a relatively sharp bend in the power channel (see Fig. 1). This bend does not create excessive turbulence because the canal velocity has been lowered locally to less than 0.5 m/s for the protection of the fishes which might otherwise get caught at the trash rack.

Table 1. Main geometric properties of the PKW at Emmenau II.

Parameter	Value m	Parameter	Value m
Length of Weir B	2.16	Base length B_b	1.08
Inlet key width W_i	0.625	Outlet key width W_o	0.500
Total width of the PKW	3.36	Total developed length L	13.53
Weir height P	1.20	Side wall thickness T_s	0.080
Parapet wall height P_{px}	0.100	crest thickness parapet T_{px}	0.080

2 PKW DESIGN

2.1 *Selection of PKW geometry*

During the detailed design process in early 2012, new information on the PKW design became available. The detailed head-discharge relationships published by Anderson (2011) and Leite Ribeiro et al. (2012) were used in the optimization process for the PKW at the Emmenau II spillway. The conceptual PKW with symmetrical upstream and downstream overhangs, i.e., a PKW Type A according to Lempérière & Ouamane (2003) was further improved by adding parapet walls and rounded crests. The final design features the following geometric properties.

The ratio of inlet to outlet key width is 1.25 and the floor slope selected was 3 to 2 (horizontal to vertical) which is in agreement with an optimum design reported by Anderson (2011). There is no fillet or nose underneath the upstream overhang. Parapet walls and side walls have been designed with a rounded crest. In accordance with the nomenclature used in the model tests by Anderson (2011), the PKW at Emmenau II can be named a $PKRR_{1.25}$, i.e., a PKW with raised and fully rounded crest.

The PKW was placed into the old powerhouse whose machine floor was opened and the former power conduit equipped with trash rack and lateral spillway. Due to the limited space available, the downstream overhang of the most upstream PKW unit had to be cut in order to fit into the old powerhouse structure.

2.2 *Hydraulic design and layout*

The small size of the PKW at Emmenau II did not warrant a model study for the verification of the hydraulic performance of the selected design. Instead, the head-discharge relationships established by Anderson (2011) in his model tests were used. The corresponding equation is as follows:

$$Q = C_d \frac{2}{3} \sqrt{2g}\, L\, H_t^{1.5} \tag{1}$$

where Q = discharge; C_d = discharge coefficient; g = acceleration of gravity; L = developed crest length; and H_t = total upstream head including approach velocity head $v^2/(2g)$.

The design flow of the PKW with a total width W = 3.36 m is 2.5 m^3/s, i.e., the existing flap gate is functioning and the PKW should evacuate only the excess flow in the canal; in overload conditions (existing flap gate blocked), the PKW has to handle a flow of 4.35 m^3/s. In the experimental study by Anderson (2011), C_d was given as a function of the ratio between total head and weir height H_t/P. Applied to the design flow and overload conditions, C_d-values of 0.50 and 0.37 respectively were used, thus giving total upstream heads of 0.24 m and 0.45 m respectively. With these values, canal flow conditions were checked; the resulting backwater profiles were found acceptable. In overload condition, a bankful flow establishes without any freeboard (see Fig. 3b).

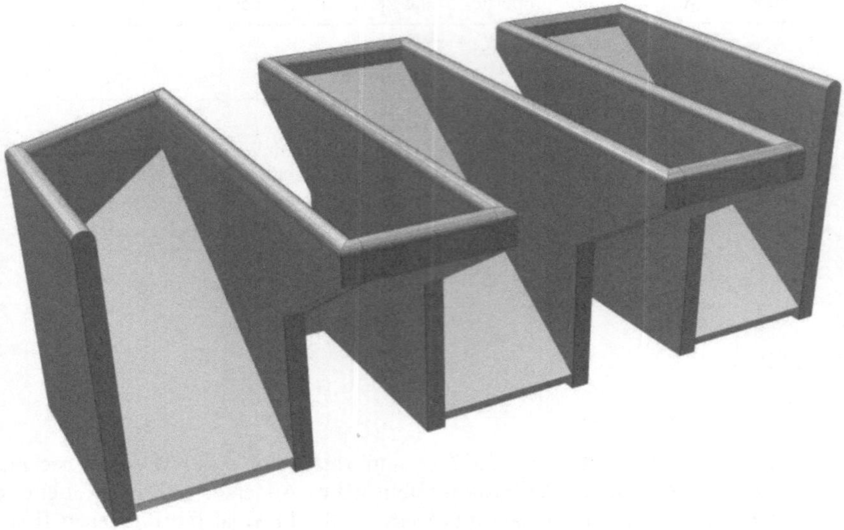

Figure 2. 3D view of the final design of the PKW at Emmenau II.

Figure 3. Diversion canal upstream of the Emmenau II plant: flowing just above full supply level (left) and bankful flow/no freeboard at 5 m^3/s through PKW, flap gate blocked (right).

2.3 *Other design requirements*

Floating debris is a problem at Emmenau every time flooding in the river occurs. In such cases, trash can be carried through the canal system and will eventually end up at the spillway. Clogging of the upstream overhang of the PK Weir was identified as a potential threat to the proper functioning of the proposed spillway.

The use of a PKW Type C with no upstream overhang was first considered but dropped later due to its insufficient hydraulic efficiency. As the size of the trash (timber logs, branches) that can reach the spillway is relatively small, the risk of a complete clogging of the PKW was considered low and the PKW design configuration according to Table 1 was retained. The only measure taken was the complete closure of the opening underneath the upstream overhang (flat "nose"), so that floating debris cannot get caught in between the precast side walls (see Fig. 4, right).

Aeration under the downstream overhang was considered unnecessary because of the short length of the selected PKW. It was argued that the overflowing nappe was sufficiently aerated from both

Figure 4. Construction of the PKW at Emmenau with element jointing using bolts and a hydro-swelling compound on the joint surfaces (left) and filling the void underneath the overhangs with concrete (right).

PKW end boundaries so that no artificial aeration using a pipe was required. The operational experience confirms this view.

3 PKW CONSTRUCTION

3.1 *Building material selection*

Initially, the PKW was designed using steel plates on a stiffening steel frame and with beveled plates at the overflow crest. However, corrosion of hydraulic steel structures has always been an issue for the plant owner and the prospect of having another item requiring a renewal of the corrosion protection coating every 20 years did not please him. The alternatives considered were concrete cast-in-place and prefabricated concrete elements. The cast-in-place alternative was soon dropped because the minimum side wall and crest thickness required for construction purposes (concrete compaction/vibration) is as large as 0.12 m. The hydraulic efficiency of this design is lower than for the steel and pre-cast options, which have crest thicknesses of 0.06 m and 0.08 m respectively. When considering life cycle costs, i.e., up-front costs plus recurring costs for maintenance and corrosion protection (discounted to present values), the pre-cast concrete option turned out to be the best alternative.

3.2 *Pre-fabrication aspects*

The pre-fabrication of the PKW elements in a factory specializing in this type of work turned out to be more difficult than initially thought because of the additional design features such as parapet wall and rounded crest. Initially, only three different items had been considered, i.e., two floors for inlet and outlet keys and a side wall. Eventually, a total number of seven different items had to be ordered to meet the requirements putting some stress on the contractor's formwork preparation.

The individual elements of 0.08 m thickness and with masses ranging from 200 kg to 325 kg were bolted together at the site using the stainless steel sockets previously cast into the elements in the factory. The elements were also bolted to the ground floor. However, stability of the PKW structure was achieved by filling the void underneath the upstream overhang with concrete. The downstream overhang was only partially filled. In addition to increasing stability, this concrete fill was used to enhance waterproofing and protection of the joints which were sealed using a hydro-swelling compound.

4 ECONOMIC CONSIDERATIONS

4.1 *BPKW costs*

The PKW costs amounted to CHF 11,750 excluding VAT corresponding to some €9,400. The distribution of these costs is depicted in Figure 5. Planning and design costs do neither include the evaluation of the spillway alternatives (flap gate) nor the selection of the most economic construction material and method (prefabrication). These costs have been attributed to the development of a new application of the PKW concept and will not occur again in similar projects in the future.

The ratio of PKW cost to its design discharge can be computed to 3.75 k€/(m^3/s). This is very similar to PKW projects implemented so far, where Vermeulen et al. (2011) stated ratios of PKW cost to PKW discharge capacity ranging from 2.9 to 4.4. Specific costs in small hydro applications are usually much higher than with comparable full scale developments. The reason for the attractive economic result at Emmenau is due to the fact that the PKW was built as part of the refurbishment and upgrade of the entire hydropower scheme with a budget of close to one million Euros. Hence, there was no need for establishing site access, crane and other contractor's requirements specifically for PKW construction.

4.2 *Economic comparison with spillway alternatives*

The cost of an alternative spillway design using a flap gate with a width of 3.36 m and a height of 0.5 m can be estimated at CHF 15,000 excluding the hydraulic pressure unit and the associated control equipment. The latter can be incorporated into the turbine-generator controls but must include a redundant circuit which opens the gate also in case of a failure of the primary controls due to lightning or similar risks.

The cost advantage of the PKW at Emmenau is more than 50% over the alternative flap gate. In addition, the PKW as a simple overfall spillway features a higher safety margin against flooding in the upstream channel than any other alternative; this advantage was not expressed in economic terms. The owner of the Emmenau plant did not require such estimates; for him, risk aversion is a major consideration in any of his decisions regarding technology selection. He might have chosen the PKW despite its economic disadvantage.

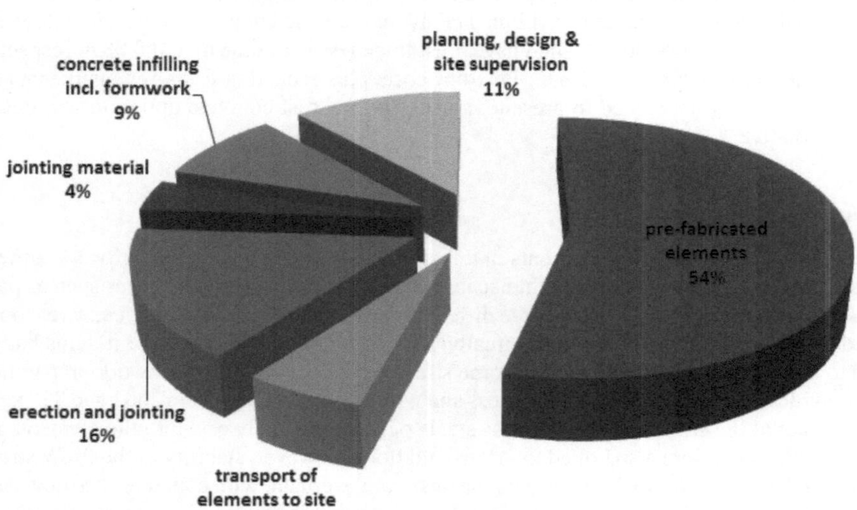

Figure 5. Distribution of PKW construction costs at the Emmenau II hydro plant.

5 OPERATIONAL EXPERIENCE

The PKW has been in commercial operation since November 2012 when the refurbished hydropower plant was commissioned. The hydraulic performance of the PKW was tested only up to its design discharge. The overload condition was checked for a relatively short duration without any measurement campaign: running the canal at full flow without freeboard for extended periods was not acceptable to the operating crew.

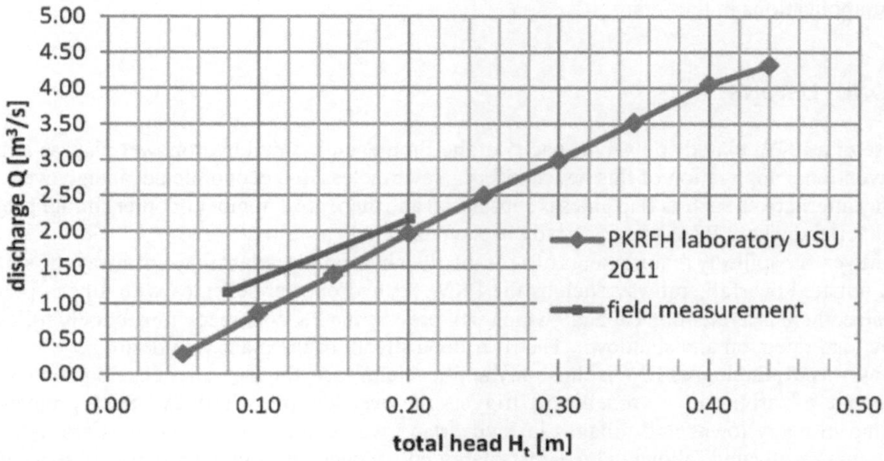

Figure 6. Hydraulic efficiency according to laboratory tests by Anderson (2011) and field measurements.

Figure 7. PKW at Emmenau in operation with a flow in the range of 2 m^3/s.

Measurements at site were difficult due to eddy development in the approach flow; accurate head measurements were impossible. These flow distortions originated from the sharp bend upstream of the PKW (see Fig. 1). In addition, flow over the PKW was determined indirectly using the power output of the turbine unit. As a result, the accuracy of the head discharge relationship thus established is low. Therefore, the comparison with the laboratory test results by Anderson (2011) at Utah State University for a model with similar geometric properties must be taken with caution. Even though Figure 6 seems to suggest that the hydraulic efficiency achieved with the prototype in Emmenau is better than the one at the lab, it is advisable to use the C_d-values from the lab for similar applications in the future.

6 CONCLUSIONS

The use of a PKW to add spillway capacity at the Emmenau II small hydropower plant is a rather unconventional application of this new concept. Nevertheless, the economic advantage over other spillway alternatives such as flap gates is substantial and there are a number of other similar projects in Switzerland where PKWs have already been considered.

However, the spillway requirements of a run-of-river hydropower plant may not always be suitable for an ungated overfall spillway such as the PKW. At hydropower schemes with a head pond of some size, there is a retention capacity which will prevent the PKW to react immediately following turbine load rejection and shutdown. The river downstream of the plant would fall dry if no gated spillway was available. A PKW is thus only applicable in fast flowing canal developments where any change of turbine flow immediately triggers the overflow over the PKW. Such projects can be found in many towns and villages in Switzerland where small hydro schemes are more and more being reactivated following the latest energy policy decisions in Switzerland. As space along the canals in these mostly historical settings is limited, the PKW can play a vital role in this niche market due to its compact size, low cost and added safety features.

REFERENCES

Anderson, R. M. (2011) Piano key weir head discharge relationships. *All Graduate Theses and Dissertations*, Paper 880. http://digitalcommons.usu.edu/etd/880/, Utah State University.

Leite Ribeiro, M., Bieri, M., Boillat, J., Schleiss, A. J., Singhal, G. & Sharma, N. (2012) Discharge capacity of piano key weirs. *Journal of Hydraulic Engineering* 138(2): 199–203.

Lempérière, F. & Ouamane, A. (2003) The piano key weir: A new cost effective solution for spillways. *The International Journal on Hydropower & Dams* (5): 144–149.

Vermeulen, J., Laugier, F., Faramond, L. & Gille, C. (2011) Lessons learnt from design and construction of EDF first piano key weirs. In S. Erpicum, F. Laugier, J. L. Biollat, M. Pirotton, B. Reverchon & A. J. Schleiss (eds.) *Proceedings of the international conference on labyrinth and piano key weirs (PKW 2011)*: 215–224. Liège, Belgium: CRC Press/Balkema.

Labyrinth and Piano Key Weirs II – PKW 2013 – Erpicum et al. (eds)
 ISBN 978-1-138-00085-8

Raising the bellmouth spillway at Black Esk reservoir using Piano Key weirs

J.C. Ackers & F.C.J. Bennett
Black & Veatch, Redhill, Surrey, UK

T.A. Scott
Black & Veatch, Glasgow, UK

G. Karunaratne
Scottish Water, Glasgow, UK

ABSTRACT: Work started in early 2013 on a scheme for enlarging Black Esk reservoir in southern Scotland. The reservoir, which was completed in about 1962, is impounded by an embankment dam with a rolled clay core and bellmouth shaft and tunnel (also known as a 'morning glory') spillway. The overflow level is being raised by 2.5 m, increasing the storage volume of 2200 Ml by about 40%, by the innovative adoption of precast piano-key (PK) weirs around the rim of the bellmouth. The new PK weirs are designed to pass the design flood (the PMF) of 183 m^3/s with a flood surcharge of 0.97 m, saving 0.64 m from the amount of dam raising that would otherwise have been required if the bellmouth rim had been simply raised. The initial hydraulic design was undertaken by adapting published empirical relationships and then refined using computational fluid dynamics (CFD) analyses.

1 BLACK ESK RESERVOIR

1.1 *Description of original reservoir*

Black Esk reservoir, which is situated in Dumfries & Galloway Region about 75 km south of Edinburgh, was completed in about 1962 to designs by Robert H. Cuthbertson & Associates. The reservoir had an original surface area of about 33 hectares and a storage capacity of about 2200 Ml. It is impounded by an embankment dam, whose crest is about 20 m above the downstream valley floor and whose crest length is about 250 m.

The original dam features a rolled clay core formed of glacial till, and the shoulders of the dam are also built with similar material, but include a chimney drain at the downstream face of the core and approximately horizontal drainage layers (Fig. 1). It was originally designed for raising by up to 9.1 m (30 feet), which would have involved building up both the upstream and downstream shoulders, together with raising of the rim of the bellmouth shaft.

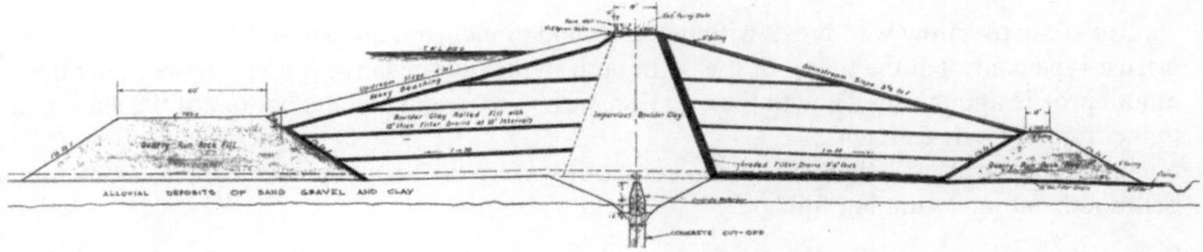

Figure 1. Indicative cross section of Black Esk dam (reduced from original record drawing).

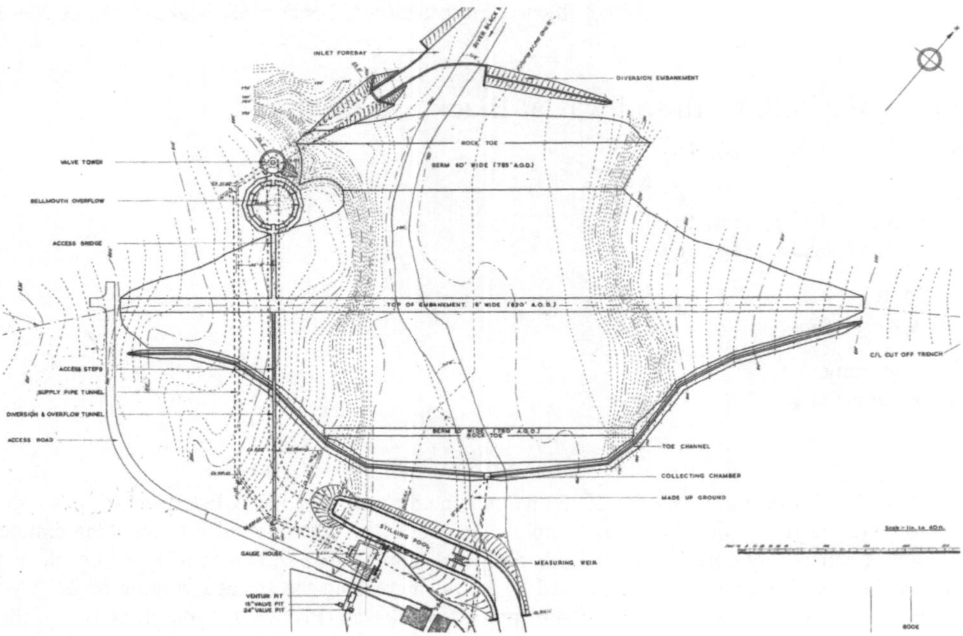

Figure 2. Overall layout of Black Esk dam (reduced from original record drawing).

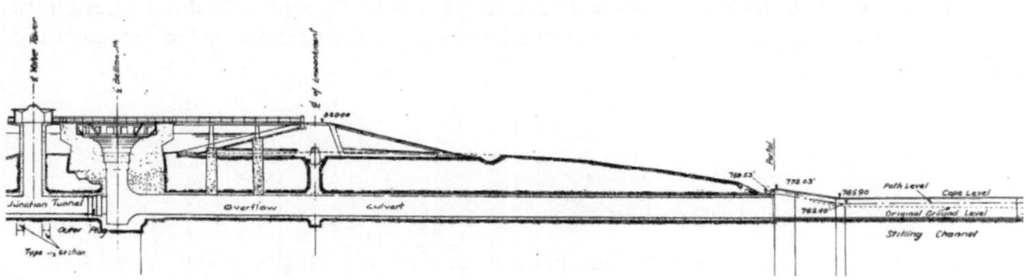

Figure 3. Long section of bellmouth shaft and spillway tunnel (reduced from original record drawing).

Figure 2 gives the overall layout of the original dam, with the spillway and drawoff works located at the southwest or right abutment. Figure 3 contains a longitudinal section of the bellmouth and spillway tunnel, and a photograph of the original bellmouth is given in Figure 4.

1.2 *Hydraulic behaviour of original bellmouth spillway*

There are, in principle, three flow modes that are possible for the bellmouth form of spillway which are, in the order of increasing flows:

- control at the overflow weir crest, with the shaft and tunnel running part-full;
- orifice-type control at the throat of the bellmouth (typically the tangent point between the bellmouth profile and the shaft), with the overflow weir submerged and the bottom of the shaft and tunnel running part-full; and
- tunnel-full control, in which the flow conditions are akin to a pipeline with a submerged bellmouth entrance running full.

For low heads, weir control always occurs, with the rim of the bellmouth forming the overflow weir from the reservoir and the shaft and tunnel flowing part-full. For higher heads, control may

Figure 4. Original bellmouth shaft with access bridges above and drawoff shaft on left.

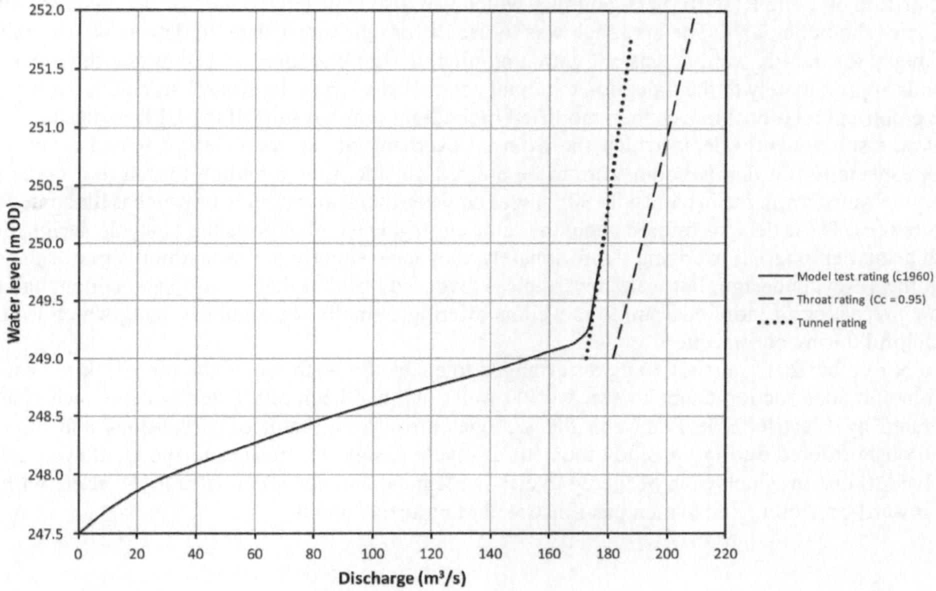

Figure 5. Original spillway rating for Black Esk reservoir.

first move to the throat of the shaft and then migrate to tunnel-full control, or alternatively there may be a direct transition from weir control to tunnel-full control.

Figure 5 illustrates the key rating relationships for the original Black Esk spillway. The spillway rating based on model testing (solid line) is believed to be the result of physical model testing undertaken around 1960 at Aberdeen University. Although a copy of the report on the model testing is no longer available, the rating was still available in 2006 and was used for a flood study that was undertaken that year as part of an inspection under Section 10 of the Reservoirs Act 1975.

In order to check the higher part of the original rating, where the weir was drowned out by downstream flow conditions in the bellmouth shaft and tunnel, we undertook analyses of the two potential flow modes that would cause this drowning:

- bellmouth throat control (dashed line), adopting a contraction coefficient of 0.95; and
- tunnel flowing full (dotted line).

The close match of the dotted line with the original rating, together with the distinct difference in gradient for the dashed line, demonstrate that hydraulic control in the original spillway shaft migrates, at a head of about 1.7 m, from the ogee weir around the bellmouth rim to tunnel full-flow control, without an intermediate region of throat (orifice-type) control.

2 PIANO KEY WEIRS

2.1 *Design development for PK weirs*

The original ogee weir around the rim of the bellmouth shaft comprised 12 straight segments, with angles of 30° between them. In order to suit this, the initial concept for the PK weir was based on accommodating a single weir cycle into each of the 12 segments, as illustrated in Figure 6(a), with each cycle of the PK weir positioned so that its outward-projecting bay (inlet key) was centred within the segment. The inward-projecting bays (outlet keys) straddled the 30° angles between the segments and were consequently truncated.

This overall design geometry for the 12-cycle PK weir was established with the aid of a spreadsheet that was compiled to follow the 'universal formula' proposed by Leite Ribeiro et al. (2011), as clarified by Leite Ribeiro (pers. comm.), which was also adjusted to take into account the truncation of the outlet keys. The approach was to use the design spreadsheet to determine a suitable geometry for the PK weir, consistent with a nominal design discharge of 180 m^3/s (which corresponds approximately to the transition to tunnel control) at a target flood surcharge of about 0.6 m. The empirical relationship was then modified in the light of the results of the CFD analyses.

It was subsequently decided that the external overhang of this weir design would be excessive, especially if a debris boom were to be erected outside it, so attention turned to a range of 24-cycle weirs, arranged in pairs with 30° angles between them, an example of which is illustrated in Figure 6(b). These designs offered about the same aggregate weir length as the 12-cycle design, but with a smaller external overhang. Unfortunately, they gave slightly poorer hydraulic performance than the 12-cycle design, but were nevertheless favoured because they would be small enough to allow precasting of individual units, as well as offering a smaller external overhang which would be helpful during construction.

In November 2011, a study tour was arranged to enable Scottish Water and the Black & Veatch design and construction teams to see two PK weirs that had been retrofitted at dams owned and operated by Electricité de France in the Limoges area. As a result of discussions and advice generously offered during the study tour, the 24-cycle design was further refined and eventually evolved to one in which each of the 24 cycles is identical and accommodates a 15° angle within the inward-projecting embayment, as illustrated in Figures 7 and 8.

(a) Full 12-cycle PK weir

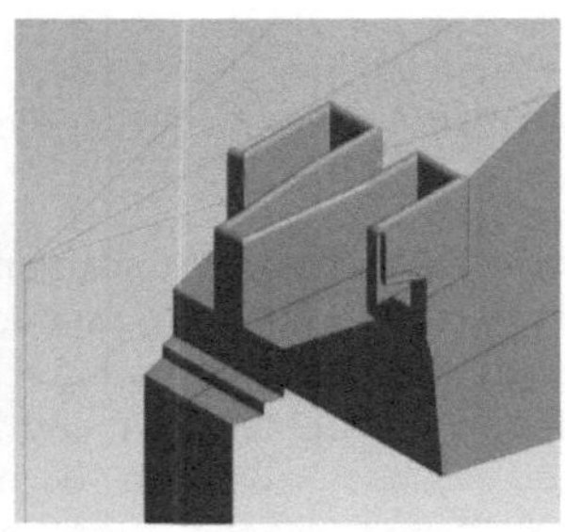

(b) 30° sector of initial design for 24-cycle PK weir

Figure 6. CFD models of alternative 12-cycle and 24-cycle PK weirs.

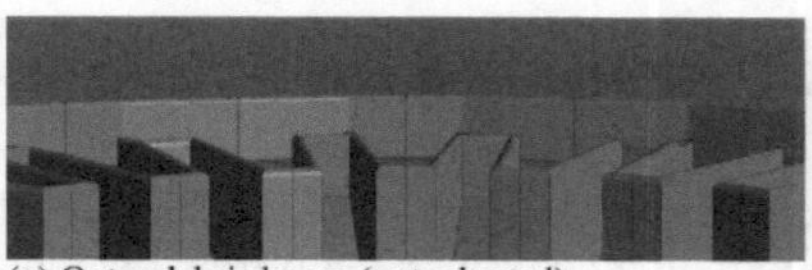

(a) Outer debris boom (not adopted)

(b) Inner debris boom (adopted)

Figure 7. CFD models of 24-cycle PK weirs with alternative outer and inner debris booms.

Two variants of the 24-cycle PK weir with equal 15° sectors were analysed by CFD, as illustrated in Figures 7(a) and 7(b):

(a) with a separate debris boom, located about 1 m outside the PK weir; and
(b) with an inner debris boom, located only between the outward-projecting embayments, leaving the weirs at the end of those embayments unprotected.

It was found that the outer debris boom gave the slightly better hydraulic performance, but the inner debris boom was favoured on the grounds of simpler and cheaper construction and also because the gaps would provide a means for ice floes that are breaking up in the thaw to escape down the spillway, in much the same way as could occur with the 'windows' through the debris boom at the original spillway (see Fig. 4). Figure 8 shows the full 24-cycle PK weir design, together with one of the constituent 15° sectors.

Table 1 lists the key design information for the Black Esk PK weirs, based mainly on the parameters and dimensions presented by Vermeulen et al. (2011) for the EDF PKW projects.

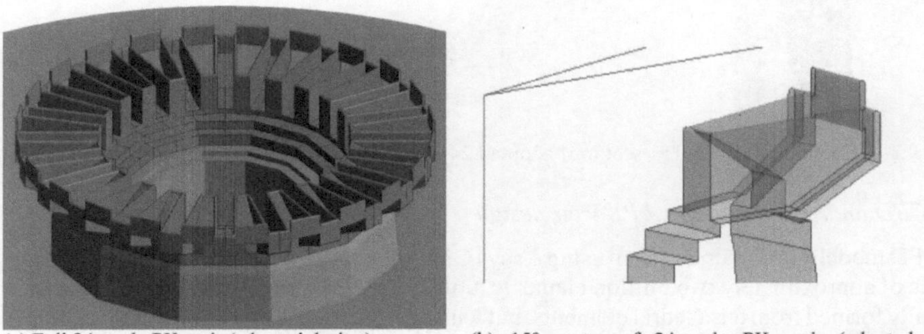

(a) Full 24-cycle PK weir (adopted design) (b) 15° sector of 24-cycle PK weir (adopted design)

Figure 8. CFD models of adopted 24-cycle PK weir with inner debris boom.

Table 1. Key parameters for Black Esk reservoir and PK weirs.

Parameter	Symbol	Description or value
Dam type		Embankment with rolled clay core
Dam height above downstream valley floor		20 m
Date of original construction		1962
PKW location		On rim of original bellmouth spillway
Reservoir design flood flow (PMF)		183 m^3/s
PKW design flow		183 m^3/s
Number of PKW cycles		24
Upstream–downstream length	B	4.4 m
Upstream crest overhang length	B_o	2.3 m
Downstream crest overhang length	B_i	0
Base length	B_b	1.9 m
Total height	P	2.1 m (internal invert to crest)
Amount by which overflow is being raised		2.5 m
Total width	W	54.0 m (based on mid-point radius = 8.6 m)
Inlet key width	W_i	Tapers 1.60 m to 0.50 m
Outlet key width	W_o	0.80 m
Sidewall thickness	T_s	0.20 m
Crest thickness at inlet key extremity	T_i	0.20 m
Crest thickness at outlet key extremity	T_o	0.20 m
Total developed length along weir centreline	L	238 m
Weir crest shape		Rounded (full semicircle)

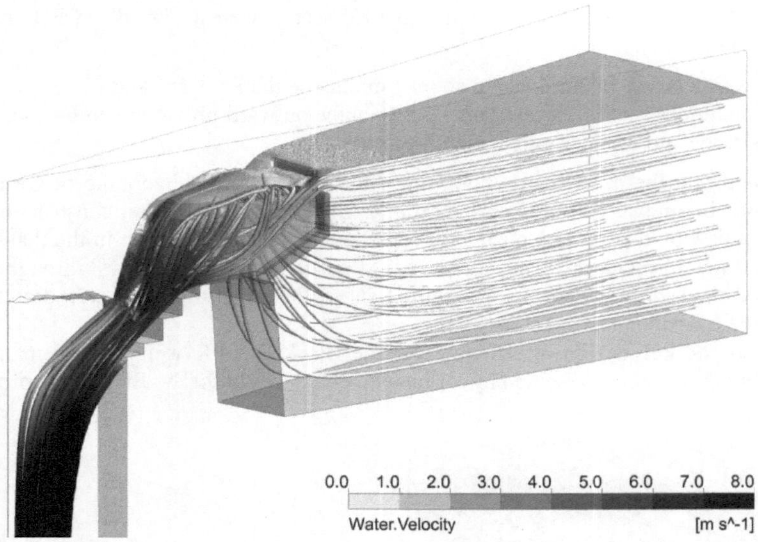

Figure 9. CFD simulation for 15° sector of adopted 24-cycle PK weir at total spillway flow of 168 m^3/s.

2.2 *CFD analyses for adopted PK weir design*

The CFD modelling was undertaken using Ansys CFX version 13, with multiphase, inhomogeneous models of approximately two million elements, utilising the SST turbulence model. The mesh was generally formed from tetrahedral elements, but with prismatic elements at the free surface upstream of the weir.

The CFD model of the PK weir actually comprises only a representative 15° sector, extending between the centres of successive inlet keys, as illustrated in Figures 8(b) and 9. It should also be noted that the CFD modelling of the PK weir does not include the shaft and tunnel, largely because it would not be possible to achieve a sufficiently fine mesh in a model that includes even half of the entire PK weir and shaft, together with half of at least the upstream end of the tunnel (with a vertical line of symmetry). In addition, it was considered preferable to obtain an undrowned weir rating, then apply an empirical treatment of the tunnel hydraulics (supported by the original model test rating) and of the drowning of the PK weir.

Figure 10 shows the rating curves for the Black Esk spillway with the adopted design of PK weirs, derived by the following methods:

- modular 24-cycle PK weir, based on adaptation of the 'universal formula' (Leite Ribeiro et al., 2011; Leite Ribeiro, pers. comm.);
- modular 24-cycle PK weir with inner boom, using CFD analyses; and
- the overall CFD-based rating curve also taking account of drowning of the PK weirs and the transition to tunnel control at about 180 m^3/s.

It may be noted that there is a good comparison between the 'universal formula' and the CFD analyses.

Also shown on Figure 10 is the rating that we assumed would have applied with simple raising of the original ogee weir, but adopting a less obstructive arrangement of debris boom than in the original scheme. This illustrates the substantial advantage in terms of flood surcharge requirements resulting from the adoption of the PK weirs.

2.3 *Designing the PK weirs for construction*

The design of the PK weirs has developed with ease of construction a significant consideration. The choice of the number of bays was determined in order to facilitate construction of the PKWs as

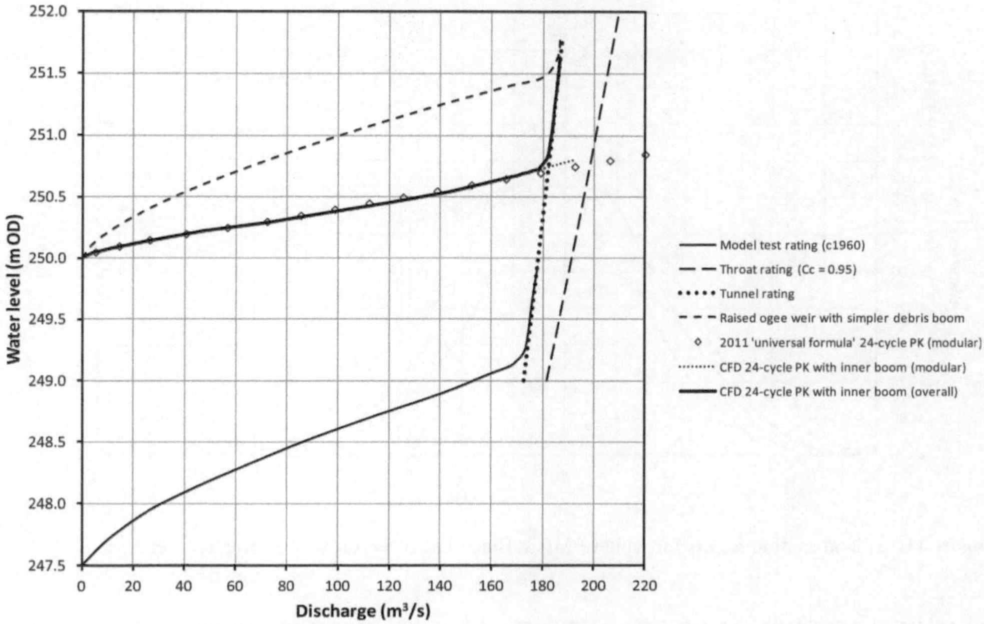

Figure 10. Original and raised spillway ratings for Black Esk reservoir.

precast concrete units, despite the slightly poorer hydraulic performance. The adoption of identical 15° units (instead of the symmetrical pairs illustrated in Figure 6(b)) allows the use of a single mould and makes sequencing of work simpler to manage; and working directly over water has been minimised by ensuring that most of the construction activities occur on the original bellmouth rim.

A crane platform adjacent to the bellmouth facilitates the extensive lifting required for removing the original debris boom and walkway, installation of the PKW units and also the raising of the existing valve tower adjacent to the bellmouth. The original debris boom and walkway were removed by wire sawing the structure into sections suitable for removal by crane. The original bellmouth had a large relatively flat weir crest onto which the foundation of the PKWs was cast.

In order to maximise the developed weir length the PKW units were pushed to the outside edge of the bellmouth. With the arrangement chosen, this sets the minimum inlet key width W_i which was chosen to allow a small section of insitu wall and base to be cast to stitch individual precast units together. The upstream overhang B_o was limited to approximately half of the whole upstream–downstream length B to ensure the stability of each precast unit while being placed and aligned, before being secured to the foundation and to ensure stability in transit and standing on site. Overall stability for the unit is not achieved until the PKW unit is anchored to the existing structure and concrete ballast blocks added to the inlet key. The ballast block also provides the hydraulic shape to the inlet key. The debris boom and ballast units are also precast.

Manufacturing the units in precast concrete allows smaller bay widths than would be feasible for an insitu construction method, as no scaffolding is required within the outlet key. Casting the PKW units inverted allows the casting of the rounded crest to be achieved with greater control than for insitu methods and also allows thinner sidewall sections to be used. The size of the units allows for two units to be carried on a standard articulated lorry without attendant pilot vehicles or other restrictions that would be required for an oversized load.

A mixture of waterproofing barriers are being employed depending on the connection detail. Between precast and insitu wall sections, waterstops have been designed to be the primary barrier. Between the PKW units and the foundation concrete, hydrophilic strips are being used in conjunction with pressure grouting to provide a double barrier.

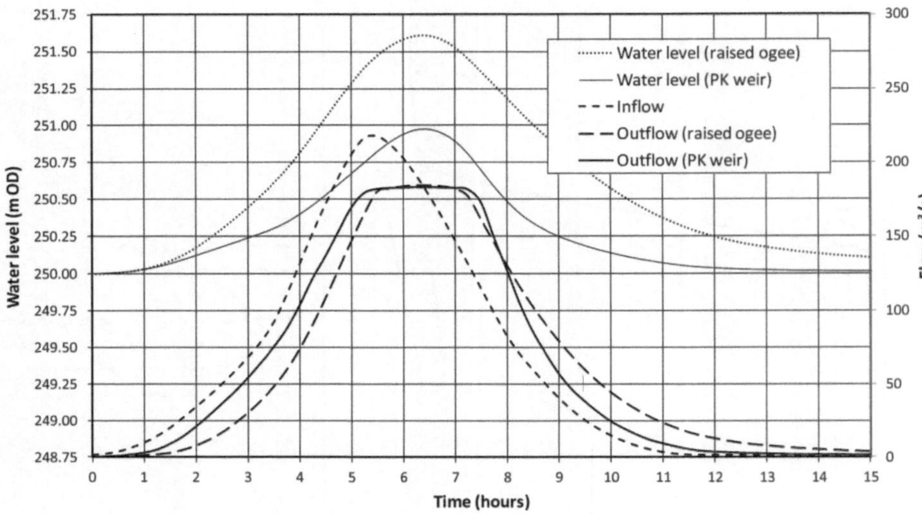

Figure 11. Flood routing results for winter PMF at Black Esk reservoir with alternative weirs.

3 FLOOD PERFORMANCE OF RAISED BELLMOUTH WITH PK WEIRS

Figure 11 shows the routing of the PMF (the probable maximum flood) through the raised reservoir, for two cases:

- the adopted design, with the 24-cycle PK weirs; and
- an alternative design, based on raising the ogee weir crest, but simplifying the debris boom arrangements to cause less obstruction.

Hydrographs are given for the inflow, for the outflows for the alternative weir designs, and for the corresponding flood surcharges.

The flood routing results in Figure 11 illustrate the reduction in flood surcharge that occurs throughout the flood, lowering the peak level by 0.64 m. During the rising limb of the flood, the outflow of course follows the inflow more closely with the PK weirs. For two to three hours around the peak of the flood, the outflow from the reservoir is governed by tunnel control, so the peak outflow with the PK weirs is marginally less than with the ogee weir.

A sensitivity test was undertaken in which the PMF flow ordinates for the PK weir case were all increased by 10%. This results in a peak flood level being raised by about 0.3 m, which would still be well within the residual freeboard being provided as part of the dam raising scheme.

REFERENCES

Leite Ribeiro, M., Boillat, J.-L., Schleiss, O.J., Le Doucen, O. & Laugier, F. 2011. Experimental parametric study for hydraulic design of PKWs. In Erpicum et al. (eds.), *Labyrinth and piano key weirs-PKW 2011*, CRC press, London, 183–190.

Leite Ribeiro, M. 2001. *Personal communication, 28 July 2011.*

Vermeulen, J., Laugier, F., Faramond, L. & Gille, E. 2011. Lessons learnt from design and construction of EDF first Piano Key Weirs. In Erpicum et al. (eds.), *Labyrinth and piano key weirs-PKW 2011*, CRC press, London, 215–224.

Labyrinth and Piano Key Weirs II – PKW 2013 – Erpicum et al. (eds)
© 2014 Taylor & Francis Group, London, ISBN 978-1-138-00085-8

Spillway capacity upgrade at Malarce dam: Construction of an additional piano key weir spillway

T. Pinchard, J.-L. Farges, J.-M. Boutet, A. Lochu & F. Laugier
EDF – Centre d'Ingénierie Hydraulique, Le Bourget du Lac, France

ABSTRACT: The Malarce dam is located on the Chassezac river in southern France. The design flood estimate was 4000 m^3/s at the construction of the dam. Recent hydrologic surveys increased the design flood up to 4600 m^3/s, resulting in a 600 m^3/s shortage for existing spillways. The optimal solution designed by EDF-CIH was to implement a Piano Key labyrinth weir (PKW) on the right side of the dam, next to the third gated weir of the dam. The use of a free-flow spillway versus a gated weir was selected to increase the safety of the dam for worst-case scenarios. Indeed, PKW requires no energy or human presence to operate. This is enhanced by the very fast flow rise that could occur during exceptional floods (1400 m^3/s/hr in case of a 1000 years return period flood).

To anchor the PKW to the dam, and to improve the dam's stability, 12 prestressed cable units were used. The concrete lining, whose function is to carry the 600 m^3/s flow down to the river bed without risk of scouring at the dam toe, is another work almost as complex as the new weir.

This paper focuses on the design of the spillway and its dissipation works, the problems encountered during construction and the instrumentation that will be integrated in the spillway to monitor it during floods.

This is the sixth PKW to be built on a dam owned by EDF, and it should overspill at least once a year at 200 m^3/s, which is about the third of its design flow (50 cm head).

The construction started in November 2011 and lasted 13 months.

1 INTRODUCTION

In 1968, Malarce dam was built on the Chassezac River, in southern France. In this area, climate is Mediterranean, which is very dry but sensitive to stormy weather, with heavy rainfall during less than a day. The catchment area, of only 500 km^2, is characterized by a very dry rocky soil, steep slopes and almost no capacity to retain water. Therefore floods are often very strong with high flowrates during short periods. The design flood in 1968 was 4100 m^3/s, but latest hydrological study revealed a 1000 year return period of 4600 m^3/s with a maximum increase gradient of 1400 m^3/s/hr, which is very high by comparison to the 14 m^3/s of the natural mean flowrate of the Chassezac.

Today the dam is equipped with 3 radial gates (14 m width × 11.5 m height, Fig. 1-left), which include a 2.2 m high flap to ensure progressive flowrate downstream for minor floods and floating debris evacuation. Malarce dam is 31.4 m high above foundations and a total retention volume of about 3.5 hm^3. Total discharge capacity of the original dam was measured on hydraulic model (Cicero et al., 2010), and was estimated to 4000 m^3/s at Maximum Water Level, revealing a discharge capacity deficit of approximately 600 m^3/s. Design studies of a new spillway started in 2008 concluding that the adjunction of a Piano Key Weir (PKW) 42.5 m wide on the right bank side of the dam (Fig. 1-right) was the best option.

Hydraulic design and stability assessment will not be discussed in this paper, but some references can be found in the literature (Pinchard et al., 2011).

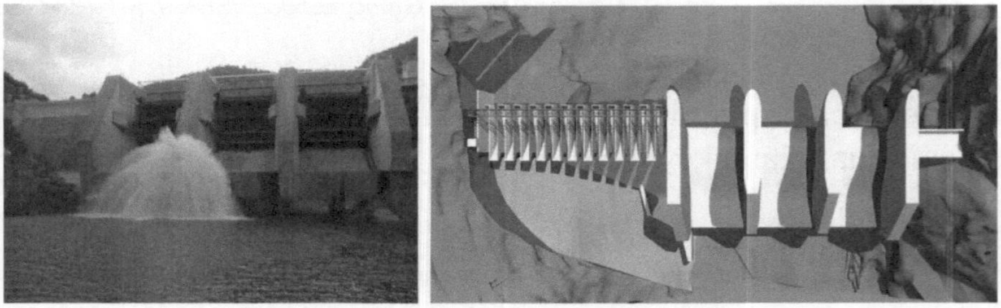

Figure 1. Downstream view of Malarce Dam (left) and 3D view of the spillway that will be implemented on the right bank side of the dam (right).

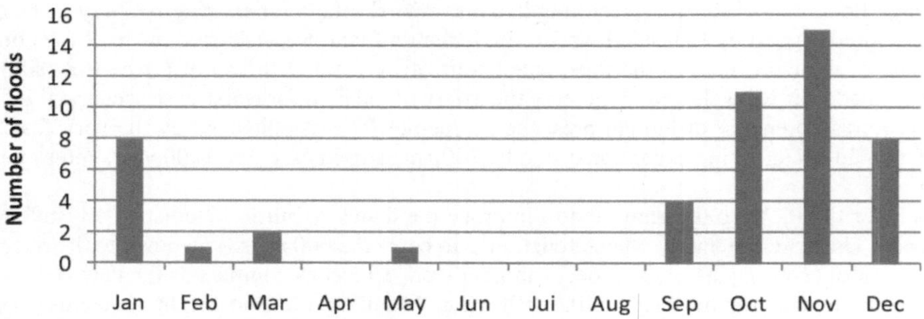

Figure 2. Distribution of the 50 major floods during 1995–2009 period.

2 PREPARATION WORKS

2.1 *Time schedule and water management*

The new spillway has been constructed on the right side of the dam where no access road exists. Therefore a backfill temporary "fuse" road had to be made, in order to cross the river at the dam toe. In case of floods, this road was designed to be easily rebuilt and the backfill materials were only natural materials from the site in order to avoid pollution of the river. Indeed, the powerhouse offered the only possible discharge evacuation. Its capacity is only 27 m^3/s, which is twice the mean flow but 71 times less than a 10-yr return period flood. Then, the probability to have to open the radial gates, causing major damages to the access road, was rather high.

During the 1995–2009 period, an average of 12 days per year had a daily average flow higher than 27 m^3/s. Considering the January–August period, this number drops to 4. In addition, major floods have been recorded on the September–December period (Fig. 2), corresponding to the rainiest months. Then, it was decided that most of the work had to be done on the January-August period, keeping September as a safety month.

Only 3 episodes of high flow causing the opening of the gates were registered during the works, with a maximum flow under 100 m^3/s each time, lasting only a few days. 2012 was indeed a very lucky year.

During works, the reservoir pool had to be lowered, in order to demolish the top of the existing dam and to reconstruct the PKW. In order to keep the reservoir level between 210 and 212 m a.s.m, the power plant was controlled automatically by the water level. Statistical analysis showed that it was possible to keep a 100 yrs flood under the demolition level any times thanks to the low risk in the spring-summer period and the high capacity of the gates.

Figure 3. Earthwork downstream in steep conditions.

Figure 4. View of the demolition process.

2.2 *Access roads and earthworks*

A fuse track was backfilled as close as possible to the dam toe, in order to protect it from the impact flow issued from the gates in case of flood and a working platform was created on the right side.

Difficult earthwork due to the steepness of the banks (Fig. 3) had to be done to prepare the upstream and downstream platform at 213–215 m a.s.l.

2.3 *Demolition of the upper part of the dam*

In order to obtain a good shape at the interface between the dam and the future PKW, and to reduce vibrations transmission, a diamond-cable saw was used to cut the dam at the base of the PKW. Then, the upper part of the dam was taken down using a 2 stages explosive demolition process (Fig. 4). Explosive ratio was about 500 gr/m^3, distributed in 4 m holes deep with a 60 × 60 cm grid. This demolition scheme allowed to limit the vibrations under the contractual threshold.

A modal-spectral analysis was performed, prior to the demolition, and determined that stresses were allowable up to a particular velocity of about 50 mm/s on the 5–10 Hz range measured at the top of the pier N°4. This pier is supporting a 14 m × 11.5 m radial gate, therefore, it was essential to respect the threshold. This velocity threshold ensures that the concrete will remain in compression.

Three firing tests allowed the contractor to adapt its demolition process in order to respect the contract.

3 CONSTRUCTION OF THE PK-WEIR

3.1 *Base preparation*

All PKW built by EDF have been designed with an aeration system, which goal is to avoid nappe vibration when the spillway is overtopped. On Malarce PKW, the first 4 inlets from the right bank are provided by a direct aeration pipe connected to the drainage gallery, the 4 in the middle and the 4 on the left are provided by 2 collector pipes connected to the drainage gallery but drowned in the PKW base (Fig. 5-left).

Numerous steel bars were anchored to the existing dam. These are of no use for the PKW stability (thanks to the prestressed cables) in the definitive state, but they were very useful to ensure the temporary stability of the PKW during construction and to fix the upstream scaffolding (Fig. 5-right) used to concrete the overhangs.

3.2 *Gallery reconstruction*

The drainage gallery of Malarce dam had only one entrance which was right in the middle of the new spillway (Fig. 6-left). Therefore, the gallery was extended to the right bank (Fig. 6-right), under the PKW, and a new 9 m deep pit was created to access the gallery.

A short grout curtain has been realized under the new gallery which is constructed directly on the rock. In addition, a drainage curtain has been drilled to limit uplift in this zone.

Figure 5. PKW-base concreting (left) and view of the upstream scaffolding (right).

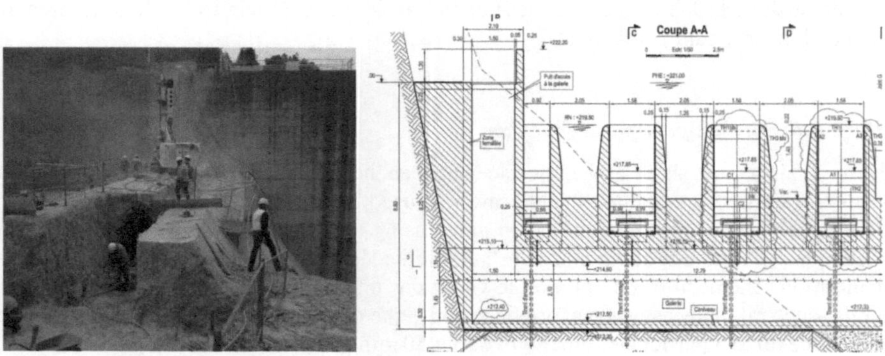

Figure 6. Existing gallery (left) and drawings of the extension (right).

3.3 *PKW erection*

The contractor decided to build the PKW in 4 steps:

- Construction of the base (approx. 60 cm high) of the PKW, which will be useful to put the upstream construction platforms
- Construction of the vertical walls
- Construction of the inlets
- Construction of the outlets

At each junction of walls and inlets/outlets, the continuity of the steel reinforcement is ensured by coupling devices. This technique allows prefabricating reinforcement cages of the vertical walls (Fig. 7-left), in a dedicated area, and putting them in the formwork (Fig. 7-right) in a single handling operation. Everyday a new wall was concreted thanks to this organization.

After the vertical walls cured sufficiently, specific formworks were hung on the walls using embedded bushings in order to concrete the inlets and outlets (Fig. 8). This was necessary to relieve the platforms from the weight of the fresh concrete of the inlets/outlets. Because of the big upstream overhangs (6.63 m), it was mandatory to build the inlets before the outlets to avoid traction on the downstream PKW toe, moving the gravity center downstream.

3.4 *Watertightness*

Specific attention was paid to the sealing of the PKW. Indeed, the new spillway is built over 3 different blocks of the concrete dam. The 3 contraction/expansion joints have been extended in the PKW with normal (Fig. 9) and L-shaped waterstop joints.

Figure 7. Prefabricated reinforcement cages (left) and formwork (right).

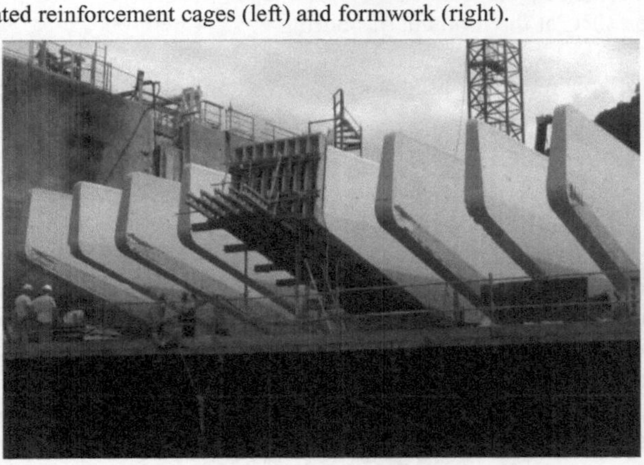

Figure 8. Formwork of the outlets.

Figure 9. Connection of the new waterstops with the existing ones.

To achieve the goal of no leakage at the PKW-dam interface, a flat waterstop embedded in the new concrete and sealed in a notch in the existing dam was used. In addition, a resin injection device was installed in case of waterstop failure. This device is still reachable and could be injected if leakage was discovered during PKW's life.

A drainage notch, connected to the elevation drainage system (bored holes every 3 m), is used to detect and measure leaks and reduce uplifts if necessary.

3.5 *Concrete*

The concrete to be used for the construction of the PKW had to meet the following requirements:

- 28d compressive strength of 35 MPa (cylindrical test)
- High manoeuverability: mean slump test 200 mm
- Small aggregates: 16 mm max because of thin elements with high rebar ratios (20–40 cm thick, 80–180 kg/m^3 rebar ratio)
- Cement capable to withstand aggressive pure water
- Watertightness
- Frost resistance obtained by the use of air entrainment
- Temperature under 30°C at the moment of pouring

Approx. 3000 m^3 of concrete were used, and the development of the concrete composition took 5 months.

3.6 *Miscellaneous*

In order to increase the dam stability and to fasten the PKW, 12 prestressed units were used (Fig. 10-left). Each one consists in an assembly of 12 cables of 7 wires, anchored 12 m deep in the foundation. Every anchor head is reachable in the outlets under a watertight hatch, for maintenance operations or further tensioning. Access to the dam is made on the left bank, and the access pit to the gallery lies on the right bank.

Then, it was necessary to cross the 42.5 m wide gap created by the PKW. A 43.5 m gangway, consisting of a 20 t steel truss was designed. This footbridge (Fig. 10-right) embeds a hauling rail and a mobile ladder to access in every outlet to carry future maintenance work on the prestressed cables if needed. Due to the high span, special care was taken to estimate wind loads.

Figure 10. View of one of the 12 prestressed cables (left) and footbridge during installation (right).

Figure 11. View of the bottom of the apron during rock preparation (left) and view of the anchors, geogrid strips and shotcrete beams (right).

4 CONSTRUCTION OF THE DOWNSTREAM APRON AND DISSIPATION STRUCTURES

A scouring study showed that the rock, due to the fractures orientation, was unable to withstand the flow issued from the PKW. Therefore it was decided to design a concrete lining and dissipation structures (flip bucket and deflector walls) to lead safely the water to the dissipation pit, 20 m downstream from the dam.

4.1 *Rock preparation*

After a difficult rock excavation due to the 75% slope, a particular attention was paid to the preparation of the rock supporting the concrete lining (Fig. 11-left).

A small geological fault was discovered during that phase, and was cleaned out by some more excavation, drainage and shotcrete.

To ensure the stability of the lining, it was essential to limit the uplifts at the under-face of the concrete skin (40–60 cm thick), and to anchor the lining in the rock (Fig. 11-right). Numerous geogrid strips were used, and the collection of leakage was performed by a 400 mm drainage pipe surrounded by 20/40 mm gravel. The anchor mesh is relatively light in the upper part of the lining, using 20 mm rebars, and quite dense at the foot of the apron, using 40 mm rebars.

4.2 *Downstream apron concreting*

In order to reduce thermal cracking in this thin shell exposed to high temperatures in summer (southern France), contraction/expansion joints have been arranged every 10–15 m, slicing the apron into 12 plane slabs, except for two of them which are twisted.

To ensure a waterproof shell, waterstop joints have been used in every contraction joint. A triangular shaped void was created under each joint to collect any leakage and avoid uplifts in case of failure.

Figure 12. Downstream apron during concreting.

Figure 13. View of the crane foot (left) and view of the dissipation structures (2 deflector walls and a flip bucket under construction – right)

Shotcreate beams following the contraction joints scheme were done during the preparation works and were used to ensure a good closure of the formwork (Fig. 12).

4.3 *Dissipation structures*

These structures were the first and last structures to be erected. Indeed, the flip bucket was divided in a 2-stage job. The first lift incorporated the bottom of the tower crane (Fig. 13-left). Indeed, the contractor decided to use this weight to withstand the 11 MN.m torque created by the crane, and it was placed to be protected in case of an important flood. Then it was necessary to wait until all other works were completed to disassemble the crane in order to complete the flip bucket with an additional mobile crane.

The triangular shaped deflector will receive an estimated 600 t thrust (Fig. 13-right) and is therefore anchored to the ground and to the dam.

5 PKW OPERATION

5.1 *Planning*

The global duration of this project was 12–13 months. The preparation of the PKW construction (access roads, earthworks, demolition, base preparation) was the most time-consuming task, requiring about 5–6 months compared to less than 3.5 months for the PKW construction itself.

Figure 14. Downstream (left) and upstream (right) views of the completed spillway.

November	2011	Upgrading dam drainage curtain
December	2011	Upgrading dam drainage curtain, preparation
January	2012	Access works
February	2012	Earthworks
March	2012	Dam demolition
April	2012	PKW base preparation (aeration pipe, surface preparation, anchoring), rock preparation for downstream works
May	2012	Downstream works in preparation, tower-crane installation, gallery extension
June	2012	PKW base concreting, anchorage and drainage for downstream apron
July	2012	Access pit for gallery, PKW vertical walls, energy dissipation structures 1st stage, anchoring and drainage for downstream apron
August	2012	PKW vertical walls, 1st bottom third of the downstream apron
September	2012	PKW inlets and outlets, 2nd third of the downstream apron, energy dissipation structures 2nd stage
October	2012	Top of the downstream apron, prestressed cables, footbridge installation
November	2012	Crane taking down, finishing, access dismantling and reservoir refilling
December	2012	Reservoir refilling

5.2 *Reservoir filling*

Because the reservoir had been kept very low during almost a year, and because the PKW construction impacted greatly the dam, a special care was taken during refilling the reservoir.

Therefore, water rise has been limited to 1 m/day, and inspections were conducted twice a day.

Displacement measures, uplifts and water seepage were absolutely normal. The spillway is now operable (Fig. 14) and will now be used for future floods before the opening of any gate up to 200 m^3/s (approx 50 cm head), then the gates will be used to regulate the water level at el. 220. It is only for extreme floods (>100 yr return period) that the PKW will be used with more than 1.5 m head.

Because Malarce PKW will be functioning approx. once a year, it is equipped with 3 dynamic pressure cells, 1 vibration sensors, 10 strain gauges and 15 temperature sensors.

6 CONCLUSION

The Piano Key Weir has proven to be a very interesting alternative to increase the discharge of existing dams without modifying too much the Normal Water Level or the Maximum Water Level and increasing the safety level with a completely autonomous solution.

Figure 15. Panoramic view of the spillway.

Figure 16. PKW with a discharge of 200 m^3/s (left) and 50 m^3/s (right).

The construction of Malarce dam is a new example where a PKW revealed as the best solution. The construction of the PKW was almost routine, thanks to the experience gained in the 5 previous PKW built by EDF, but the downstream apron was the real challenge of this project.

By the middle of 2013, the PKW has been overtopped already twice, the first time with a maximum discharge of 50 m^3/s and the second time with a maximum discharge of 200 m^3/s (Fig. 16). The PKW behaved as expected, suffering no damage, and the downstream flow was very similar to the simulated flow observed on the physical model. Sensors were active and recorded data concerning downstream pressures and vibrations. These values are currently under analysis.

REFERENCES

Cicero, G.-M., Guene, C., Luck, M., Pinchard, T., Lochu, A. & Brousse, P.-H. 2010. Experimental optimization of a piano key weir to increase the spillway of the Malarce dam. *in proceedings of 1st IAHR European Congress*, Edimburg, United Kingdom.

Lempérière, F. & Ouamane, A. 2003. The piano keys weir: a new cost-effective solutions for spillways. *Hydropower & Dams* 10 (5), 144–149.

Pinchard, T., Boutet, J.-M. & Cicero, G.-M. 2011. Spillway capacity upgrade at Malarce dam: Design of an additional Piano Key Weir Spillway. *Labyrinth and piano key weirs-PKW 2011,* CRC Press, London, 233–240.

Pralong, J., Montarros, F., Blancher, B. & Laugier, F. 2011. A sensitivity analysis of Piano Key Weirs geometrical parameters based on 3D numerical modeling. *Labyrinth and piano key weirs-PKW 2011,* CRC Press, London, 133–139.

Truong Chi Hien, Huynh Thanh Son & M. Ho Ta Kanh, M. 2006. Results of some "Piano Keys" Weir hydraulic model tests in Vietnam. *Proceedings of the 22nd ICOLD Congress*. (Q87, R39). Barcelona, Spain.

Vermeulen, J., Faramond, L. & Gille, C. 2011. Lessons learnt from design and construction of EDF first Piano Key Weirs. *Labyrinth and piano key weirs-PKW 2011,* CRC Press, London, 215–224.

F. Future developments

Labyrinth and Piano Key Weirs II – PKW 2013 – Erpicum et al. (eds)
© 2014 Taylor & Francis Group, London, ISBN 978-1-138-00085-8

Effect of driftwood on hydraulic head of Piano Key weirs

M. Pfister & A.J. Schleiss
Laboratory of Hydraulic Constructions (LCH), Ecole Polytechnique Fédérale de Lausanne (EPFL), Lausanne, Switzerland

B.P. Tullis
Utah State University, Logan, Utah, USA

ABSTRACT: Driftwood is often transported by a river course during flood events and can be problematic where it accumulates at hydraulic structures. Near obstacles, driftwood can block and thus reduces the open flow area for water passage, resulting in a lowered discharge capacity combined with increasing flow depths in the upstream region. The blocking of driftwood at weirs and spillway inlets is of particular interest, as it may affect the safety of such structures by drastically reducing the maximum spilling capacity. The high discharge efficiency of Piano Key weirs (PKWs), relative to linear weirs, may influence driftwood collection. The flow depth over a PKW for a given discharge will be less than would be required by linear structures, thus providing less flow momentum per unit weir length for flushing driftwood. If the catchment is forested, the potential effects of driftwood on the PKW head-discharge relation should be considered as part of the hydraulic design process. This paper describes systematic model tests considering the effect of driftwood on the PKW discharge efficiency. Trunks and rootstocks were supplied upstream of different laboratory-scaled PKW configurations. The test series closely approximated real-world conditions by investigating the nature of driftwood accumulations on the PKW discharge efficiency.

1 INTRODUCTION

1.1 *Driftwood*

Catchment areas of Mountain Rivers are often forested. During extreme flood-generating rain events, riverbank erosion and landslides can develop. It is common for both processes to introduce trees and brush into the watercourse. In steeper boulder-lined rivers, trees are quickly reduced to smaller fractured woody debris elements during transport. In lower zones of a catchment, infrastructure [Bezzola and Hegg 2007 (weirs); Bezzola et al. 2002, Schmocker and Hager 2011 (bridge decks); Melville and Dongol 1992, Pagliara and Carnacina 2010 (bridge piers)] along a river often limits driftwood transport capacity; urban waste can add to the problem. Near such structures, driftwood often blocks and thereby reduces the open flow area for water passage, resulting in a lowered discharge capacity combined with an increased upstream flow depth. The blocking generated by driftwood at weirs and spillway inlets is of particular interest, as it may affect the safety of such structures. Namely, the maximum spilling capacity is reduced due to obstruction.

Determination of the potential driftwood volume that might be generated by a catchment represents an important input parameter for the analysis. The nature and volume of the driftwood can vary significantly based on terrain, geographical location, season, and climate. Rickenmann (1997), however, observed a correlation between the catchment surface area and the effective driftwood production, based on several filed surveys following floods, particularly in Switzerland. The catchment area seems a suitable first-order indicator, as it can be used to approximate the size of the forested surface and the hydrologic peak flood discharge. For the catchment areas evaluated, Rickenmann (1997) proposed the following pragmatic estimate for potential driftwood volume V [m^3]:

$$V = 45A^{0.67} \quad (1)$$

where A = catchment surface in [km^2]. A significant spread was observed for small catchment areas (<1 km^2), whereas the equation overestimated the value V for large catchments (> 1,000 km^2). Alternative methods to derived the driftwood volume at a given river section are provided by Uchiogi et al. (1996), Rimböck (2003), and Lange and Bezzola (2006).

1.2 *Piano Key weirs*

Piano Key weirs (PKWs) are a hydraulically attractive alternative to linear overflow weirs, increasing the unit discharge at the unregulated spillway inlet for given upstream heads and spillway widths. This advantage allows for operation of reservoirs at higher supply levels (less reservoir volume set aside for flood routing) thus providing an increased retention volume. In the last decade, the implementation of PKWs in practice has increased significantly, particularly in France (e.g., Laugier 2007, Laugier et al. 2009) and presently under construction in Vietnam (Ho Ta Khan et al. 2011). Lempérière et al. (2011) and Schleiss (2011) presented historical reviews on the evolution from Labyrinth weirs to PKWs, with a focus on the hydraulic characteristics of PKW operation. Note that the standard naming convention for PKWs as introduced by Pralong et al. (2011) is applied herein.

1.3 *Driftwood at Piano Key weirs*

PKW research to date has primarily focused on hydraulic efficiency (Kabiri-Samani and Javaheri 2012; Leite Ribeiro et al. 2012; Machiels 2012; Anderson and Tullis 2012a, 2012b, 2013), typically ignoring a possible negative effect of driftwood on the rating curve. Ouamane and Lempérière (2006) were among the first to conduct preliminary model tests with driftwood. They observed that no driftwood was trapped under the apex overhangs during the reservoir filling process. As soon as the PKW started to spill, however, driftwood got stuck in the inlet key at small discharges, resulting in a slight reduction in discharge efficiency. A discharge coefficient reduction of about 10% was reported for $H/P < 0.5$, with H = total upstream head measured relative to the PKW crest elevation, and P = weir height. As the discharge increased, the driftwood was eventually washed downstream.

LCH (2007) investigated the effects of floating wood on the St. Marc Dam (France) emergency PKW spillway and reported no driftwood-associated reduction in discharge capacity when the gates of the adjacent principal spillway were fully open, attracting a significant portion of the driftwood.

Laugier (2007) reported systematic model tests with driftwood of the emergency PKW spillway of Goulours Dam (France), concluding that 1) increasing head tended to remove previously blocked driftwood, 2) the principal gated spillways attracted driftwood before the PKW started operation, 3) most driftwood passed the PKW for flow depths at the PKW larger than 1 m, 4) most streamlines passed below the driftwood of a fully blocked PKW entering the inlet key, so that the water flowed "around" the blockage, and 5) the residual discharge capacity of a blocked PKW was more than 80% of the driftwood-free condition.

SOGREAH (2011) conducted model tests of the Luziere Dam (France) emergency PKW spillway. It was observed that 1) driftwood was mainly attracted by the principal spillway inlet, as it generated higher approach flow velocities than the PKW, 2) the volume of blocked driftwood decreased with increasing discharge, 3) the density of the driftwood accumulations were compact for high heads and loose for small heads, and 4) a relative head increase of some 10% occurred for a flood, with an initial head of 1.0 m.

Despite of a lack of systematic tests, one may summarize that PKWs seem prone to blockage with driftwood, particularly when a gated primary spillway isn't available to attract the majority of the floating debris. Although the effect occurs within certain limits, an increased hydraulic head was observed, particularly for small specific discharges. This has to be considered for the design of such structures.

2 PHYSICAL MODEL TESTS

2.1 *Piano Key weir configurations*

Systematic physical model tests were conducted at the Laboratory of Hydraulic Constructions (LCH) of EPFL, Switzerland. Three different PKW configurations (Table 1 and Fig. 1) were tested at the downstream end of a 2 m wide and 10 m long channel, with an approach flow depth on the order of 0.8 m. The PKWs were installed in the opening of a vertical-walled dam (contracted-weir configuration) as shown in Fig. 1. With linear weir width W to upstream channel width ratios of 0.27 to 0.33 and the base of each PKW installed 0.5 m above the channel invert, the upstream approach flow condition approximated that of a reservoir. Additionally, no tailwater submergence occurred. The water level in the channel was measured using a point gauge (±0.5 mm). The latter measurement provided H, as the measurement was taken in a zone with stagnant water. Generally, the channel flow velocities tended to zero because the tested discharges Q were small. The latter was measured with a magnetic inductive flow meter (±0.5% at full span).

To better generalize the results, three different PKW configurations (all Type-A) were tested as detailed in Table 1. They included variations in cycle number, characteristic lengths, and abutment wall detail. The physical models were built with a geometrical scale factor of 1:30 (relative to their respective prototype structures) and operated under the similitude of Froude. Herein, only prototype values are mentioned.

2.2 *Driftwood*

The tested driftwood was chosen based on two approaches, namely: A large spectrum of trunk lengths to find sensitive dimensions and to allow for a general application of the results; and to

Table 1. Characteristics of tested PKW configurations A to C, with W = linear transversal width, W_i = inlet key width, W_o = outlet key width, T_s = side wall thickness, B = stream-wise length, B_i = inlet key overhang length, B_o = outlet key overhang length, $P_i = P_o$ = inlet or outlet key weir height, R = parapet wall height, and W_u = cycle width.

	A	B	C	
W	16.35	16.65	19.95	[m]
W_i	2.46	1.32	1.20	[m]
W_o	1.50	1.17	0.99	[m]
T_s	0.25	(0.03)	0.30	[m]
B	9.75	5.70	9.75	[m]
B_o	3.54	1.65	3.00	[m]
B_i	2.70	1.50	2.25	[m]
Crest type	half rounded	sharp	half rounded	
$P_i = P_o$	2.88	2.52	4.50	[m]
R	0.00	0.50	0.60	[m]
W_u	4.44	2.55	2.79	[m]
L/W	5.20	5.50	7.80	[–]
W_i/W_o	1.64	1.13	1.21	[–]
B/P	3.39	2.26	2.17	[–]
B_i/B	0.28	0.26	0.23	[–]
B_o/B	0.36	0.29	0.31	[–]

(a)

(b)

Figure 1. Photos of PKW configurations (a) A, and (b) C (Table 1).

cover common flood-related driftwood element sizes as determined by Bezzola and Hegg (2007), based on survey after major flood events in Switzerland. Herein, seven trunk length T classes were derived from the latter reference, namely $T = 1.0$, 3.0, 5.0, 7.0, 9.0, 11.0, and 13.0 m. Small values T represent tree branches, and large values entire trunks not yet fractured by the transport processes. The ratio between individual trunk length and diameter D was constant as $T/D = 20$. The trunk lengths compared with the cycle width W_u of the herein investigated PKWs (Table 1) ranges between $0.23 \leq T/W_u \leq 5.10$. Furthermore, rootstocks were included, with a trunk length of 5.0 m and a root-diameter of 2.5 m, defined perpendicular to the trunk axis.

For the tests, ten different driftwood batches (i.e., accumulations) were prepared, which included 19 to 22 trunk elements of varied sizes and two rootstocks (RS) (Table 2). Further, the trunk dimensions within a driftwood batch were chosen in an effort to be consistent with the natural driftwood size distribution statistics presented by Bezzola and Hegg (2007). The ten driftwood batches were added sequentially at random stations across a common reservoir cross-section located 2 m upstream of the PKW, for a constant Q. Without removing the driftwood trapped on the PKW from previous batches, subsequent driftwood batches were introduced upstream of the flume; this was repeated for all ten driftwood batches. All three PKW configurations (Table 1) were tested under four different discharge rates (discharge was held constant during a test) spanning a dimensionless upstream head spectrum of $0.07 \leq H_i/P \leq 0.49$, where H_i = initial upstream head prior to adding driftwood. For each discharge, the ten batches were added sequentially and their effect on H measured. Thus, 40 values of H resulted per PKW geometry, giving a total of 120 data points for all PKWs collectively.

Table 2 indicates the supplied integral driftwood volume V supplied according to the individual batch number. In total, approximate driftwood volume associated with the 10 batches was 144 m^3. This volume describes exclusively the effective wood volume, assuming the trunks and rootstocks to be linear cylinders. The total volume of a loosely formed batch (low heads) including void spaces exceeded the related wood volume.

Table 2. Batch characteristics, indicating number of elements per trunk class.

T [m]→	13	11	9	7	5	3	1	RS	Els. per batch	V [m^3]
Batch 1	1	2	2	4	7	4	2	2	24	18
Batch 2	1	1	1	4	6	4	2	2	21	31
Batch 3	1	1	2	5	7	5	0	2	23	47
Batch 4	1	1	1	4	6	4	2	2	21	60
Batch 5	1	1	2	5	7	4	2	2	24	76
Batch 6	1	1	1	5	7	5	0	2	22	90
Batch 7	1	1	2	4	6	4	2	2	22	105
Batch 8	1	1	1	5	7	4	2	2	23	119
Batch 9	0	1	2	5	7	4	0	2	21	130
Batch 10	0	2	2	5	6	4	2	2	23	144

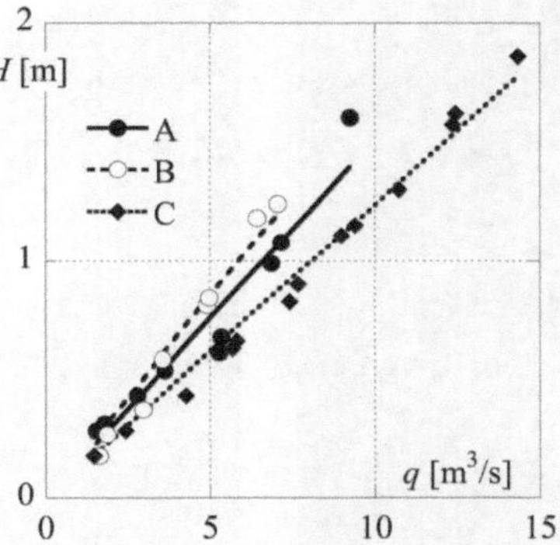

Figure 2. Rating curves of PKW configurations A to C.

2.3 *Test range*

In order to generalize the resulting observations, some key figures are given as semi-normalized values. The specific discharge ranged from $1.63 \leq q = Q/W \leq 12.45$ m^3/sm. The corresponding head-specific discharge curves are shown in Fig. 2, along with some higher discharges for validation. Configuration C had the highest discharge efficiency of the three PKWs tested, followed by B then A. This order is in agreement with L/W given in Table 1, which represents the dominant parameter regarding discharge efficiency (Leite Ribeiro et al. 2012). Similarly, the specific driftwood supply is defined as $V_s = V/W$, with V from Table 2. The parameter V_s describes thus the driftwood volume arriving at the PKW per weir width. The first batch of configuration C ($V_s = 0.88$ m^3/m) represented the minimum specific driftwood supply; the tenth batch of configuration A ($V_s = 8.79$ m^3/m) represents the maximum value tested herein.

3 RESULTS

Figure 3 shows the blockage process of driftwood at two different PKW configurations (left configuration B, right configuration C), subjected to two different specific discharges (left $q = 2.96$ m^2/s, right 9.39 m^2/s), as a function of the specific driftwood supply (increasing supply from Fig. 3a to 3e).

Figure 3 indicates that:

- the driftwood volume trapped at the PKW increases with increasing supplied driftwood volume V_s.

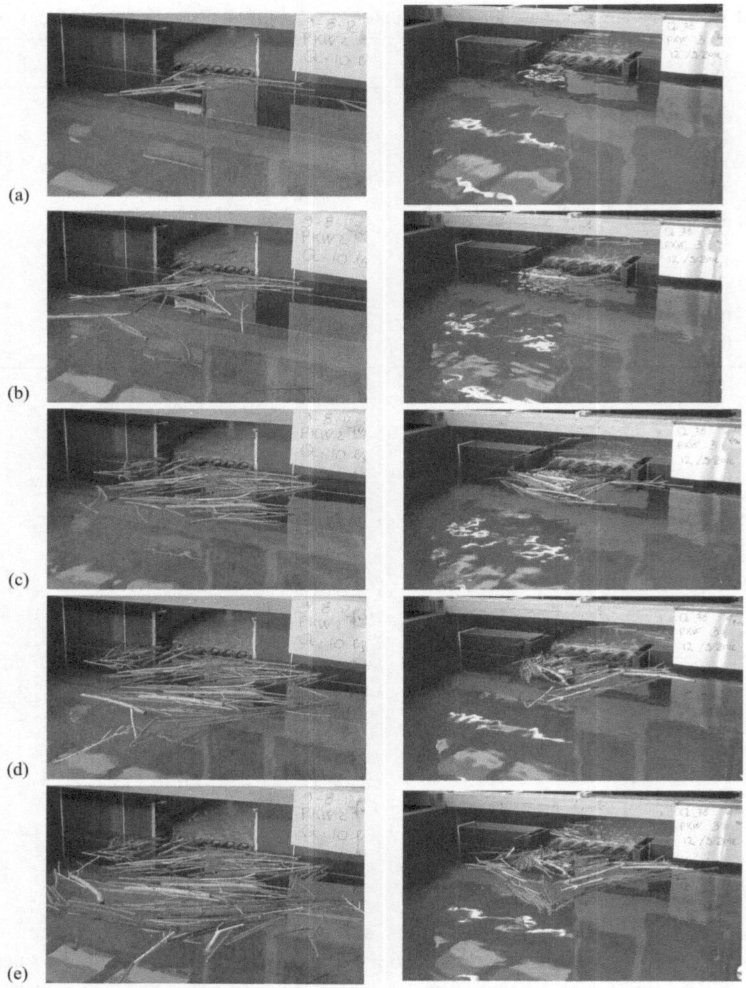

Figure 3. Progressive driftwood blockage at PKW configuration B for $q = 2.96$ m^2/s (left), and configuration C for 9.39 m^2/s (right), after supply of batch (Table 2) (a) 2, (b) 4, (c) 6, (d) 8, and (e) 10.

- large specific discharges (right) generate a smaller blockage than small values (left). Or, vice-versa, the probability to be washed over the PKW increases for large specific discharges.
- for small specific discharges, most of the trunks from the initial batches get trapped (Fig. 3b, left); for larger specific discharges, most of the trunks from the initial batches passed (Fig. 3b, right), given that no blockage was present at their arrival.
- Generally, the accumulations appear loose for small specific discharges (due to the reservoir approach flow type), and denser for higher specific discharges. For the latter, the flow momentum is comparably higher, supporting the passage of trunks as well as the tighter "packing" of trapped trunks.
- The trunks in direct contact with the PKW weir crest are mostly orientated parallel to the flow (perpendicular to the weir axis), whereas those further upstream were arranged more tangentially.

Figure 4a shows the relative head increase H/H_i versus the driftwood supply V_s. Generally, the relative head increased with increasing V_s, with few exceptions for small specific discharges.

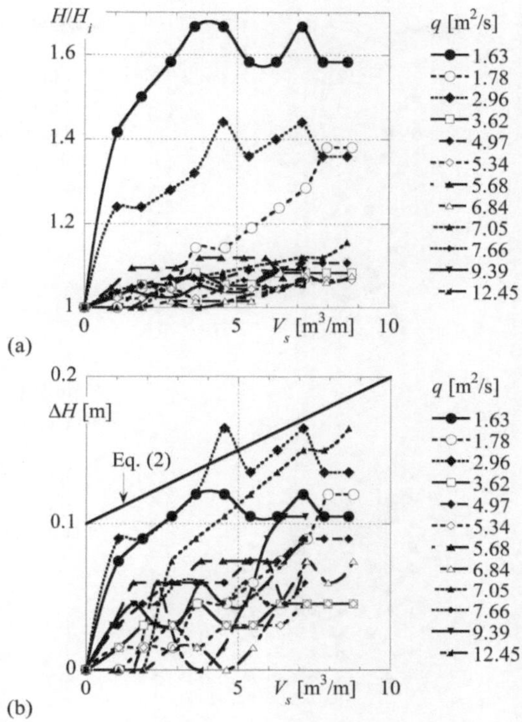

Figure 4. Effect of V_s and q on (a) relative H/H_i, and (b) absolute ΔH head increase.

For these cases, the arrival of additional driftwood at the PKW provided sufficient momentum to remobilize some of the previously trapped debris clusters and pass them over the PKW. As the trapped debris was liberated, H/H_i subsequently decreased. Furthermore, smaller discharges ($q < 3.0\,\mathrm{m^2/s}$) produced larger relative head increase ($1.4 < H/H_i < 1.7$), whereas large specific discharges above the aforementioned limit resulted in $H/H_i < 1.2$. The latter values apply for V_s between 4 to 9 m³/m approximately. However, the conclusion that a large value q is less sensitive regarding the effect of driftwood on H might be misleading, as Fig. 4b indicates. There, the absolute head increase $\Delta H = H - H_i$ is given as a function of q and V_s.

An envelope of the data in Fig 4b indicates that the maximum hydraulic head ΔH_M in [m] additionally generated by driftwood can be expressed pragmatically as

$$\Delta H_M = E + 0.01V_s \tag{2}$$

where V_s = specific driftwood volume transported to the considered section in in [m³/m]. Note that the limits of the model tests apply, namely related the PKW configurations, $q \leq 12.45\,\mathrm{m^2/s}$ and $V_s \leq 8.79\,\mathrm{m^3/m}$. The variable E covers the initial and pronounced head increase occurring as soon as driftwood arrives at a PKW, and is $E = 0.10$ for $q \leq 3\,\mathrm{m^2/s}$, and $E = 0.05$ for $q > 3\,\mathrm{m^2/s}$. In Fig. 4b, Eq. (2) is shown with $E = 0.10$. For a design discharge of $q_D \leq 3\,\mathrm{m^2/s}$, the ΔH_M from Eq. (2) with $E = 0.10$ represents the additional reservoir freeboard required to account for debris accumulation. For larger design discharges, the larger value ΔH_M from Eq. (2) either with $E = 0.10$ or with $E = 0.05$ has to be provided. This is necessary, as PKWs are unregulated weir structures, imperatively spilling small discharges at the first phase of every flood passage.

For PKWs, the hydraulic head is frequently expressed relative to the weir height as H_i/P. Figure 5 includes basically the same information as Fig. 4, i.e., shows that the relative head increase H/H_i

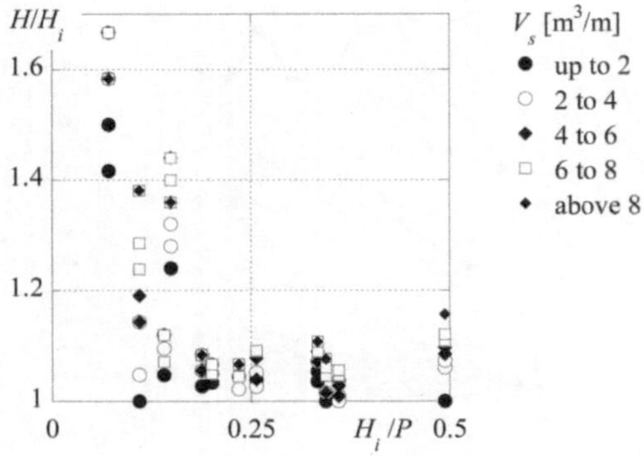

(a)

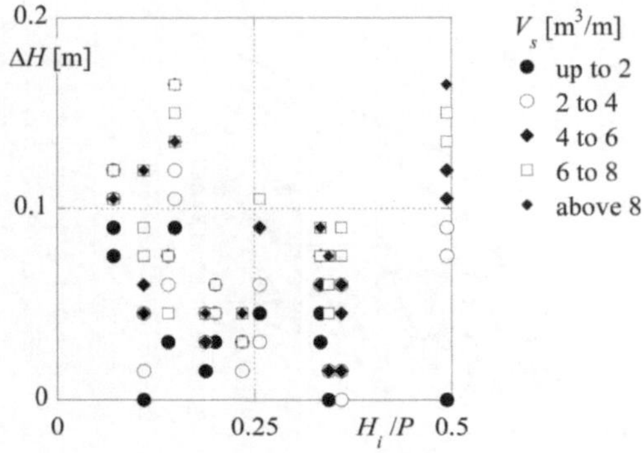

(b)

Figure 5. Effect of H_i/P and V_s on (a) relative H/H_i, and (b) absolute ΔH head increase.

generated by driftwood is significant (H/H_i up to 1.7) for $H_i/P < 0.2$, and small ($H/H_i < 1.2$) for $0.2 < H_i/P < 0.5$. As for the absolute head increase ΔH (Fig. 5b), the trend is again less significant, however, the maximum additional head becomes $\Delta H_M < 0.2$ m.

The tests were conducted under a constant discharge q, and the increased head H as a function of the driftwood supply was measured. Formulating the related discharge coefficients C_d gives

$$\frac{q}{q_i} = \frac{C_d\sqrt{2gH^3}}{C_{di}\sqrt{2gH_i^3}} = 1 \tag{3}$$

where the subscript i refers to initial conditions prior to adding driftwood. Equation (3) can be reformulated as

$$\frac{C_d}{C_{di}} = \left(\frac{H}{H_i}\right)^{-3/2} \tag{4}$$

Considering the herein measured maximum $H/H_i = 1.7$ results is $C_d/C_{di} = 0.45$, and $H/H_i = 1.1$ in $C_d/C_{di} = 0.87$. If $q > 3.0$ m^2/s then generally $H/H_i < 1.2$ (Fig. 4a), so that minimally $C_d/C_{di} = 0.76$. For the latter (typical) situation, one may thus assume a reduction of the discharge coefficient up to $C_d = (3/4)C_{di}$, serving as basis for a rough estimation.

The hydraulic load on the PKW increases due to the presence of driftwood. For the reservoir approach flow type as investigated herein, mainly the augmented hydrostatic pressure (by ΔH) seems relevant, whereas the dynamic impact of the trunks is small due to negligible approach flow velocities.

4 CONCLUSIONS

To develop a better understanding of the potential impact of driftwood accumulation upstream of PKWs, laboratory-scale testing was conducted featuring three different in-reservoir PKW geometries and varying sizes of driftwood trunks consistent those commonly found in alpine and pre-alpine catchments (Bezzola and Hegg 2007). Each PKW was tested at four different specific discharges; each test included the addition of ten sequential batches of tree trunks of distributed diameter and length upstream of the PKW. The driftwood elements were allowed to pass or collect at the PKW as flow conditions dictated. Trapped driftwood elements were not removed prior to the addition of the subsequent batches. The changes in the upstream reservoir level were noted with the addition of each batch. Based on the results of these tests, the following conclusions are made:

1. In general, the amount of driftwood collecting at the PKWs increased with decreasing unit discharge (q). In some low specific discharge cases, however, the momentum provided by the later driftwood batch additions, along with the increased reservoir level (ΔH) caused by the driftwood accumulation, was sufficient to liberate some of the previously trapped trunks and pass them over the PKW.
2. The driftwood volume trapped at the PKW increases with increasing supplied driftwood volume (V_s). The density of the trapped driftwood debris field increased with increasing q, as shown in Fig. 3.
3. The probability of driftwood trunks being washed over the PKW increased with increasing q.
4. The trunks in direct contact with the PKW weir crest were mostly orientated parallel to the flow (perpendicular to the weir axis); trapped trunks located further upstream in the reservoir were arranged more parallel to the PKW axis.
5. Eq. 2 provides a first-order approximation of the increase in upstream reservoir head (ΔH) caused by driftwood accumulation at a PKW. The nature and volume of the driftwood will likely vary significantly, however, with terrain, geographical location, season, and climate.
6. The discharge coefficient (C_d) decreases due to driftwood collection at the PKW. It is typically around (3/4) of the value without drift wood occurrence for specific discharges (q) exceeding some 3 m^2/s.

Future research areas regarding driftwood and PKW interaction should include characterizing the influence of adjacent gated spillway operation as a means of controlling driftwood collection at PKWs following flood events. Additionally, the influence of the crest shape is of interest. Besides, Eq. (2) has to be expressed dimensionless, based on a dimensional analysis.

ACKNOWLEDGMENTS

The authors thank Mr. Damiano Capobianco, MSc Student at University of Cassino (Italy), for his excellent support during the model tests.

REFERENCES

Anderson, R.M. and Tullis B.P. 2012a. Piano Key Weirs: Reservoir versus Channel Applications. *J. Irrig. Drain. Eng*. 138(8): 773–776.

Anderson, R.M. and Tullis B.P. 2012b. Comparison of Piano Key and Rectangular Labyrinth Weir Hydraulics. *J. Irrig. Drain. Eng*. 138(6): 358–361.

Anderson, R.M. and Tullis B.P. 2013. Piano Key Hydraulics and Labyrinth Weir Comparisons. *J. Hydraulic Eng*. 139(3): 246–253.

Bezzola, G.R. and Gantenbein, S., Hollenstein, R. and Minor, H.-E. 2002. Verklausung von Brückenquerschnitten [Blocking of bridge cross sections]. VAW *Mitteilung* 175, H.-E. Minor, ed., ETH Zurich, 87–97 (in German).

Bezzola, G.R. Hegg, C., eds. 2007. *Ereignisanalyse Hochwasser* 2005 [2005 Flood event analysis]. Bundesamt für Umwelt & Eidgenössischen Forschungsanstalt für Wald, Schnee und Landschaft, Bern, Switzerland (in German).

Ho Ta Khan, M., Truong, C.H. and Nguyen, T.H. 2011. Main results of the P.K. weir mode tests in Vietnam (2004 to 2010). *Proc. Intl. Conf. Labyrinth and Piano Key Weirs*, Liège B, 191–198. CRC Press, Boca Raton FL.

Kabiri-Samani, A. and Javaheri, A. 2012. Discharge coefficient for free and submerged flow over Piano Key weirs. *J. Hydraulic Res.* 50(1): 114–120.

Lange, D. and Bezzola, G.R. 2006. Schwemmholz: Probleme und Lösungsansätze [Driftwood: Problems and methods of resolution]. VAW *Mitteilung* 188, H.-E. Minor, ed., ETH Zurich (in German).

Laugier, F. 2007. Design and construction of the first Piano Key Weir spillway at Goulours Dam. *Hydropower & Dams* 13(5): 94–100.

Laugier, F., Lochu, A., Gille, C., Leite Ribeiro, M. and Boillat, J.-L. 2009. Design and construction of a labyrinth PKW spillway at St-Marc Dam, France. *Hydropower & Dams* 15(5): 100–107.

LCH 2007. Barrage de Saint-Marc: Etude sur modèle des évacuateurs de cure [Saint-Marc Dam: Spillway model study]. *Report* LCH 10/2006, EPFL Lausanne, Switzerland (in French, not published).

Leite Ribeiro, M., Pfister, M., Schleiss, A.J. and Boillat, J.-L. 2012. Hydraulic design of A-type Piano Key Weirs. *J. Hydraulic Res.* 50(4): 400–408.

Lempérière, F., Vigny, J.P. and Ouamane, A. 2011. General comments on Labyrinth and Piano Key Weirs: The past and present. Proc. Intl. Conf. *Labyrinth and Piano Key Weirs*, Liège B, 17–24. CRC Press, Boca Raton FL.

Machiels, O., Erpicum, S., Dewals, B.J., Archambeau, P. and Pirotton, M. 2011. Experimental observation of flow characteristics over a Piano Key Weir. *J. Hydraulic Res.* 49(3): 359–366.

Machiels, O. 2012. Experimental study of the hydraulic behaviour of Piano Key Weirs. PhD *Thesis* ULgetd-09252012-224610, Univerity of Liège (B).

Melville, B.W. and Dongol, D.M. 1992. Bridge pier scour with debris accumulation. *J. Hydraulic Eng.* 118(9): 1306–1310.

Ouamane, A. and Lempérière, F. 2006. Design of a new economic shape of weir. *Proc. Intl. Symp. Dams in the Societies of the 21st Century*, Barcelona, 463–470.

Pagliara, S. and Carnacina, I. 2010. Temporal scour evolution at bridge piers: effect of wood debris roughness and porosity. *J. Hydraulic Res.* 48(1): 3–13.

Pralong, J., Vermeulen, J., Blancher, B., Laugier, F., Erpicum, S., Machiels, O., Pirotton, M., Boillat, J.L., Leite Ribeiro, M. and Schleiss, A.J. 2011. A naming convention for the piano key weirs geometrical parameters. *Proc. Intl. Conf. Labyrinth and Piano Key Weirs*, Liège B, 271–278. CRC Press, Boca Raton FL.

Rickenmann, D. 1997. Schwemmholz und Hochwasser [Driftwood and floods]. *Wasser Energie Luft* 89(5/6): 115–119 (in German).

Rimböck, A. 2003. *Schwemmholzrückhalt in Wildbächen* [Driftwood retention in torrents]. PhD Thesis, TU Munich (in German).

Schleiss, A.J. 2011. From labyrinth to piano key weirs: A historical review. *Proc. Intl. Conf. Labyrinth and Piano Key Weirs*, Liège B, 3–15. CRC Press, Boca Raton FL.

Schmocker, L. and Hager W.H. 2011. Probability of drift blockage at bridge decks. *J. Hydraulic Eng.* 137(4): 470–479.

SOGREAH 2011. Modèle physique du barrage de Luzieres [Physical model of the Luzieres Dam]. *Report* 136 0 886, SOGREAH (Now: The ARTELIA Engineering Company), France (in French, not published).

Uchiogi, T., Shima, J., Tajima, H. and Ishikawa, Y. 1996. Design methods for woody-debris entrapment. *Interpraevent*, Gramisch-Partenkrichen 5:279–288.

Labyrinth and Piano Key Weirs II – PKW 2013 – Erpicum et al. (eds)

Experimental and numerical study of the hydraulic performance of a trapezoidal Piano Key weir

G.M. Cicero & J.R. Delisle
EDF – National Hydraulics and Environment Laboratory, Chatou, France

V. Lefebvre & J. Vermeulen
EDF – Hydro Engineering Center, Le Bourget-du-Lac, France

ABSTRACT: For ten years, many experimental and numerical studies have been carried out on the Piano-Key Weir (PKW) to better understand the influence of numerous geometric parameters on hydraulic performance. The classical PKW has a rectangular shape in plan view (sidewall angle as 0); the purpose of this study is to examine the effect of a sidewall angle on hydraulic performance. Two trapezoidal PKWs, similar to a rectangular type A, were designed with either the same developed length (trapezoidal 1) or the same upstream-downstream length (trapezoidal 2). Both the rectangular and the trapezoidal 1 PKWs were tested in a 2 m wide channel to measure the discharge characteristics under free flow conditions, and in both positions for the trapezoidal (non symmetrical in plan view). Then the tests performed on both geometries were simulated by Flow-3D® and the numerical results were in rather good agreement (i.e., <8%) with the measured discharge characteristics. The numerical model so validated was then used to predict hydraulic performance of the trapezoidal 2. In the position where the inlet is larger than the outlet, both trapezoidal PKWs were more efficient than the rectangular one, with a gain in discharge that should be greater than 5%.

1 INTRODUCTION

The Piano Key Weir (PKW) is a type of labyrinth weir with efficient hydraulic performance especially at low heads. Since the first tests performed for Hydrocoop in 1998 at the National Hydraulics and Environment Laboratory (LNHE) in Chatou (Lemperiere et al. 2011), the laboratory experiments conducted in channels by Anderson (2011), Ho Ta Khanh et al. (2011), Ouamane (2011), Leite Ribeiro et al. (2012), Kabiri-Samini & Javaheri (2012), Machiels (2011, 2012) contributed to improve the knowledge on the hydraulic behaviour of PKWs under free-flow or submerged conditions. In addition, the numerical modeling studies (Erpicum et al. 2011, Pralong et al. 2011a) were complementary to better understand the flow behaviour and the relevant parameters on hydraulic performance. The primary parameters influencing hydraulic performance were identified such as the width ratio of the inlet and outlet keys, the developed length to width ratio, and the weir height. However, some secondary parameters like the dam height, the noses under the outlet keys, and the crest shape also have an influence on hydraulic performance.

In 2011, Electricité De France (EDF) started a new experimental program at LNHE in Chatou to expand the knowledge on some geometric parameters effects such as the overhangs (Cicero et al. 2012), the crest shape, the dam height, and the noses under the outlet keys. This program includes the testing of a trapezoidal PKW design with sidewalls angles, and the testing of all geometries under submerged conditions. This program will also provide a useful database to validate 1D or 3D numerical modeling and empirical correlations.

The idea of designing a trapezoidal PKW appeared thanks to the study of Falvey (2003) concerning triangular and trapezoidal labyrinth weirs. He revealed that a sidewall angle can improve the discharge capacity and defined an optimum around 8°.

The present article is devoted to the analyses of the free-flow hydraulic performance of two trapezoidal PKWs designed with similar dimensions as a classical rectangular PKW of type A. The first section of this article presents the geometry of the rectangular PKW, the design principles, and the resulting geometries of the tested trapezoidal PKWs. The second part is related to the experimental results of the trapezoidal 1 and the rectangular PKWs. The last part presents the numerical simulations with the validation on the rectangular and trapezoidal 1 PKWs, and the predictions on the trapezoidal 2.

2 GEOMETRIES OF TRAPEZOIDAL PK-WEIRS

We used the standard notations defined by Pralong et al. (2011) for rectangular PKW, and the specific notations defined by Figure 1 for trapezoidal PKW.

The objective of this study is to isolate the effect of a sidewall angle (α) on the hydraulic efficiency of a PKW. The design of the trapezoidal PKW was based on a classical rectangular one (i.e. $\alpha = 0$), keeping its main geometric parameters. We chose the geometry of a type A recommended by Lemperiere et al. (2011), (i.e. $W_i = W_o = 0.6P$, $B_b = 1.2P$, $B = 2.4P + 2T_s$), with a weir height $P = 0.222$ m and a sidewall thickness $T_s = 20$ mm. The crest thicknesses of the inlet (T_i) and outlet (T_o) keys were $T_i = T_o = 20.6$ mm. The PKWs had a flat-topped crest.

Then two trapezoidal PKWs were designed that maintained P, B_b, T_s, T_i, T_o and the widths of the inlet ($W_{i,d}$) and outlet ($W_{o,d}$) keys at the downstream edge, (i.e. $W_{i,d} = W_i$, and $W_{o,d} = W_o$). The resulting geometric parameters of Table 1 were calculated by Equations 1 to 5, according to the design constraints. The first design constraint (trapezoidal 1) was to maximize the sidewall angle keeping constant the ratio L_u/W_u which is a relevant parameter on hydraulic performance of PK-Weirs and labyrinth weirs (Tullis et al. 1995). The maximum sidewall angle was found to be $\alpha = 5°$.

The second design constraint was to maintain the upstream-downstream length (B) which is a parameter influencing the building cost of PKWs. So, the inlet (B_o) and outlet (B_i) overhang lengths were the same as the rectangular type A. Although the maximum sidewall angle was found to be $\alpha = 6°$, we kept $\alpha = 5°$ to allow comparisons with the trapezoidal 1.

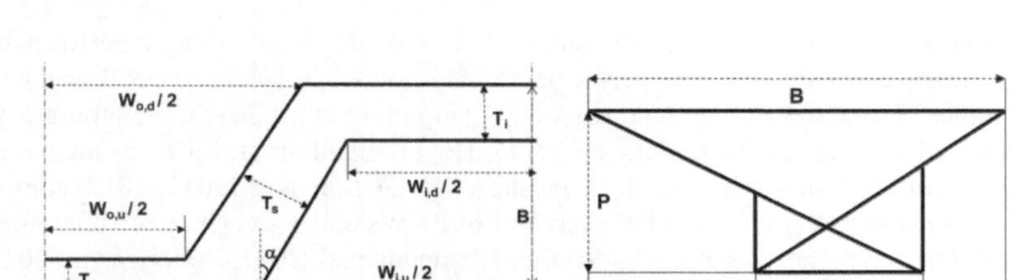

Figure 1. Geometric parameters of trapezoidal PKW. Half-element unit in plan view (a) and elevation (b).

Table 1. Computed dimensions (S.I.) of the trapezoidal PKWs compared to the rectangular PKW.

PKW	α	$W_{i,u}$	$W_{o,u}$	B_i	B_o	B	L_u	W_u	L_u/W_u
Rectangular	0°	.133	.133	.154	.154	.573	1.411	.306	4.61
Trapezoidal 1	5°	.237	.029	.175	.175	.617	1.395	.302	4.61
Trapezoidal 2	5°	.230	.036	.154	.154	.573	1.315	.302	4.35

The PKWs width was the total channel width (~2 m) to maximize the number of PKW units and to avoid side effects. Thus, the number of PKW units was $N_u = 6.5$ with half an inlet key and half an outlet key on each end.

$$tg\alpha = \frac{W_{i,u} - W_{i,d}}{2(B - T_i)} = \frac{W_{o,d} - W_{o,u}}{2(B - T_o)} \tag{1}$$

$$B_i = B_o = \frac{1}{2}\left(B - B_b\right) \tag{2}$$

$$B_h = \frac{2B - T_o - T_i}{2\cos\alpha} \tag{3}$$

$$W_u = W_{i,u} + W_{o,u} + \frac{2(T_s - T_o \sin\alpha)}{\cos\alpha} = W_{i,d} + W_{o,d} + \frac{2(T_s - T_i \sin\alpha)}{\cos\alpha} \tag{4}$$

$$L_u = W_u + 2B_h(1 - \sin\alpha) \tag{5}$$

3 EXPERIMENTAL RESULTS

3.1 *Testing apparatus*

The PKW testing was conducted in a 2 m wide by 1 m deep by 25 m long rectangular channel devoted to the discharge capacity measurements up to 500 l/s.

The weir discharge (Q) was measured by an electromagnetic flow-meter of +/−1% accuracy and the piezometric level (h) was measured at ~6 m upstream of the PKW with an accuracy of +/−0.18 mm. The discharge and the upstream level were recorded for a 3 min period at a 0.5 Hz frequency. As a result, the discharge and head measurements used for the analyses were the average and standard deviation of 90 temporal values.

The total upstream head (H) was calculated by adding the velocity head corresponding to the average cross-sectional velocity at the location of the piezometric level measurement.

3.2 *Experimental program*

The objective of the experimental program was to compare the hydraulic efficiency of the classical type A to the trapezoidal 1 under free-flow conditions. Since a trapezoidal PKW is not symmetric in plan view, it was tested in the "design position" where the inlet keys are larger than the outlet keys at the upstream edge (Fig. 2a), and in the "reverse position" where the outlet keys are larger than the inlet keys (Fig. 2b).

The PKWs were installed on top of a 0.361 m platform in concrete, so that the dam height was $P_d = 1.63P$. They were tested without noses under the upstream overhangs.

3.3 *Experimental results*

For both positions of the trapezoidal 1, the rating curves $H(Q)$ were measured in 50 l/s increments from 50 l/s to 450 l/s. Figure 3a shows that the trapezoidal shape is hydraulically more efficient than the rectangular shape in the design position, and became less efficient in the reverse position. These results are consistent with the behaviour of the classical PKW, which is known to be more efficient when the cross-section is greater in the inlet keys than in the outlet keys (i.e. the design position).

Figure 3b compares the rating curves $C_w(H/P)$ when the discharge coefficient C_w is calculated by the classical formula:

$$C_W = \frac{Q}{W\sqrt{2g}H^{1.5}} \tag{6}$$

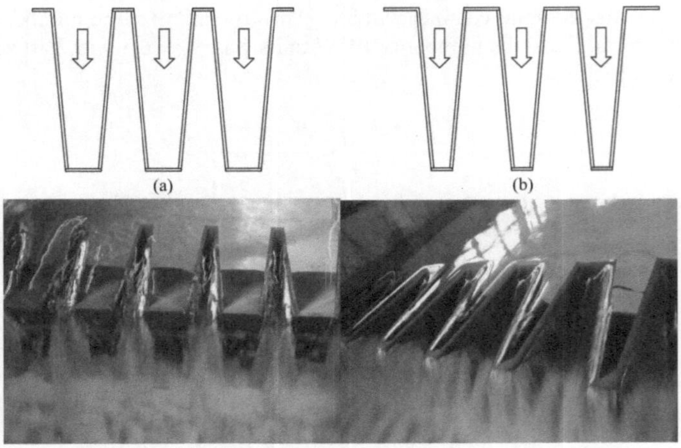

Figure 2. Views of the PKW trapezoidal 1 in the design position (a) and in the reverse position (b).

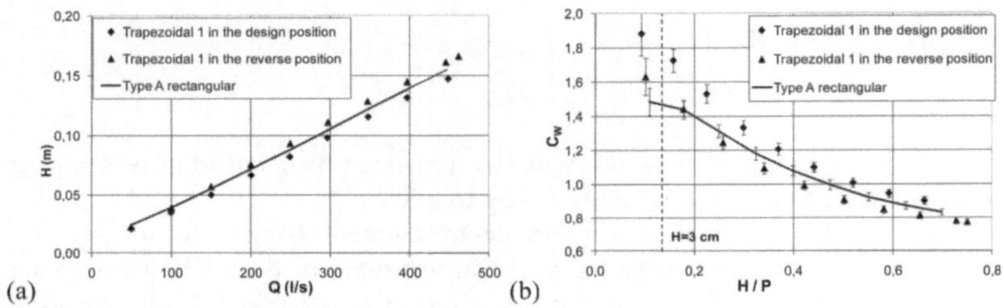

Figure 3. Experimental rating curves of the rectangular, and the trapezoidal 1 in design and reverse positions.

The measurement error on C_W was estimated by using the measurements accuracy of Q and h and the standard deviations (σ_Q and σ_h) on the 90 temporal values by:

$$\frac{\Delta C_W}{C_W} = \frac{\Delta Q + \sigma_Q}{Q} + 1{,}5\frac{\Delta H}{H} \tag{7}$$

$$\Delta H = \Delta h + \sigma_h + \frac{Q^2}{gW^2(h+P+P_d)^2}\left(\frac{\Delta Q + \sigma_Q}{Q} + \frac{\Delta h + \sigma_h}{(h+P+P_d)}\right) \tag{8}$$

with $\Delta Q/Q = 0.01$ and $\Delta h = 0.18$ mm. For the 3 rating curves (rectangular, and trapezoidal 1 in both positions), the average and minimal measurement uncertainty $\Delta C_w/C_w$ was 3% and 2% respectively.

The measured rating curves $C_W(H/P)$ were correlated by regression equations to compare the discharge efficiencies of the trapezoidal and rectangular PKWs at the same upstream head. (Note also that $P_d/P = 1.6$).

For the rectangular type A PKW, the regression Equation 9 ($R^2 = 0.9962$), allowed to predict the measured discharge with a maximum error of 2.0% for $0.1 < H/P < 0.7$:

$$C_W\left(\frac{H}{P}\right) = 2.71\left(\frac{H}{P}\right)^3 - 2.27\left(\frac{H}{P}\right)^2 - 0.85\left(\frac{H}{P}\right) + 1.62 \tag{9}$$

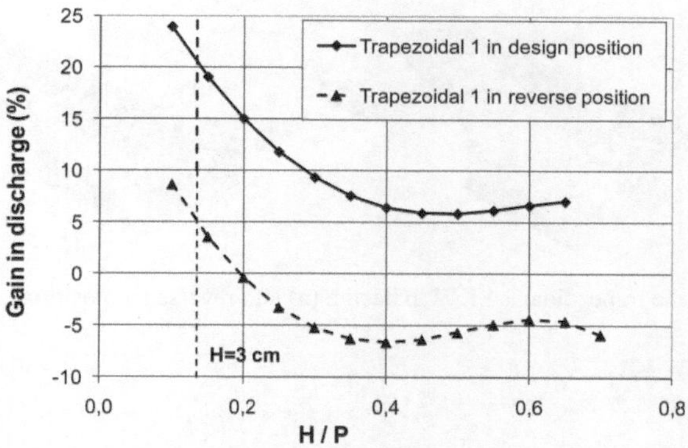

Figure 4. Gain in discharge of the trapezoidal 1 (in both positions) relative to the rectangular PKW.

For the trapezoidal 1 in the design position, the regression Equation 10 ($R^2 = 0.9987$) allowed to predict the measured discharge with a maximum error of 1.4 % for $0.1 < H/P < 0.66$:

$$C_W\left(\frac{H}{P}\right) = 0.07\left(\frac{H}{P}\right)^3 + 2.47\left(\frac{H}{P}\right)^2 - 3.64\left(\frac{H}{P}\right) + 2.22 \qquad (10)$$

For the trapezoidal 1 in the reverse position, the regression Equation 11 ($R^2 = 0.9995$) allowed to predict the measured discharge with a maximum error of 1.0 % for $0.1 < H/P < 0.75$:

$$C_W\left(\frac{H}{P}\right) = -1.65\left(\frac{H}{P}\right)^3 + 4.21\left(\frac{H}{P}\right)^2 - 3.86\left(\frac{H}{P}\right) + 1.99 \qquad (11)$$

Figure 4 shows that the gain in discharge of the trapezoidal 1 relative to the rectangular PKW, decreased with the upstream head. Since the measurement uncertainty $\Delta C_w / C_w$ is greater than 2%, the gain (or loss) in discharge relative to the rectangular PKW become significant if it is greater than 4%. In the design position, the trapezoidal PKW was from 5 to 25% more efficient than the rectangular one. For $H > 3$ cm, where the scale effects due to viscosity and surface tension became negligible (Leite Ribeiro et al. 2012), the trapezoidal PKW was from 20 to 5% more efficient than the rectangular PKW. In the reverse position, except at low heads ($H/P < 0.2$), the trapezoidal PKW became less efficient (5 to 7%) than the rectangular.

4 NUMERICAL SIMULATIONS

4.1 *Description of the model*

Flow-3D® is a CFD software developed by Flow Science based on the original and true form of the Volume-of-Fluid technique. It numerically simulates the hydraulic behaviour of the rectangular and trapezoidal PKWs in order to compare the results with the physical model experimentations.

The three spillway geometries (rectangular, trapezoidal 1 and trapezoidal 2) were drawn using AutoCAD® and implemented on Flow-3D® with an optimized mesh. The flow was assumed to be symmetrical, only half an inlet and half an outlet were represented with one sidewall. The mesh (about 200,000 cells in total, 6 mm size on the PKW) extended 1 m upstream and 0.5 m downstream of the PKW to respect approach conditions and free flow evacuation. Symmetrical conditions (velocity derivative equal to zero at the limit) were used for lateral boundaries and specific fluid elevations were defined for upstream and downstream limit (the fluid elevation was assumed equal to the hydraulic head measured on the physical model in order to include entrance velocity). The solver uses first order momentum advection and no turbulence model is activated.

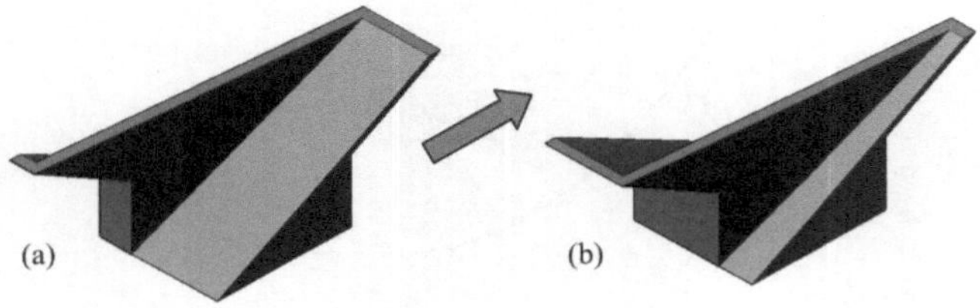

Figure 5. 3D views of the trapezoidal 1 PKW in design (a) and reverse (b) positions.

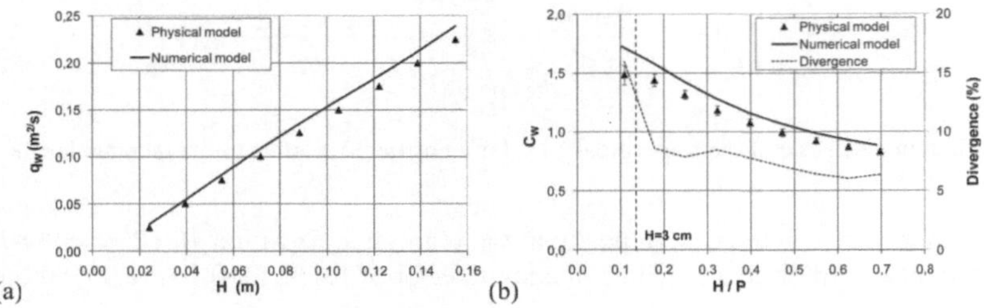

Figure 6. Specific discharge (a), and discharge coefficient (b) of the Rectangular PKW on both models.

Flow is simulated on 30 seconds to ensure its stability, the mass flow rate converges after 10 seconds (also verified with mass-averaged fluid mean kinetic energy convergence).

The simulations were conducted by setting up the upstream hydraulic heads measured on the physical model and computing the total discharge over the spillway through control sections. Five sets of ten simulations were performed to plot the rating curves $H(Q)$ of each geometry in different positions.

4.2 *Validation on the rectangular and trapezoidal 1 experiments*

Results of the numerical simulation can be interpreted as those of the physical models. The following figures show either the specific discharge or the discharge coefficient computed as before (Equation 6). Graphs for physical and numerical models are drawn together to compare them and the divergence between the numerical and the physical model is added on the right axis.

First of all, we can see that the qualitative behaviour of the rating curves is respected by Flow-3D®. The physical and numerical model curves are almost stacked for the trapezoidal 1 in the design position (Fig. 7a), and rather close for the rectangular PKW (Fig. 6) and the trapezoidal 1 in reverse position (Fig. 7b).

The numerical model overestimates the discharge coefficients measured on the physical model. Most of the divergence between both models varies from 5% to 10%. The average deviations are about 8% for the rectangular (Fig. 6a), 2% for the trapezoidal in design position (Fig. 7a), and 8% in reverse position (Fig. 7b).

The software results are in rather good agreement with the physical model outcomes and it can be considered as valid for further studies on trapezoidal PKWs.

4.3 *Predictions on the trapezoidal 2 PKW*

The trapezoidal 2 PKW is slightly different form the trapezoidal 1 (Table 1). The trapezoidal 1 keeps the parameter L/W equal to the rectangular one, whereas the trapezoidal 2 focuses on the structural parameter B (upstream-downstream length) to keep it constant since it is often a building constraint for the project. Then, the trapezoidal 2 and the rectangular PKW have the same global dimensions (overhangs are not extended as in the trapezoidal 1).

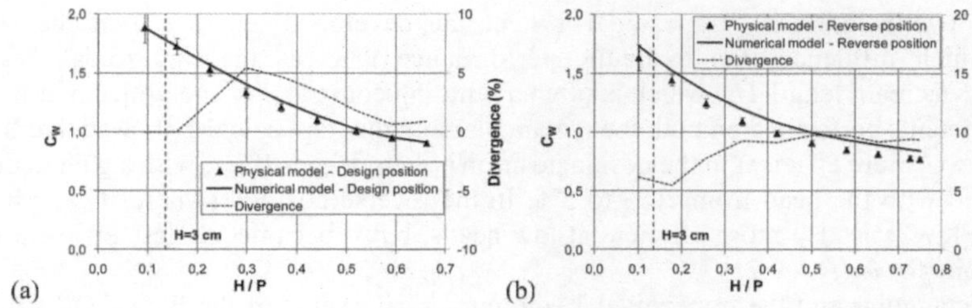

Figure 7. Discharge coefficient for both models of trapezoidal 1, in design (a) and reverse (b) positions.

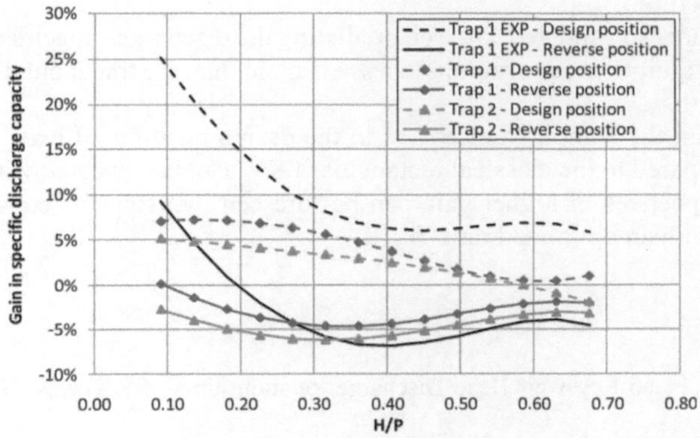

Figure 8. Gain in specific discharge capacity for trapezoidal PKWs 1 and 2, in both positions: experimental and numerical results for Trap 1, and numerical results for Trap 2.

The following figure shows the gain in terms of specific discharge relatively to the rectangular PKW for trapezoidal 1 and 2 in both the design and reverse positions. Gains are computed relative to the rectangular results for experimental and numerical models respectively. Therefore, the rating curves $C_w(H/P)$ computed by Flow-3D® were also correlated by regression equations to calculate the discharges of the trapezoidal and rectangular PKWs at the same upstream head.

Once again, we can see that the design position (dotted lines) is the most efficient orientation. The trapezoidal configuration in the reverse position (solid lines) is less efficient than the rectangular geometry. This behaviour coincides with the classical rectangular PKW where a W_i/W_o ratio larger than 1 is preferable to increase the discharge capacity. It is the case at the upstream edge in the design position, and the opposite in the reverse position. Actually, the inlet volume is enlarged in the design position which induces a better inlet of the PKW.

Globally, the trapezoidal 2 configuration is less efficient than the trapezoidal 1 for all heads, since the developed crest length L_u is shortened to keep a constant value for B (Table 1). For the trapezoidal 1, Flow-3D® underestimates the gain in discharge by more than 5%, at all heads in the design position, and at low heads in the reverse position. This is mainly due to the computation uncertainties (or divergence) which depends on the head and on the numerical model mesh.

Finally, taking into account the computations errors, in the design position, at least 5% of gain in discharge compared to the classical rectangular PKW can be expected with both trapezoidal configurations.

5 CONCLUSIONS

In this study, a physical and a Flow-3D® numerical model were used to compare the free-flow hydraulic performance of two trapezoidal PKWs similar to a rectangular PKW of type A (with symmetrical overhangs). The trapezoidal 1 kept the developed length (L) of the rectangular PKW which influences the hydraulic performance whereas the trapezoidal 2 kept the upstream-downstream length (B) which is often a building constraint of the project.

The experiments performed on both the rectangular and the trapezoidal 1 showed that the trapezoidal shape was more efficient in the design position (where $W_{i,u} > W_{o,u}$), with a gain in discharge that decreased with the head from 20% to 5%. In the reverse position (where $W_{i,d} < W_{o,d}$), the trapezoidal PKW was also more efficient at low heads, but it became 5% less efficient than the rectangular one for $H/P > 0.2$.

For the rectangular and the trapezoidal 1, the numerical results of the Flow-3D® simulations were in rather good agreement with the physical model outcomes: the mean deviation between both models was about 8% for the rectangular PK, 2% for the trapezoidal 1 in the design position, and 8% in the reverse position.

The Flow-3D® model so validated, allowed predicting the discharge capacities of the trapezoidal 2. Globally, this configuration was around 2% less efficient than the trapezoidal 1 due to its shorter developed crest length.

Taking into account the computation errors, in the design position, at least 5% of gain in discharge capacity compared to the classical rectangular PKW can be expected with both trapezoidal configurations. Perspectives of higher gain can be foreseen by using the concept of trapezoidal scheme on a PKW with an optimized ratio W_i/W_o.

REFERENCES

Anderson, R.M. 2011. Piano Key Weir Head Discharge Relationships. M.S. Thesis, Utah State University, Logan, Utah.

Cicero, G.M., De Miranda D.A. & Luck, M. 2012. Assessment of the code Wolf 1D PKW for predicting the hydraulic behaviour of PK-Weirs. *Congrès SHF- 33èmes journées de l'hydraulique "Grands aménagements hydrauliques 2012", Paris, France.*

Erpicum, S., Machiels, O., Archambeau, P., Dewals, B. & Pirotton, M. 2011. 1D numerical modeling of the flow over a Piano Key Weir. *Proc. International Workshop on Labyrinths and Piano Key Weirs PKW 2011, CRC Press,* 151–158.

Falvey, H.T. 2003. Hydraulic Design of Labyrinth Weirs, ASCE Press, Reston, Virginia.

Ho Ta Khanh, M., Chi Hien, T. & Thanh Hai, N. 2011. Main results of the P.K. weir model tests in Vietnam (2004 to 2010). *Proc. International Workshop on Labyrinths and Piano Key Weirs PKW 2011, CRC Press,* 191–198.

Kabiri-Samani, A. & Javaheri, A. 2012. Discharge coefficient for free and submerged flow over Piano Key Weirs. *Journal of Hydraulic Research,* 50(1): 114–120.

Leite Ribeiro, M., Pfister, M., Schleiss, A.J. & Boillat, J.L. 2012. Hydraulic design of A-type Piano Key Weirs. *Journal of Hydraulic Research,* 50(4): 400–408.

Lemperiere, F., Vigny, J.P. & Ouamane, A. 2011. General comments on Piano Kew Weirs: The past and present. *Proc. International Workshop on Labyrinths and Piano Key Weirs PKW 2011, CRC Press,* 17–25.

Ouamane, A. 2011. Nine years of study of the Piano Key Weir in the university laboratory of Biskra "lessons and reflections". *Proc. International Workshop on Labyrinths and Piano Key Weirs PKW 2011, CRC Press,* 51–58.

Machiels, O., Erpicum, S., Dewals, B.J., Archambeau, P. & Pirotton, M. 2011. Experimental observation of flow characteristics over a Piano Key Weir. *Journal of Hydraulic Research,* 49(3): 359–366.

Machiels, O. 2012. Experimental study of the hydraulic behaviour of Piano Key Weirs. M.S. Thesis, Université de Liège, Liege, Belgium.

Pralong, J., Montarros, F., Blancher, B. & Laugier, F. 2011a. A sensitivity analysis of Piano Key Weirs geometrical parameters based on 3D numerical modeling. *Proc. International Workshop on Labyrinths and Piano Key Weirs PKW 2011, CRC Press,* 133–140.

Pralong, J., Vermeulen, J., Blancher, B., Laugier, F., Erpicum, S., Machiels, O., Pirotton, M., Boillat, J.L., Leite Ribeiro, M. & Schleiss, A.J. 2011b. A naming convention for the piano key weirs geometrical parameters. *Proc. International Workshop on Labyrinths and Piano Key Weirs PKW 2011, CRC Press,* 271–278.

Tullis, J.P., Armanian, N. & Waldron, D. 1995. Design of labyrinth spillways. *Journal of Hydraulic Engineering,* 121(3): 247–255.

Labyrinth and Piano Key Weirs II – PKW 2013 – Erpicum et al. (eds)
© 2014 Taylor & Francis Group, London, ISBN 978-1-138-00085-8

Overview of Piano Key weir prototypes and scientific model investigations

M. Leite Ribeiro
Stucky Ltd., Renens, Switzerland

M. Pfister & A.J. Schleiss
Laboratory of Hydraulic Constructions (LCH), Ecole Polytechnique Fédérale de Lausanne (EPFL), Lausanne, Switzerland

ABSTRACT: The interest in Piano Key weirs (PKW) was increasing over the past years, both in terms of application and hydraulic research. Several systematic model test series were performed and published so far, originating from different hydraulic laboratories around the world. Therein, the key geometric parameters dominating the hydraulic capacity of PKWs were identified as developed crest length, transversal weir width, height of the walls and key widths. The present paper compares the available data-sets provided by several laboratories and as constructed on prototypes, partially including a systematic model testing of the relevant parameters. The present overview is based on the geometrical dimensions of the investigated and designed PKWs, without giving any hydraulic examination.

1 INTRODUCTION

Piano Key weirs, representing an alternative to the traditional Labyrinth weirs, are used as efficient spillway inlet on dam rehabilitation projects (Leite Ribeiro et al. 2007, Laugier et al. 2009, Pinchard et al. 2011, Erpicum et al. 2011), as well as for new dam projects (Lempérière et al. 2011, Ho Ta Khanh 2011a). In addition to the increase in discharge capacity as compared to traditional Labyrinth weirs (Blancher et al. 2011, Anderson and Tullis 2011, 2012), PKWs have typically techno-economic advantages as: (i) minor impact on dam structure; (ii) minor or negligible impact on existing gates and other components, (iii) reliability of free surface overflow inlets, requiring little maintenance, (iv) low cost solution involving the construction of repeating concrete units (Lempériere and Ouamane 2003, Laugier 2007, Leite Ribeiro et al. 2009, Vermeulen et al. 2011), and (v) small sensitivity to debris accumulation (Pfister et al. 2013).

Compared to a linear weir, the efficiency of a PKW is mainly a consequence of its longer developed crest. However, with increasing upstream head, this advantage gets lost (Leite Ribeiro et al. 2012a, c, Machiels et al. 2011b, Ouamane and Lempérière 2006). As a consequence, PKWs are typically designed to operate under relative moderate heads, nevertheless resulting in a significant discharge capacity.

Given that many prototypes were erected in the past years, mainly designed using physical model studies, and as a consequence of several systematic research studies, including physical and numerical modeling, extensive hydraulic data related to hydraulic features of PKWs are today available. These contribute to the increase of comprehension regarding to the hydraulic behavior of PKWs. Nowadays, the key geometrical parameters influencing the hydraulic capacity of a PKW, namely the developed crest length, L, the transversal width W, the width of the inlet and outlet keys W_i and W_o, and the height of the inlet key P_i, were determined (Figure 1).

The discharge capacity of a PKW is affected influenced by further geometrical parameters, as the upstream approach flow channel conditions, the height and position of eventual parapet walls

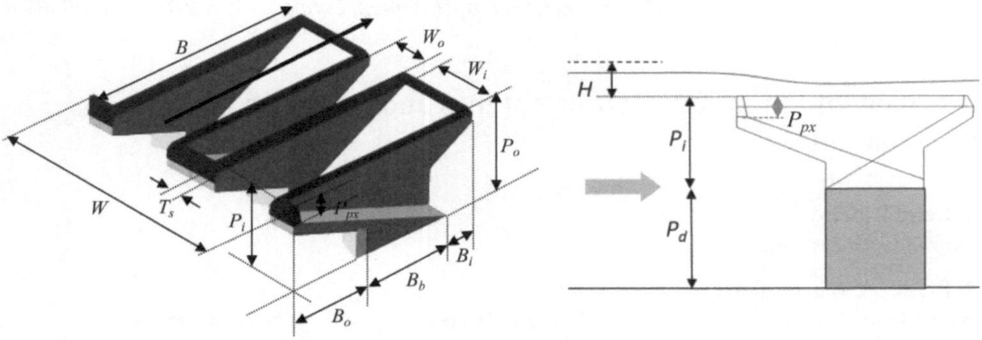

Figure 1. PKW Nomenclature according to Pralong et al. (2011).

P_{px}, the crest shape, the shape of the upstream diffuser, the height of the foundation P_d, the up- and downstream overhang lengths B_o and B_i, and the tailwater conditions. However, the influence of these parameters can be considered as secondary in comparison with the influence of the key parameters as described above. Some attention has been recently directed to the thickness of the walls (Laugier et al. 2011). As PKWs typically operate under low heads, low values of the ratio H/T_s can decrease the discharge coefficient of the walls, then operating as broad-crested weirs instead of sharp-crested weirs.

The present paper assembles the – so far available – data-sets regarding PKW models tested in numerous laboratories under free overflow conditions (i.e., no submergence from the tailwater was observed). Both, scientific model investigations to systematically vary the dominant parameters (typically sectional models mounted in a tested in a linear channel) as well as prototype-related physical model investigations (typical full models with a reservoir as upstream condition) are considered. The aim of the paper is to summarize and interpret the geometry of the different PKW models already tested, and to compare these models to existing PKW prototypes, without giving a detailed hydraulic investigation. This comparison might be interesting to compare the outcome of the different investigations, and to find gaps for further research.

2 EXISTING DATABASES

2.1 *Prototype PKW*

Table 1 lists several built PKWs with their main hydraulic and geometrical characteristics. Data that were not accessible in literature are indicated with N/A.

The use of PKWs as a solution for dam rehabilitation has started in France, at EDF dams (Laugier 2007). Until 2013, five PKWs were built, namely at Goulours Dam, St-Marc Dam, Gloriettes Damand Etroit Dam (Vermeulen et al. 2011), and Malarce Dam. In addition, supplementary PKWs were tested in the frame of model investigations and will presumably be constructed in the next years (namely at Gage II Dam and Raviège Dam). In the EDF projects, the maximum design discharge over the PKWs is $Q = 525$ m^3/s (Malarce Dam), which corresponds to a specific discharge ($q = Q/W$) equal to 12.3 m^2/s.

More recently, several PKWs were built, are under construction or planned in other countries, such as Algeria, Burkina Faso, India, Switzerland and Vietnam. For some cases, the design discharge exceeds $Q = 1{,}000$ m^3/s, reaching up to 8,700 m^3/s at Van Pongh Dam. The specific discharges are also considerable, reaching up to $q = 40$ m^2/s at of Vinh Son 3 Dam.

PKWs are typically provided to operate under low heads, taking advantage of their over-proportional developed crest length. As can be seen in Table 1, most of the PKWs have a ratio H_d/P_i around 0.3, where H_d is the upstream design head. Some exceptions are the PKWs of Sawaa Khuddu Dam (India) and Ouldjet Mellegue Dam (Algeria), where $H_d/P_i = 0.56$ and 0.66 respectively.

Table 1. Main hydraulic and geometrical characteristics of existing and planned PKWs.

Project	Country	Q_d [m^3/s]	$q(Q_d/W)$ [–]	H_d/P_i [–]	L/W [–]	P_i/W_i [–]	W_i/W_o [–]	S_i [–]	P_i/T_s [–]	P_{pxi}/P_i [–]	Main reference
Bambakari	Burkina Faso	1000	5.0	0.27	6.0	2.67	0.77	N/A	N/A	0.0	Lempérière et al. (2011)
Gage II Dam	France	400	11.8	0.25	7.8	3.75	1.23	0.67	15	0.0	Dugué et al. (2011)
Raviege Dam	France	400	10.4	0.32	6.8	2.08	1.50	0.51	13	0.2	Erpicum et al. (2011)
Malarce	France	525	12.3	0.34	8.1	2.67	1.04	0.64	22	0.4	Pinchard et al. (2011)
Goulours	France	68	5.7	0.31	4.9	1.15	1.80	0.53	16	0.0	Leite Ribeiro et al. (2009)
St-Marc	France	138	7.7	0.32	4.3	1.35	1.41	0.48	14	0.0	Leite Ribeiro et al. (2009)
Etroit	France	82	7.0	0.18	6.7	1.96	1.54	0.58	N/A	0.1	Leite Ribeiro et al. (2009)
Gloriettes	France	90	4.9	0.27	4.7	1.30	1.53	0.48	10	0.0	Leite Ribeiro et al. (2009)
Lhasi	India	115	1.0	0.13	6.0	2.17	1.25	N/A	N/A	0.0	Lempérière et al. (2011)
Van Pongh Dam	Vietnam	8700	28.9	N/A	5.8	2.09	1.19	N/A	22	N/A	Ho Ta Khanh et al (2011)
Dakmi 2	Vietnam	500	6.7	N/A	5.0	N/A	N/A	N/A	N/A	0.0	Ho Ta Khanh et al (2011)
Ngan Truoi	Vietnam	1560	16.3	N/A	5.0	2.04	1.35	N/A	16	N/A	Ho Ta Khanh et al (2011)
Vinh Son 3	Vietnam	4000	40.0	N/A	5.0	N/A	N/A	N/A	N/A	N/A	Ho Ta Khanh et al (2011)
Sawaa Kuddu	India	5240	38.0	0.56	4.9	1.33	1.00	N/A	N/A	0.0	Das Singhal and Sharma (2011)
Ouldjet Mellegue	Algeria	5240	0.7	0.66	4.8	N/A	1.37	N/A	N/A	N/A	Pfister et al. (2012)

2.2 *Scientific model PKW*

According to present literature, most of the systematic research PKW model studies were conducted in eight laboratories around the world. Table 2 lists the key research projects in the field of experimental PKW investigations. In total, almost 300 PKW model configurations were reported.

3 OVERVIEW AND COMPARISON

Table 3 gives an overview of the tested range of six dimensionless geometric parameters of so far tested PKWs, including all listed investigations, i.e., prototype PKWs as well as scientific models (being the sum of Tables 1 and 2). It can be seen that – with the exception of P_{pxi}/P_i, – all prototypes values were also studied in the scientific models. Or, in other words, the scientific investigations include larger limits than so far applied on prototypes.

The dominant dimensionless parameters influencing the discharge capacity of PKWs are the ratios L/W and H_d/P_i. For the experiments performed at EPFL (Leite Ribeiro et al. 2012c), exclusively these two terms are sufficient to predict the rating curve of all tested PKWs under all discharges with a maximum error of only ±17%. Herein, in order to focalize on the geometric parameters of the PKWs, the parameter H_d/P_i is not further considered. However, most of the

Table 2. Main current players of the PKW investigation.

Laboratory	Country	Number of PKW models	Main references
Laboratory of Hydraulic Developments and Environment – Briska University	Algeria	83	Ouamane and Lempérière (2006); Ouamane (2011)
Ho Chi Minh City University of Technology (HCM)	Vietnam	5	Truong Chi Hien et al. (2006) HMC Internal reports
Water Resources Development & Management Department – Indian Institute of Technology Roorkee	India	28	Das Singhal (2010)
Laboratory of Hydraulic Constructions – Ecole Polytechnique Fédérale de Lausanne (LCH – EPFL)	Switzerland	50	Leite Ribeiro et al. (2011) Leite Ribeiro et al. (2012b)
National Hydraulic and Environment Laboratory – Electricité de France (LNHE – EDF)	France	9	De Miranda (2011); Cicero and Delisle (2013)
Utah Water Research Laboratory – Utah State University	USA	13	Anderson (2011)
Hydraulics in environmental and civil engineering – Liège University (HECE – ULg)	Belgium	52	Machiels (2012)
Department of Civil Engineering – Isfahan University of Technology	Iran	33	Kabiri-Samani and Javaheri (2012)

Table 3. Limits of dimensionless geometrical ratios, comparison of scientific models with prototype PKWs.

		L/W	W_i/W_o	P_i/W_i	P_i/T_s	S_i	P_{pxi}/P_i
Scientific model	Maximum	8.5	2.45	4.71	300	2.00	0.43
	Minimum	2.5	0.00	0.59	5	0.25	0.00
Prototype	Maximum	8.1	1.80	3.75	22	0.67	0.38
	Minimum	4.3	0.77	1.15	10	0.48	0.00

scientific models include a test range of $0.1 < H_d/P_i < 3.0$, largely exceeding that of the prototypes (Table 1).

Hereafter, some dimensionless geometrical ratios of the numerous studies are visually compared.

Figure 2 plots P_i/W_i versus L/W for both, prototype and the scientific models. Generally, P_i/W_i increases in parallel with augmenting ratios L/W. This is explained with the fact that, for a given total PKW width W, an increase of the ratio L/W is typically linked to increasing the number of cycles and consequently decreasing the width of the inlet key W_i.

Another possibility to increase the ratio L/W is the augmentation of the streamwise PKW length B (particularly B_i and B_o). However, this impacts the slope S_i of the keys. According to Figure 3, this option was rarely exhausted, as most tests range around slopes of $S_i = 0.5$ to almost 1.0 m/m.

Concerning the scientific model tests, the major part of them was conducted with slopes between $S_i = 0.3$ and 1.0 m/m. Leite Ribeiro et al. (2011) have suggested that the influence of the slopes disappears when the upstream hydraulic head H_d is normalized with the height of the inlet key P_i for a range of slopes between 0.3 and 0.8. Results presented by Machiels (2012) and Machiels et al. (2011) mention an upper limit in around $S_i = 0.75$.

Although the ratio between the inlet and outlet key widths W_i/W_o is associated to the discharge efficiency of a PKW, its influence has been considered as secondary (Leite Ribeiro et al. 2012c). The scientific model tests varied this ration in large ranges to find optimal configurations (roughly

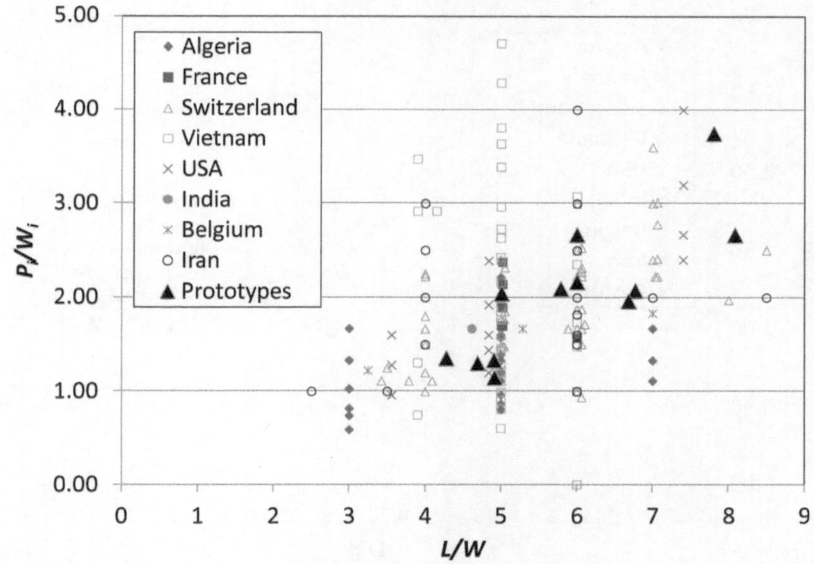

Figure 2. P_i/W_i as a function of L/W for all herein listed PKWs (Tables 1 and 2).

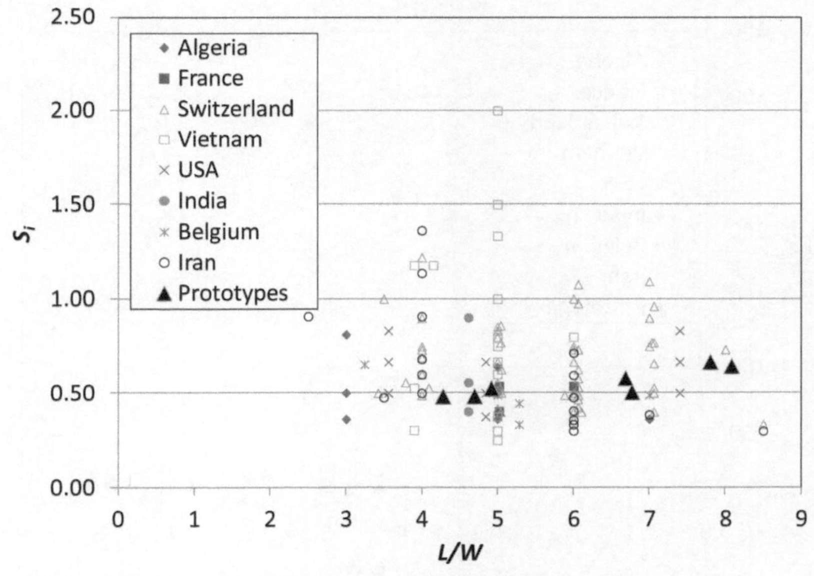

Figure 3. S_i as a function of L/W for all herein listed PKWs (Tables 1 and 2).

between $2.5 < W_i/W_o < 0.5$), with a typical optimal value around $W_i/W_o = 1.5$ (Leite Ribeiro 2012c, Machiels et al. 2010, Ouamane and Lempérière 2006).

According to a 3D numerical study performed by Laugier et al. (2011), the thickness of the side walls has an important impact on the PKW discharge capacity (up to 20%), especially for low heads. For the existing PKW prototypes, the vertical walls of each key are made of reinforced concrete, with thickness between 0.2 and 0.3 m. Considering the height of the walls (P_i) as reference, it leads

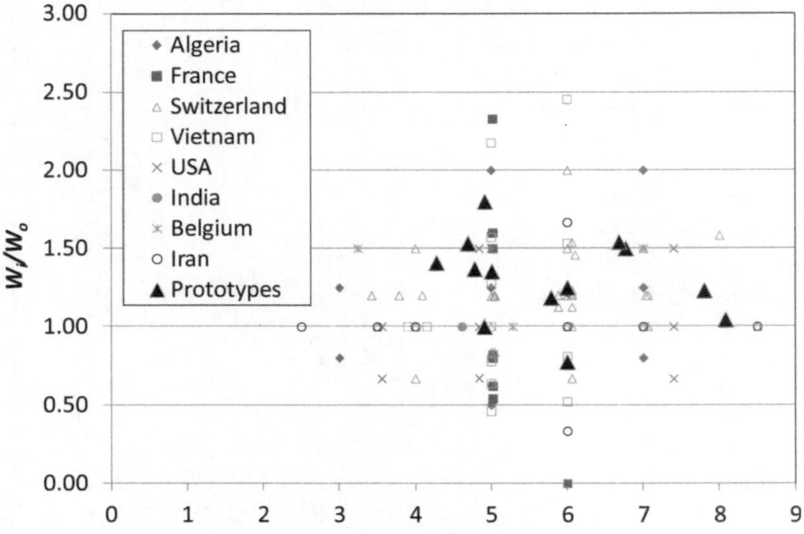

Figure 4. W_i/W_o as a function of L/W for all herein listed PKWs (Tables 1 and 2).

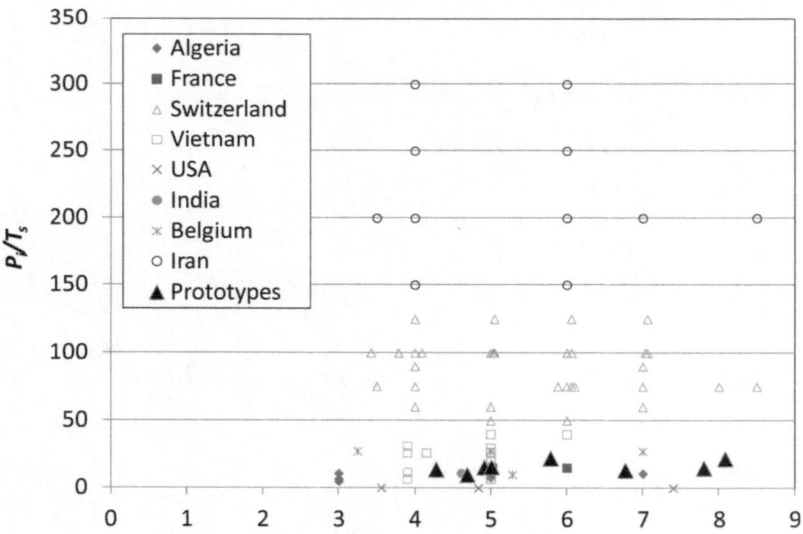

Figure 5. P_i/T_s as a function of P_i/W_i for all herein listed PKWs (Tables 1 and 2).

to ratios P_i/T_s between 10 and 22 for the existing PKWs. As shown in Figure 5, most of the scientific model tests were performed with P_i/T_s is in the range of those considered in the prototypes. As suggested by Laugier et al. (2011), the PKW discharge capacity may be increased by reducing the thickness of the walls and consequently increasing the ratio P_i/T_s. Such configurations were tested in Algeria and Iran.

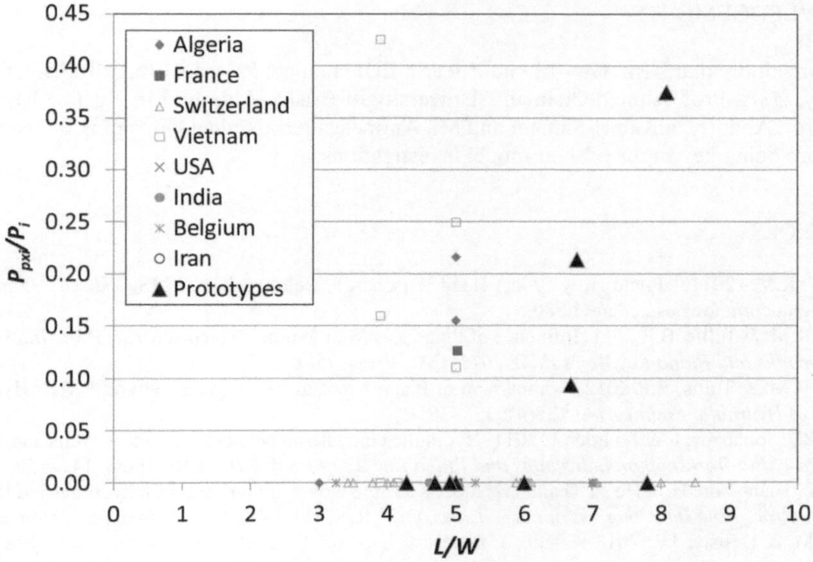

Figure 6. P_{pxi}/P_i as a function of L/W for all herein listed PKWs (Tables 1 and 2).

Until today, parapet walls were studied in three PKWs of EDF, namely at Etroit and Malarce Dams (built in both inlet and outlet keys) and in Raviège Dam (studied in both inlet and outlet keys and also in the outlet keys only). According to the results of Leite Ribeiro et al. (2012b) and Machiels et al. (online), the use of the parapet walls in the inlet key acts as an increase of the PKW height P_i. Therefore, the differences between the scientific tests performed in Switzerland, USA and Belgium and the two prototype models shown in Figire 6 are not important in terms of representativeness of the tests. However, as demonstrated by Leite Ribeiro et al. (2012b), parapet walls are efficient if applied in the outlet key.

4 CONCLUSIONS

The present paper gives a rough overview on so far performed model tests related to PKW, including investigations related to specific prototypes as well as research models to conduct systemic parameter analyses.

The main outcomes are:

- The application of PKW as an alternative for dam rehabilitation and new spillway project increased during the last years. In parallel, an important database of systematic laboratory tests was created.
- The key parameters controlling the PKW discharge capacity are L, W, P_i and H. These have already been studied in detail by different authors based on systematic tests performed in laboratory channels.
- The studies in the next years regarding PKWs operating under free flow conditions will probably concentrate on the analysis of the influence of the secondary parameters, such as the approach conditions, the shape and the thickness of the crest, the shape of the deflectors below the outlet keys, the length of the overhangs, and the presence of parapets walls in the outlet key.

Nevertheless, one has to notice that the efficiency of PKWs in terms of their rating curve may be optimized thereby maximally by few percent. As compared to the uncertainty related to the derivation of the design flood, such an optimization process is to question.

ACKNOWLEDGEMENTS

The authors kindly thank Mr. Guy-Michel Cicero, EDF, France; Prof. Dr. Blake Tullis, Utah State University, USA; Prof. Ahmed Ouamane, University of Briska, Algeria; Mr. Ho Ta Khanh, EDF, France; Prof. Abdorreza Kabiri-Samani and Mr. Amir Javaheri, Isfahan University of Technology, Iran, for providing key limits of their model investigations.

REFERENCES

Anderson, R.M. 2011. Piano Key Weir Head Discharge Relationships. MSc *Thesis*. Paper 880. http://digitalcommons.usu.edu/etd/880

Anderson, R.M. & Tullis, B.P. 2011. Influence of Piano Key Weir geometry on discharge. *Proc. Intl Workshop on Labyrinths and Piano Key Weirs PKW 2011*, CRC Press, 75–80.

Anderson, R.M. & Tullis, B.P. 2012. Comparison of Piano Key and Rectangular Labyrinth Weir Hydraulics. *Journal of Hydraulic Engineering* 138(4): 358–361.

Blancher, B., Montarros, F. & Laugier, F. 2011. Hydraulic comparison between Piano Key Weirs and labyrinth weirs. *Proc. Intl Workshop on Labyrinths and Piano Key Weirs PKW 2011*, CRC Press, 141–150.

Chi Hien, T., Thanh Son, H. & Ho Ta Khanh, M. 2006. Results of some 'piano keys' weir hydraulic model tests in Vietnam. *Proc. 22nd Intl Congress of Large Dams,* Q87, R39, Vol. IV, ICOLD, Barcelona, Spain, 581–595.

Cicero, G.M. & Delisle, J.R. 2013. Effects of the crest shape on the discharge efficiency of a Type A PKW. *This issue.*

Da Singhal, G. & Sharma, N. 2011. Rehabilitation of Sawara Kuddu Hydroelectric Project – Model studies of Piano Key Weir in India. *Proc. Intl Workshop on Labyrinths and Piano Key Weirs PKW 2011*, CRC Press, 241–250.

De Miranda, D.A. 2011. Etude expérimentale sur modèle réduit de la débitance des seuils PK weir. MSc *Thesis, Laboratoire de constructions hydrauliques (LCH/EPFL), Switzerland, and Laboratoire d'Hydraulique St-Venant (EDF), France*. June 2011. In French

Dugué, V., Hachem, F., Boillat, J.-L., Nagel, V., Roca, J.-P. & Laugier, F. 2011. PK Weir and flap gate spillway for the Gage II Dam. *Proc. Intl Workshop on Labyrinths and Piano Key Weirs PKW 2011*, CRC Press, 35–42.

Erpicum, S., Nagel, V. & Laugier, F. 2011. Piano Key Weir design of Raviege dam. *Proc. Intl Workshop on Labyrinths and Piano Key Weirs PKW 2011*, CRC Press, 43–50

Ho Ta Khanh, M., Sy Quat, D. & Xuan Thuy, D. 2011a. P.K. weirs under design and construction in Vietnam. *Proc. Intl Workshop on Labyrinths and Piano Key Weirs PKW 2011*, CRC Press, 225–232.

Ho Ta Khanh, M., Chi Hien, T. & Thanh Hai, N. 2011b. Main results of the P.K. weir model tests in Vietnam (2004 to 2010). *Proc. Intl Workshop on Labyrinths and Piano Key Weirs PKW 2011*, CRC Press, 191–198.

Kabiri-Samani, A. & Javaheri, A. 2012. Discharge coefficient for free and submerged flow over Piano Key Weirs. In *Journal of Hydraulic Research* 50(1): 114–120.

Laugier, F. 2007. Design and construction of the first Piano Key Weir (PKW) spillway at the Goulours dam. *Hydropower & Dams* 13(5): 94–100.

Laugier, F., Lochu, A., Gille, C., Leite Ribeiro, M. & Boillat, J.-L. 2009. Design and construction a labyrinth PKW spillway at St-Marc dam, France. *Hydropower & Dams* 15(5): 100–107.

Laugier, F., Pralong, J. & Blancher, B. 2011. Influence of structural thickness of sidewalls on PKW spillway discharge capacity. *Proc. Intl Workshop on Labyrinths and Piano Key Weirs PKW 2011*, CRC Press, 159–165

Leite Ribeiro, M., Bieri, M., Boillat, J.-L., Schleiss, A., Delorme, F. & Laugier, F. 2009. Hydraulic capacity improvement of existing spillways – Design of piano key weirs. *23rd Congress of Large Dams Q90, R43*, 25–29. Brasilia, Brazil.

Leite Ribeiro, M., Bieri, M., Boillat, J. L., Schleiss, A.J., Singhal, G. & Sharma, N. 2012a. Discharge capacity of Piano Key Weirs. *Journal of Hydraulic Engineering* 138(2): 199–203.

Leite Ribeiro, M., Boillat, J.-L. & Schleiss, A.J. 2011. Experimental parametric study for hydraulic design of PKWs. *Proc. Intl Workshop on Labyrinths and Piano Key Weirs PKW 2011*, CRC Press, 183–190.

Leite Ribeiro, M., Boillat, J.-L., Schleiss, A., Laugier, F. & Albalat, C. 2007. Rehabilitation of St-Marc Dam - Experimental Optimization of a Piano Key Weir. *Proc. 32nd Congress IAHR*, Venice, Italy.

Leite Ribeiro, M., Pfister, M., Boillat, J.-L., Schleiss, A.J. & Laugier, F. 2012b. Piano Key Weirs as efficient spillway structure. *24th Congress of Large Dams Q94 R2,* Kyoto, Japan.

Leite Ribero, M., Pfister, M., Schleiss, A.J. & Boillat, J.L. 2012c. Hydraulic design of A-type Piano Key Weirs. *Journal of Hydraulic Research* 50(4): 400–408.

Lempérière, F. & Ouamane, A. 2003. The Piano Keys weir: a new cost-effective solution for spillways. *Hydropower & Dams* 9(5), 144–149.

Lempérière, F., Vigny, J.P. & Ouamane, A. 2011. General comments on Piano Kew Weirs: The past and present. *Proc. Intl Workshop on Labyrinths and Piano Key Weirs PKW 2011*, CRC Press, 17–25.

Machiels, O. 2012. Experimental study of the hydraulic behaviour of Piano Key Weirs, *PhD Thesis,* Faculty of applied science, University of Liège, Belgium.

Machiels, O., Erpicum, S., Archambeau, P., Dewals, B. & Pirotton, M. 2010. Analyse expérimentale de l'influence des largeurs d'alvéoles sur la débitance des déversoirs en touches de piano. *La Houille Blanche* (2), 22–28.

Machiels, O., Erpicum, S., Archambeau, P., Dewals, B. & Pirotton, M. 2011a. Influence of Piano Key Weir height on its discharge capacity. *Proc. Intl Workshop on Labyrinths and Piano Key Weirs PKW 2011*, CRC Press, 59–66

Machiels, O., Erpicum, S., Dewals, B. Archambeau, P. & Pirotton, M. 2011b. Experimental observation of flow characteristics over a Piano Key Weir, *Journal of Hydraulic Research* 49(3): 359–366.

Machiels, O., Erpicum, S., Archambeau, P., Dewals, B. & Pirotton, M. (online). Parapet wall effect on Piano Key Weirs efficiency. *Journal of Irrigation and Drainage Engineering* doi:10.1061/(ASCE)IR.1943-4774.0000566.

Ouamane, A. 2011. Nine years of study of the Piano Key Weir in the university laboratory of Biskra. *Proc. Intl Workshop on Labyrinths and Piano Key Weirs PKW 2011*, CRC Press, 51–58.

Ouamane, A. & Lempérière, F. 2006. Design of a new economic shape of weir. *Proc. Intl Symposium on Dams in the Societies of the 21st Century*, Barcelona, Spain, 463–470.

Pfister, M., Erpicum, S., Machiels, O. Schleiss, A.J. & Pirotton, M. 2012. Discussion Discharge coefficient for free and submerged flow over Piano Key weirs, *Journal of Hydraulic Research* 50(6), 642–643.

Pfister, M., Capobianco, D., Tullis, B. & Schleiss, A.J. 2013. Debris blocking sensitivity of Piano Key weirs under reservoir type approach flow. *Journal of Hydraulic Engineering* [in press].

Pinchard, T., Boutet, J.-M. & Cicero, G.M. 2011. Spillway capacity upgrade at Malarce dam: Design of an additional Piano Key Weir spillway. *Proc. Intl Workshop on Labyrinths and Piano Key Weirs PKW 2011*, CRC Press, 233–240.

Pralong, J., Vermeulen, J., Blancher, B., Laugier, F., Erpicum, S., Machiels, O., Pirotton, M., Boillat, J.L., Leite Ribeiro, M. & Schleiss, A.J. 2011. A naming convention for the Piano Key Weirs geometrical parameters. *Proc. Intl Workshop on Labyrinths and Piano Key Weirs PKW 2011*, CRC Press, 271–278.

Vermeulen, J., Laugier, F., Faramond, L. & Gilles, C. 2011. Lessons learnt from design and construction of EDF first Piano Key Weirs. *Proc. Intl Workshop on Labyrinths and Piano Key Weirs PKW 2011*, CRC Press, 215–224.

Labyrinth and Piano Key Weirs II – PKW 2013 – Erpicum et al. (eds)
© 2014 Taylor & Francis Group, London, ISBN 978-1-138-00085-8

Effective spillways: Harmonizing labyrinth weir hydraulic efficiency and project requirements

B.M. Crookston, G.S. Paxson & D.B. Campbell
Schnabel Engineering, Inc., West Chester, Pennsylvania, USA

ABSTRACT: Due to the excellent hydraulic performance and cost-effectiveness of labyrinth weirs, these structures have been implemented in new projects and for rehabilitation of aging infrastructure not meeting current design flood requirements. The hydraulic design of a labyrinth spillway requires the consideration of complex flow characteristics and optimization of multiple geometric parameters in an iterative process. Moreover, site characteristics, construction costs, and project operational requirements moderate hydraulic optimization in order to maximize project effectiveness, as economic performance in meeting owner needs controls project decisions. Application of academic research to develop effective labyrinth spillway design and the results of recent physical and numerical modeling are discussed. Key research elements examined for practicality are the headwater ratio, cycle width ratio, and staging of labyrinth weirs. This paper highlights mutually beneficial relationships between practitioners and researchers in advancing hydraulic understanding of labyrinth weirs, with applied research used to develop design solutions responsive to site and operational constraints.

1 INTRODUCTION

1.1 *Background*

As in many other countries, dams continue to be built and the risks existing dams pose are increasing throughout the USA (ASCE 2013). Populations continue to grow and development downstream of dams spreads. Water supply demands, risk, and dam safety requirements heighten. A rise in heavy precipitation events has been documented; as a consequence, probable maximum precipitation estimates have gone up by as much as 20 percent (Easterling et al. 2013). And, the average age of water infrastructure continues to increase. There are more than 84,000 dams in the USA and the Association of State Dam Safety Officials estimates a required investment of $US 21 billion to repair those in most urgent need of rehabilitation, which is only about two percent of all USA dams (ASCE 2013). As a result, greater emphasis is being placed on spillways that effectively meet project requirements.

1.2 *Labyrinth spillways*

Labyrinth weirs are well suited for both dam rehabilitation and new dams. The geometric shape (see Figure 1) can convey more flow at a given surcharge pool elevation than a linear overflow structure. Indeed, the spillway layout can be adapted to site-specific requirements to provide flood control, additional reservoir storage, and greater spillway capacity. Additional advantages include energy dissipation, aeration, operation (passive-control), reliability, inherent stability, and aesthetics. However, the hydraulic and geometric complexities of labyrinth weirs can present design challenges. Furthermore, the hydraulic optimization of a spillway is often moderated by economics, constructability and, at times, the potential incremental increase in downstream flooding (Paxson et al. 2011, Dabling & Crookston 2012).

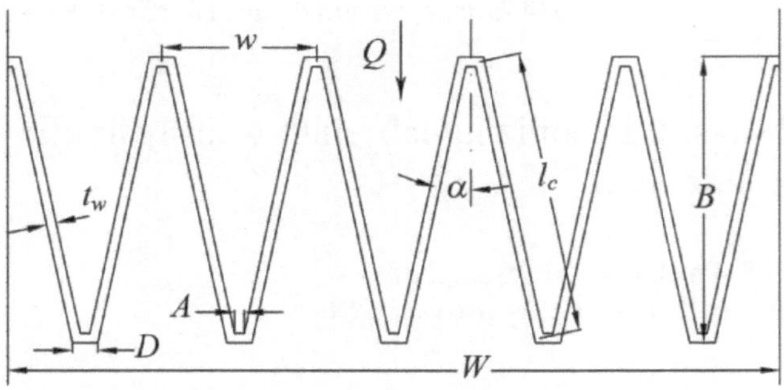

Figure 1. A labyrinth spillway (plan) with geometric parameters.

Physical modeling has been used over the past 50-years to develop labyrinth spillway designs (e.g., Darvas 1971, Houston 1982,1983, Cassidy et al. 1983, Magalhães & Lorena 1989, Tullis 1993, Quintela et al. 2000, Tullis & Young 2007, Tullis & Crookston 2008). In general, these models are less than 0.3 m in height and have geometric scaling ratios ranging from 1:10 to 1:80. The collaboration of practitioners and researchers has greatly increased understanding of labyrinth weir flow characteristics and hydraulic behaviors. Methodical labyrinth weir research efforts (e.g., Taylor 1968, Houston 1982, Crookston 2010) in controlled laboratory environments (e.g., USBR, LNEC, UWRL) have produced numerous technical articles and labyrinth weir hydraulic design methods (e.g., Hay & Taylor 1970, Hinchliff & Houston 1984, Magalhães & Lorena 1989, Tullis et al. 1995, Melo et al. 2002, Tullis et al. 2007, Crookston & Tullis 2013a,b). In more recent years, physical modeling has been complemented by computational fluid dynamics and commercially available solvers, a tool regularly used for design and research (e.g., Savage et al. 2004, Cordero-Page et al. 2007).

Labyrinth spillways have been placed in a variety of channel and reservoir applications and in diverse geological environments. Linear cycle configurations are most common; however, numerous arced labyrinth weirs (Crookston & Tullis 2012a) have also been constructed. Labyrinth weirs have featured a wide variety of crests, including sharp, flat, quarter-round, half-round, elliptical, truncated ogee, and WES crest shapes. Labyrinth weirs have also incorporated multiple crest elevations to generally match existing spillway outflow hydrographs for relatively frequent return-period storms. Projects have included various appurtenant structures (foot and vehicle bridges, piers, base-flow notches, cool-water release, low-level outlets) and a variety of downstream channels and energy dissipation structures (mild or steep chutes, curved channels, multiple steps, SAF stilling basins, endsills). In truth, the labyrinth weir concept originally explored by Gentilini in 1940 has been used with great success and has also lead to the development of labyrinth fuse-gates (Falvey & Trielle 1995) and the relatively new piano key weir (Lempérière & Ouamane 2003, Laugier 2007).

This paper presents the application of academic research to develop effective labyrinth spillway designs. Several constructed labyrinth projects are discussed to illustrate how site conditions and key labyrinth weir geometric features have resulted in research studies and consequently, additional insight into the hydraulic performance of these structures. The text also highlights the valuable and mutually beneficial relationship that can exist between researchers and practitioners.

2 HYDRAULIC DESIGN

2.1 *Previous physical model and research studies*

The following discussion highlights only a small selection of key labyrinth spillway projects, located in the USA, which contributed to additional research and information regarding these weirs. The authors gratefully acknowledge that many case and research studies have been conducted throughout the world that have contributed knowledge regarding these weirs.

Table 1. Labyrinth spillway layouts.

	Plan	Profile		$\alpha°$ (°)	N	P (m)	w (m)	B (m)	$L_{c\text{-}cycle}$ (m)	Crest
Charter Colony LTC 20/25	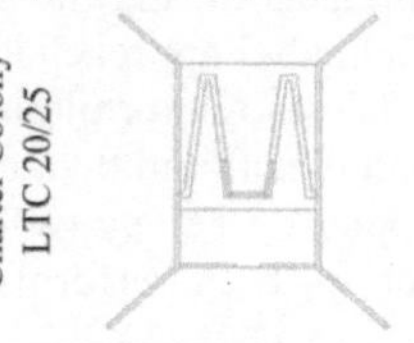	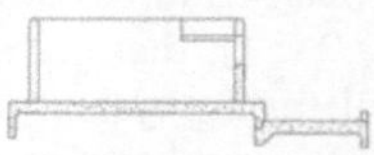 (3 stages*) $Q = 36$ m³/s $H_T = 0.5$ m	Low Stage	7.0	0.3	3.3	3.6	10.4	21.0	Quarter-round
			High Stage (*notch length = 2.4 m)	7.0	1.7	4.1	3.6	10.4	21.0	Half-round
Dog River	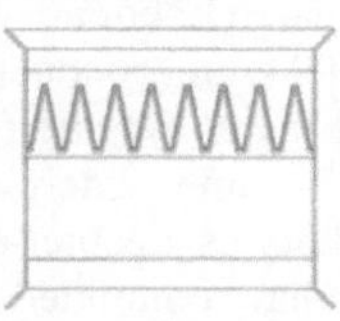	 $Q = 1{,}500$ m³/s $H_T = 2.7$ m	Original	11.3	8	4.6	9.1	16.7	36.6	Quarter-round
			Raise	11.3	8	7.6	9.1	16.7	36.6	Quarter-round
Lake Townsend	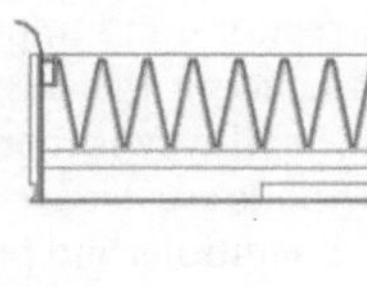	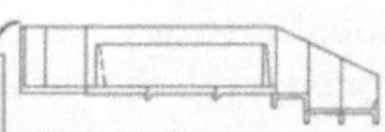 (2 stages) $Q = 4{,}050$ m³/s $H_T = 4.6$ m	Low Stage	11.4	2	5.8	12.8	26.7	40.5	Ellipse
			High Stage	11.4	5	6.1	12.8	26.7	40.5	Ellipse
Leaser	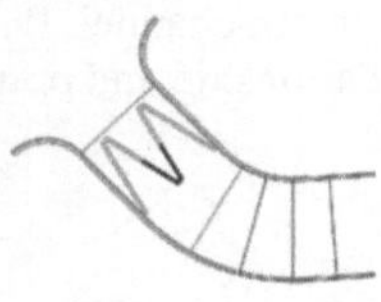	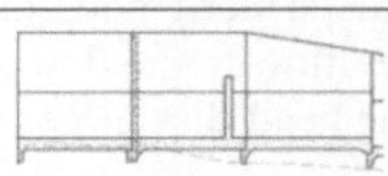 (2 stages) $Q = 300$ m³/s $H_T = 2.4$ m	Low Stage	12.3	0.5	3.7	7.6	12.8	27.2	Quarter-round
			High Stage	12.3	1.5	4.1	7.6	12.8	27.2	Quarter-round
Ute		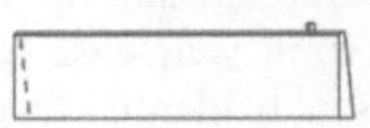 $Q = 15{,}206$ m³/s $H_T = 5.8$ m		12.2	14	9.1	18.3	34.0	73.2	Quarter-round

Ute Dam, see Table 1, is located in New Mexico, USA and original construction was completed in 1962. The original 36.9-m high earth embankment included a 256-m wide free over flow spillway, for which 8.2-m high gates were never installed. In order to meet future water demands, the US Bureau of Reclamation (USBR) was contracted to design and install the gates or develop a more economical alternative. A 10-cycle labyrinth spillway was designed based upon the design methodology of Hay and Taylor (1970) and estimated to be the most effective spillway option for the project. It should be noted that in order to meet project requirements, the design curves of Taylor (1968) for a sharp crested labyrinth weir were extrapolated and that USBR engineers had never before constructed this type of spillway. Therefore, the designers requested Kathy Frizell (Houston) to perform a physical model study and confirm that the 9.1-m high labyrinth weir would indeed pass the inflow design flood (IDF) of $Q = 16{,}040$ cms.

The study showed the spillway to pass considerably less discharge than estimated, which lead to additional systematic testing of 2-cycle sharp crested labyrinth weirs and the development of USBR design curves, presented in Hinchliff and Houston (1984). In fact, research was expanded to explore the influence of the approaching flow and various geometric parameters (including the cycle width ratio, w/P). The discrepancy between the USBR Method and the Hay and Taylor Method was in part due to their definition of the upstream head not including velocity head.

Schnabel Engineering designed the labyrinth spillway at Dog River Dam (Table 1), with construction completed in 1992. Approximately 15 years later, an expansion was designed to increase the storage capacity of the reservoir by raising the spillway by 3 m to an overall weir height (P) of 7.6 m. Available design methods did not specifically address any impacts raising the spillway crest (increasing P) would have on discharge ($w/P < 2$); therefore, Schnabel approached Bruce Savage for numerical modeling and Kathy Frizell for a physical model study. Fortuitously, the USBR hydraulic laboratory had a flume available; the results of both studies found that, contrary to estimations using the Lux (1989) Method or a spreadsheet developed by Falvey (2003) that utilized the Tullis et al. (1995) Method, the raise had minimal impact on spillway performance for the project prescribed range of H_T/P.

Paul Tullis and several of his graduate students at the Utah Water Research Laboratory performed research studies on labyrinth spillways (e.g., Amanian 1987, Waldron 1994) and conducted physical model studies for Standley Lake (1993) and the arced labyrinth spillway at Weatherford (1992). The Tullis et al. (1995) Method was a significant contribution that became widely used in the USA, in part due to Falvey (2003), and established Utah State University as a center of expertise regarding labyrinth spillways. Blake Tullis (son of Paul) and his students have continued researching labyrinth weirs and similar overflow structures. A combination of additional labyrinth spillway physical models and discerning research studies identified several corrections and improvements to the Tullis et al. (1995) Method that were adopted by Schnabel, including the design of the spillways at Leaser Lake and Lake Townsend (Table 1). The Leaser Lake spillway is representative of many rehabilitation projects where published research and design methods are used to formulate an effective 'fix' that does not require a physical model study. The site conditions and preliminary design for Lake Townsend, on the other hand, did indeed merit modeling owing to the cycle geometry (e.g., P/t_w), backwater effects, and steps to convey flow to the downstream channel. Physical and numerical modeling of the spillway determined the head-discharge relationship and optimized the step configuration at the base of the weir.

2.2 *Recent collaborative research*

Lake Townsend served as a stepping stone for additional labyrinth weir research at the UWRL. Crookston (2010) reviewed many published labyrinth weir case-studies and contacted multiple practitioners, including Hank Falvey and Greg Paxson, to identify facets of labyrinth weir hydraulics that would benefit from additional research. See Crookston (2010) and other sited references for information regarding research study experimental setups and testing procedures.

2.2.1 *Hydraulic performance*

Total discharge over a labyrinth spillway is a function of the flow conditions and weir geometry, expressed as:

$$Q = f(g, \nu, H_T, H_d, L_c, \alpha, A_c, l_c, P, B, w, t_w, \text{crest, approach, nappe}) \quad (1)$$

where Q = total discharge; g = acceleration constant of gravity; ν = kinematic viscosity of water; H_T = total upstream head measured relative to the weir crest, where $H_T = V^2/2g + h$; V = average cross-sectional velocity of flow; h = piezometric head upstream of the weir (relative to the weir crest elevation). H_d = downstream tailwater; L_c = total centerline length of the weir crest, where $L_c = 2N(A_c + l_c)$; N = number of labyrinth cycles; A_c = centerline length of the apex; l_c = the centerline length of the sidewall; α = sidewall angle; P = weir height; B = cycle depth; w = cycle width; t_w = wall thickness at the crest. Crest = the influence the crest shape has on discharge efficiency (Crookston & Tullis 2013a). Approach = the influence of the abutments, weir geometric layout (projecting, flush, etc.), topography, and the upstream velocity field on discharge (Crookston & Tullis 2012b). Nappe = hydraulic behavior of the nappe. The collision of adjacent nappes influences discharge capacity. As head increases, the collision and interaction of the nappes produce local submergence, wakes, and standing waves that reduce flow efficiency and could be

Table 2. Curve-fit coefficients for labyrinth weirs.

	quarter-round crest				half-round crest			
	a	*b*	*c*	*d*	*a*	*b*	*c*	*d*
6°	0.02623	−2.681	0.3669	0.1572	0.009447	−4.039	0.3955	0.1870
8°	0.03612	−2.576	0.4104	0.1936	0.017090	−3.497	0.4048	0.2286
10°	0.06151	−2.113	0.4210	0.2030	0.029900	−2.978	0.4107	0.2520
12°	0.09303	−1.711	0.4278	0.2047	0.030390	−3.102	0.4393	0.2912
15°	0.10890	−1.723	0.5042	0.2257	0.031600	−3.270	0.4849	0.3349
20°	0.11130	−1.889	0.5982	0.2719	0.033610	−3.500	0.5536	0.3923
35°	0.03571	−3.760	0.7996	0.4759	0.018550	−4.904	0.6697	0.5062

viewed as decreasing the effective length of the weir (Crookston & Tullis 2012c). The aeration condition (clinging, aerated, partially aerated, drowned) and behavior (nappe vibration, nappe instability) should also be considered in design (Crookston & Tullis 2013b).

As mentioned previously, empirical relationships have been developed to estimate the head-discharge relationship of these folded weirs. Crookston (2010) selected Equation 2, which is a common form of the general weir equation (Henderson 1966):

$$Q = \frac{2}{3} C_{d(\alpha°)} L_c \sqrt{2g} H_T^{3/2} \quad (2)$$

where $C_{d(\alpha°)}$ = experimentally determined dimensionless discharge coefficient that adjusts the theoretical discharge to that of reality. $C_{d(\alpha°)}$ includes the influences noted in Equation 1 that are not accounted for in Equation 2 and has been presented as either families of α curves as a function of the headwater ratio (H_T/P) as with Crookston and Tullis (2013a), or as families of H_T/P curves as a function of the magnification ratio ($L_{c\text{-}cycle}/w$, where $L_{c\text{-}cycle}$ = crest length per labyrinth cycle). Care must be taken to select $C_{d(\alpha°)}$ values that are representative of the labyrinth spillway design. Linear interpolation may be appropriate to estimate $C_{d(\alpha°)}$ values between α curves.

2.2.2 *Discharge rating curves*

Site constraints (e.g. PMP, rock excavation) can result in maximum H_T/P ratios greater than 1. The curve-fit equations presented by Crookston and Tullis (2013a) to predict $C_{d(\alpha°)}$ as a function of H_T/P and α are presented as:

$$C_{d(\alpha°)} = a \frac{H_T}{P}^{b\left(\frac{H_T}{P}\right)^c} + d \quad (3)$$

where a, b, c, and d, are curve-fit coefficients for quarter- and half-round crest shapes for $6° \leq \alpha \leq 35°$ (Table 2). The form of Equation 3 was selected for data representation ($R^2 \geq 0.99$) and it is well behaved when extrapolated for $H_T/P > 1$ (beyond the laboratory data), as presented through a comparison to the Tullis et al. (1995) polynomial curve-fit equations shown in Figure 2. The $\alpha = 15°$ data has been validated to a H_T/P ratio of 2.1 by Crookston et al. (in press) through physical and numerical modeling, increasing design flexibility of the Crookston & Tullis Method.

2.2.3 *Reservoir applications*

Research performed at the USBR for labyrinth prototypes prompted additional applied research on labyrinth spillways located in a reservoir application and the effects of several abutments (Crookston & Tullis 2012b). The geometric layouts tested are presented in Figure 3. The study documented hydraulic performance including flow characteristics and discharge.

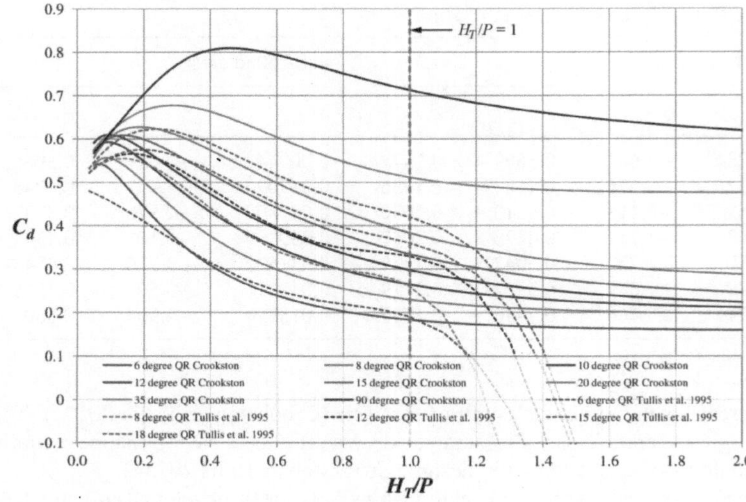

Figure 2. Crookston & Tullis (2013a) and Tullis et al. (1995) quarter-round $C_{d(\alpha°)}$ curve-fit equations.

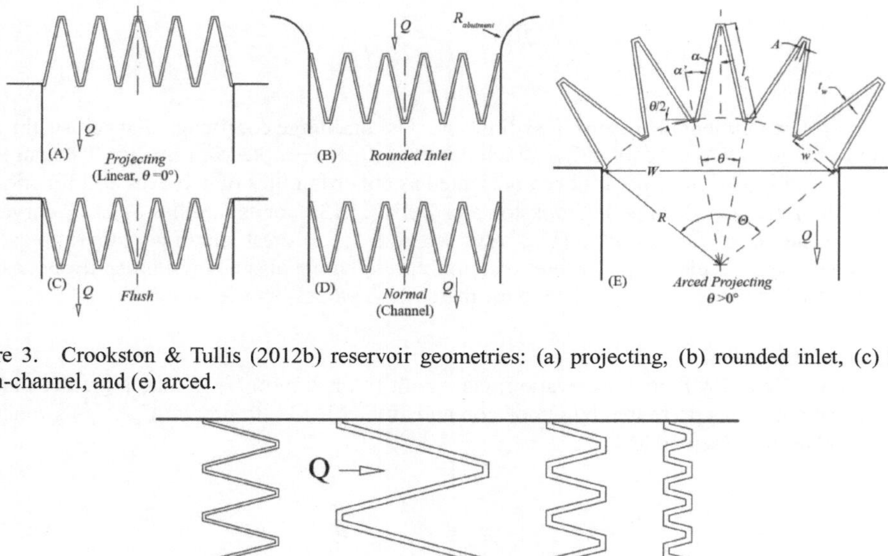

Figure 3. Crookston & Tullis (2012b) reservoir geometries: (a) projecting, (b) rounded inlet, (c) flush, (d) in-channel, and (e) arced.

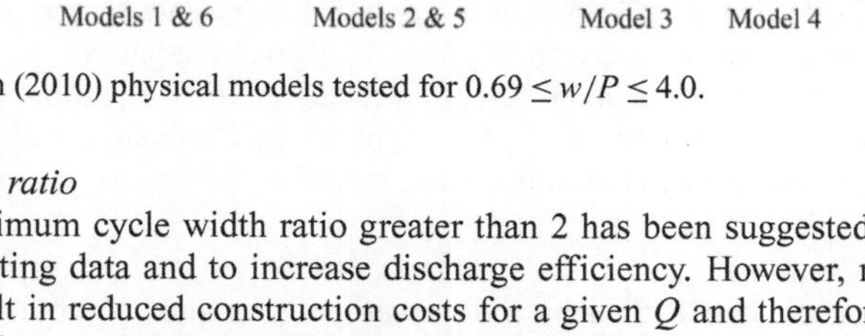

Figure 4. Crookston (2010) physical models tested for $0.69 \leq w/P \leq 4.0$.

2.2.4 *Cycle width ratio*

Historically, a minimum cycle width ratio greater than 2 has been suggested in design methods based upon supporting data and to increase discharge efficiency. However, reductions in w and w/P typically result in reduced construction costs for a given Q and therefore increase spillway effectiveness, despite reduced flow efficiency (Paxson et al. in press). In light of this, research performed by Crookston (2010) explored the influence of w and P on discharge, as identical w/P ratios may exist for different values of w and P. Tested models are presented in Figure 4.

2.2.5 *Staged labyrinth weirs*

Labyrinth spillways can pass larger flows at a given reservoir pool elevation than linear overflow spillways. Potential effects of the increased hydraulic capacity include a decrease in reservoir attenuation and an increase in peak outflows, which could potentially increase downstream flooding for lesser return period storms (Paxson et al. 2011). Some site conditions may require that a replacement spillway generally matches the outflow hydrograph of the existing spillway. As a result, a variety of spillway types, including labyrinth weirs, have been designed with multiple crest elevations; the labyrinth spillway Charter Colony LTC 20/25 was designed with three crest elevations because it was required to match specific peak discharges for the 2-, 10- and 100-year peak outflows (Table 1).

The only systematic study of staged labyrinth weir hydraulics known to the authors was Dabling and Crookston (2012), which presented preliminary experimental results from the UWRL. This study investigated multiple staged geometries with $\alpha = 15°$ to provide new insights and design guidance for practitioners. The final results (Dabling et al. in press) include staged labyrinth weir discharge coefficients and curve-fit equations, validation of a superposition method, influence of stage height (vertical difference between crest elevations) on discharge, and laboratory observations regarding flow distribution and outlet cycle hydraulics.

3 CONCLUSIONS

This paper presents the application of academic research to develop effective labyrinth spillway designs. Several constructed labyrinth projects are discussed to illustrate how site conditions and key labyrinth weir geometric features have resulted in research studies and consequently, additional insight into the hydraulic performance of these structures. Although researchers and practitioners may focus on different aspects of hydraulic structures (e.g. generating knowledge, effective spillway projects), it is the authors' experience that invaluable and mutually beneficial relationships can and do develop from applied research collaborations.

ACKNOWLEDGEMENTS

The authors wish to recognize the following colleagues: Blake Tullis, PhD (Utah State University), Mitch Dabling (UWRL Research Assistant), Bruce Savage, PhD, PE (Idaho State University), Kathy Frizell (retired, USBR), and Joseph Monroe, PE (Schnabel Engineering).

REFERENCES

Amanian, N. (1987). "Performance and design of labyrinth spillways." M.S. Thesis, Utah State University, Logan, Utah, USA.

American Society of Civil Engineers (ASCE) (2013). 2013 report card for America's infrastructure. <http://www.infrastructurereportcard.org/> (April 9, 2013).

Cassidy, J.J., Gardner, C.A., & Peacock, R.T. (1983). "Labyrinth-crest spillway – planning, design, and construction." *Proceedings of the International Conference of the Hydraulic Aspects of Flood and Flood Control,* London, England.

Cordero-Page, D., García, V., & Nonot, C. (2007). "Aliviaderos en laberinto. Presa de María Cristina." *Ingeneiería Civil,* 146, 5–20 (in Spanish)

Crookston, B.M. (2010). Labyrinth weirs. Ph.D. Dissertation. Utah State University, Logan, Utah.

Crookston, B.M., Paxson, G.S., Savage, B.M., & Tullis, B.P. (accepted April 2013). "Increasing hydraulic design flexibility of labyrinth spillways." *ICOLD 2013 International Symposium.* Seattle, Wash.

Crookston, B.M. & Tullis, B.P. (2012a). "Arced labyrinth weirs." *J. of Hydr. Engrg., ASCE,* 138(6). 555–562.

Crookston, B.M. & Tullis, B.P. (2012b). "Discharge efficiency of reservoir-application-specific labyrinth weirs." *J. Irrig. Drain. Engr., ASCE,* 138(6). 773–776.

Crookston, B.M. & Tullis, B.P. (2012c). "Labyrinth weirs: Nappe interference and local submergence." *J. Irrig. Drain. Engr., ASCE,* 138(8), 757–765.

Crookston, B.M. & Tullis, B.P. (2013a). "Hydraulic design and analysis of labyrinth weirs. Part 1: Discharge relationships." *J. Irrig. Drain. Engr., ASCE,* 139(5), 363–370.

Crookston, B.M. & Tullis, B.P. (2013b). "Hydraulic design and analysis of labyrinth weirs. Part 2: Nappe aeration, instability, and vibration." *J. Irrig. Drain. Engr., ASCE,* 139(5), 371–377.

Dabling, M.R. & Crookston, B.M. (2012). "Staged and Notched Labyrinth Weir Hydraulics." 4th International Junior Researcher and Engineer Workshop on Hydr. Structures. B. Tullis and H. Chanson (eds.).

Dabling, M.R., Tullis, B.P., & Crookston, B.M. (2013). "Staged labyrinth weir hydraulics." *J. Irrig. Drain. Eng.,* posted ahead of print May 22, 2013.

Darvas, L. (1971). "Discussion of performance and design of labyrinth weirs, by Hay and Taylor." *J. of Hydr. Engrg., ASCE,* 97(80), 1246–1251.

Easterling, D., Kunkei, K., & Yin, X. (2013) Observed increases in probable maximum precipitation over global land areas. *Proceedings of 25th Conference on Climate Variability and Change,* Austin, TX, USA.

Falvey, H. & Trielle (1995). "Hydraulics and design of fusegates." *J. of Hydr. Engrg., ASCE,* 121(7), 512–518.

Frizell, K. (2003). Dog River Dam Hydraulic Model Study Results. *USBR Water Resources Research Laboratory Report,* Denver, Colo.

Gentillini, B. (1940). "Stramazzi con cresta a planta obliqua e a zig-zag." Memorie e Studi del Instituto di Idraulica e Construzioni Idrauliche del Regil Politecnico di Milano, No. 48 (in Italian).

Hay, N., & Taylor, G. (1970). "Performance and design of labyrinth weirs." *J. of Hydr. Engrg., ASCE,* 96(11), 2337–2357.

Henderson, F.M. (1966). "Open Channel Flow." MacMillan Company, New York, USA.

Hinchliff, D., & Houston, K. (1984). "Hydraulic design and application of labyrinth spillways." *Proc. of 4th Annual USCOLD Lecture.* Dam safety and Rehabilitation, Bureau of Reclamation, U.S. Dept. of the Interior, Washington, DC, USA.

Houston, K. (1982). "Hydraulic model study of Ute Dam labyrinth spillway." *Report No. GR-82-7,* US Bureau of Reclamation, Denver, Colo., USA.

Houston, K. (1983). "Hydraulic model study of Hyrum Dam auxiliary labyrinth spillway." *Report No. GR-82-13,* US Bureau of Reclamation, Denver, Colo., USA.

Laugier, F. (2007). "Design and construction of the first piano-key weir (PKW) spillway at the Goulours Dam." *Hydropower and Dams,* Issue 5.

Lempérière, F., & Ouamane, A. (2003). "The PK Weir: a new cost-effective solution for spillways." *Hydropower & Dams, Issue 5.*

Magalhães, A., & Lorena, M. (1989). "Hydraulic design of labyrinth weirs." *Report No. 736,* National Laboratory of Civil Engineering, Lisbon, Portugal.

Melo, J., Ramos, C., & Magalhães, A. (2002). "Descarregadores com soleira em labirinto de um ciclo em canais convergentes. Determinação da capacidad de vazão." *Proc. 6° Congresso da Água,* Porto, Portugal. (in Portuguese).

Paxson, G., Campbell, D., & Monroe, J. (2011). "Evolving design approaches and considerations for labyrinth spillways." *Proc. 31st Annual USSD Conf.,* San Diego, CA, USA. CD-ROM.

Paxson, G., Monroe, J., Crookston, B.M., & Campbell, D. (in press). "Balancing site considerations with hydraulic efficiency for labyrinth spillways." *ICOLD 2013 International Symposium.* Seattle, Wash.

Quintela, A., Pinheiro, A., Afonso, J., & Cordeiro, M. (2000). "Gated spillways and free-flow spillways with long crests. Portugese dams experience." *20th ICOLD Q79-R12,* Bejing, China. 171–89.

Savage, B., Frizell, K., & Crowder, J. (2004). Brian versus brawn: The changing world of hydraulic model studies. *Proc. of the ASDSO Annual Conference,* Denver, CO, USA, CD-ROM.

Taylor, G. (1969). "The performance of labyrinth weirs." PhD Thesis, University of Nottingham, Nottingham, England.

Tullis, B. & Crookston, B.M. (2008). "Lake Townsend Dam spillway hydraulic model study report." Utah Water Research Laboratory, Logan, Utah, USA.

Tullis, B., Young, J., & Chandler, M. (2007). "Head-discharge relationships for submerged labyrinth weirs." *J. of Hydr. Engrg., ASCE,* 133(3), 248–254.

Tullis, J.P. (1992). "Weatherford spillway model study." *Hydraulic Report No. 311,* Utah Water Research Laboratory, Logan, Utah, USA.

Tullis, J.P. (1993). "Standley Lake service spillway model study." *Hydraulic Report No. 341,* Utah Water Research Laboratory, Logan, Utah, USA.

Tullis, J.P., Amanian, N., & Waldron, D. (1995). "Design of labyrinth weir spillways." *J. of Hydr. Engrg., ASCE,* 121(3), 247–255.

Waldron, D. (1994). "Design of labyrinth spillway." M.S. Thesis, Utah State University, Logan, Utah, USA.

Labyrinth and Piano Key Weirs II – PKW 2013 – Erpicum et al. (eds)
© 2014 Taylor & Francis Group, London, ISBN 978-1-138-00085-8

Author index